Published on behalf of the Institute of Management Services

BY THE SAME AUTHOR
Work Study and Related Management Services
Measurement and Control of Indirect Work
Management of Motivation and Remuneration
Management Science
Management for Administrators

WORK MEASUREMENT

DENNIS A. WHITMORE
B.Sc., M.Tech., Ph.D., C.Eng., M.I.E.E., F.B.I.M.

SECOND EDITION

HEINEMANN : LONDON

William Heinemann Ltd
10 Upper Grosvenor Street, London W1X 9PA

LONDON MELBOURNE
JOHANNESBURG AUCKLAND

© The Institute of Management Services 1975, 1987

First published 1975
Second edition 1987

British Library Cataloguing in Publication Data
Whitmore, Dennis A.
    Work measurement. - 2nd ed.
    1. Work measurement
    I. Title
    658.5′421   T60.2
    ISBN 0 434 92238 2

Printed and bound in Great Britain by
Redwood Burn Limited, Trowbridge

Foreword

by H.R.H. THE DUKE OF EDINBURGH

Formerly President of the Institute of Management Services

There are two ways of looking at productivity: its value to the company through more efficient production and its value to the individual through more efficient output.

Attitudes to work have changed dramatically since the introduction of mass industrial production. People were prepared to put up with almost any conditions in return for considerably better income and regular work. Indeed the economic advantages of productive industry both to management and to the workers on the shop floor were so great and so obvious that industry became an end in itself.

Things have changed and we are beginning to realize that industry is only a means to an end. Where people used to be expected, and were even content, to serve industry, it is becoming more apparent that industry should be serving people. This means that the emphasis in work study projects should be on its value to the individual operative, including the individual office worker and the individual manager. There can be little doubt that a company will be prosperous if every individual in the company knows that his time and his talents are being used as efficiently as work study techniques can make possible.

I believe this book will be warmly welcomed by wide sections of industry and commerce, but I hope it will penetrate further afield. There are very few occupational or vocational activities including the self-employed which could not be made more efficient by the use of work study techniques.

Preface

The ability to change and adapt, in order to adjust properly to environmental differences stemming from advancing progress, is perhaps the most significant attribute of mankind.

It is solely by exercise of this ability that the human race has assumed, and ruthlessly maintained, sovereignty over all known living creatures. The natural process of evolution, however, is normally slow and quietly remorseless; until man becomes sufficiently perceptive and introspective to begin to want to question the natural order of things, and with his innate impatience, to strive to accelerate the rate of change.

The results of these very human desires bring many conflicts in their train—most easily perceived perhaps in a study of the sciences, where the resistance of the mass of contemporary scientific opinion so often in history has marshalled and aligned itself against the would-be pioneer.

Work study as such, and work measurement perhaps in particular, is no exception to this general situation; and I believe that the time is long overdue for the profession to challenge some of its long-held tenets objectively and critically. This book has been conceived and written by the joint and sterling efforts of many of my friends and colleagues in the Institute of Practitioners in Work Study, Organisation and Methods, who have made sincere efforts to throw off preconceived ideas and dogmas, to which all of us have been exposed and are naturally subject, and to put together an objective authoritative and forward-looking work.

It is regrettably true to say that work study as a comparatively new and integral part of management had an uneven history in its early years; the profession as a whole in the United Kingdom has only lately begun to come together as a body, uniting to a common purpose. In recent years it has become increasingly obvious to many of the more far-sighted practitioners that if there is any one profession which should be clear and of one single mind in pursuing its chosen broad objectives of raising the standards of living of the peoples of the world, then it should be beyond reproach in organising its own affairs in terms of technique, research, standards of training, and application in practice.

A very pertinent example of the random and uncontrolled development of particular techniques is surely seen in the field of predetermined motion time systems.

For many reasons there has come into being a vast complexity of systems: each additional sibling appropriating and detracting from the basic patrimony of our profession. It would be too obvious a pitfall in a work of this nature to begin to attempt to try to deal with all such systems.

It is equally true that as management obtains, through increasing experience, a deeper insight into the proper role of work measurement in the total affairs of the business, coupled with a better understanding of its proper relationship to and its order of importance with other productivity-improvement techniques, that long-established concepts may change.

Many large companies are creating productivity-services or management-services organisations which embrace some or all of the present known management techniques; a number of companies are proceeding to establish forms of measurement in important areas of their businesses other than in the manpower areas specifically. It was felt necessary for the sake of completeness to make mention of these potentially far-reaching developments. We apply our work measurements, by custom and practice, basically only to men and machines—why not to other industrial and commercial areas also where equally large or better profit-improvement potentials exist?

PETER BURMAN
Chairman, Textbook Committee,
Institute of Practitioners in Work Study,
Organisation and Methods
(now the Institute of Management Services)

Author's Note to the First Edition

In the 1960s the Institute of Practitioners in Work Study, Organisation and Methods commissioned the writing of two basic textbooks on work study. The first, called *Problem Solving—A New Look at Method Study*, was published in 1970 by Business Books Ltd. In 1968 a Work Measurement Textbook Committee was established under the Chairmanship of Mr Peter M. Burman, Manager of Productivity Services, B.I.C.C. Members of the Committee, whose names appear in the Acknowledgements, were all specialists, prominent in their particular fields. In 1970 I accepted an invitation to write this book under the supervision of the Committee, whose members approved the final draft.

I am most grateful to H.R.H. The Duke of Edinburgh, President of the Institute, for most graciously agreeing to introduce the book in a special foreword which expresses concern for the people whose jobs are being measured. Although a large part of this book is devoted to the techniques and mechanics of work measurement, prominence is given to the needs and feelings of people. However, the size of any particular section of the book should not be taken as a measure of its relative importance in work measurement; most experienced practitioners of work study appreciate the need for good human relations and communication. I sincerely hope that the techniques described in this book will be used for the betterment of man.

The book is designed to meet the needs of students of work study and productivity sciences, whose syllabuses are demanding a much deeper study of the disciplines upon which work measurement is based. It is also intended as a reference book to satisfy the experienced practitioner who requires a comprehensive manual of all current techniques and practices. The sciences of mathematics and statistics are referred to throughout the book, and I make no apology for this because I believe a thorough knowledge of these disciplines to be essential to today's work-study practitioner.

A full list of Acknowledgments is given elsewhere in the book, but I would specially like to thank the Textbook Committee, and particularly its Chairman, for guiding the project over the years. I am very grateful to Jim Francis for his kind help during the last months before publication, and to his colleagues of the Institute of Practitioners in Work Study, Organisation and Methods, who were good enough to read through the manuscript and make helpful criticisms and suggestions.

Finally my thanks go to the staff at William Heinemann Ltd, who, once again, have so effectively transformed one of my manuscripts into a printed book.

DENNIS A. WHITMORE

Author's Note to the Second Edition

Since this textbook was first published in 1975 there has been little change in the well-tried and tested methods of work measurement. However, the world has seen a rapid development in the power, and drastic reduction in the costs of microcomputers, which are now available to all practitioners to help in the practice of their art. Microcomputers have been built into study boards to reduce the time spent on data collection and analysis. These aspects of new technology have been incorporated in this revised text. This updated text is being used as the material for the open/distance learning package produced by the Institute of Management Services.

I am grateful to the many readers throughout the world who have found the book of value: indeed it has even been translated into Mandarin Chinese, to be published in Beijing (Peking), People's Republic of China, as a standard text.

Acknowledgments

H.R.H. The Duke of Edinburgh, K.G., for graciously consenting to introduce this book in a special Foreword.

Peter Burman, for chairing the Textbook Committee, and for writing the Preface, Chapter 1, and Section 7.3.

Fred Evans, William Jones, Roy Shepherd, and Geoffrey Slater who were the members of the Textbook Committee and criticized and approved the draft.

The British Standards Institution, for permission to use quote definitions from their *Glossary of Terms Used in Work Study* (B.S.I. 3138), from whom the full Glossary is available.

British Insulated Callender's Cables Ltd, whose management services staff contributed the material for Sections 8.5, 8.6, and 8.7.

D. J. Desmond, and Professor N. A. Dudley, for Figure 5.8 and the theory on which the accompanying text was based.

A. Flowerdew, and P. Malin, for material upon which Section 10.33 was based (which originally appeared in *Work Study and Management Services*, December 1963).

F. Logan, Partner, Logan Associates, for approving Section 17.3 on the Wofac System, Wocom.

McGraw-Hill Book Co., for permission to reproduce Figure 5.4.

H. B. Maynard & Co., for the diagram on pp. 344–5.

M. T. M. Association Ltd, for Figure 15.6.

A. Minter, for advice on Section 5.61.

Joseph H. Quick, former Chairman of Wofac Company, who approved the sections on Work-Factor, Mento-Factor, VeFAC Programming, and Wocom (Chapters 8, 15, and 17).

J. Sainsbury Ltd (Mr G. Slater), for Table 12.1.

W. D. Scott & Co. Ltd, for the example from their *Scott-Mulligan Manual of Standard Times*, Figure 15. 15.

G. Slater, former Management Services Manager, J. Sainsbury Ltd, for supplying the material for Chapter 9, and for Table 12.1.

United Carr Ltd, and D. J. Harris, for providing the illustrations for Figures 14.3, and 14.10 to 14.14 inclusive.

Contents

Foreword	v
Preface	vii
Author's Note to the First Edition	vii
Author's Note to the Second Edition	ix
Acknowledgments	xi

1 INTRODUCTION TO WORK MEASUREMENT
- 1.1 Introduction — 1
- 1.2 Management ratios — 5
- 1.3 Measurement of work — 6
- 1.4 Manpower-productivity measurement and evaluation — 7
- 1.5 Summary — 11

2 HISTORICAL DEVELOPMENT OF WORK MEASUREMENT
- 2.1 The philanthropists — 12
- 2.2 Origins of work measurement — 13
- 2.3 Timing — 14
- 2.4 Electronics and work measurement — 20
- 2.5 Predetermined motion–time systems — 21

3. APPLICATION OF WORK MEASUREMENT
- 3.1 The relationship between work and time — 24
- 3.2 Units of work and performance ratios — 25
- 3.3 Application of work units and performance indices — 29

4. THE PRACTITIONER AND HIS APPROACH
- 4.1 The role of psychology — 33
- 4.2 Needs and drives — 34
- 4.3 Consultation — 36

5 BASIC CONCEPTS OF MEASUREMENT
- 5.1 Objectives — 40
- 5.2 Stages of measurement — 40
- 5.3 Allowances — 42
- 5.4 Outline of work measurement — 43
- 5.5 Rating — 47
- 5.6 Rating errors — 57

6 THE MATHEMATICS OF MEASUREMENT
6.1 Fundamental concepts	66
6.2 Applications	74
6.3 The binomial distribution	77
6.4 The straight line	79
6.5 Introduction to multiple-regression analysis	86
6.6 M.R.A. in work measurement	88
6.7 Linear programming (L.P.)	94

7 ANALYSIS STAGE
7.1 Basic procedure	98
7.2 Defining the job	100
7.3 Selecting the most appropriate technique	102
7.4 Element breakdown	105
7.5 Determining the number of cycles to study	109
7.6 Compilation of the basic time	116
7.7 Background to activity sampling	118
7.8 Some mathematical concepts	121
7.9 The exercise in concise form	139

8 PREDETERMINED MOTION–TIME SYSTEMS
8.1 Introduction	143
8.2 Developing the systems	148
8.3 Characteristics	151
8.4 Elements and analyses	158
8.5 Methods–time measurement—MTM-1	161
8.6 MTM-2	176
8.7 MTM-3	192
8.8 P.M.T.S. for mental processes	198
8.9 Higher levels of data	205

9 TIME STUDY
9.1 Introduction to time study	207
9.2 Timing procedure and practice	209
9.3 Taking a time study	220
9.4 A practical example	225
9.5 Planning the study	228
9.6 At the workplace	230
9.7 The study extension	240
9.8 Use of data recorded	245

10 RATED ACTIVITY SAMPLING
10.1 Background	246
10.2 Procedure for making a study	247

	10.3	Mathematical aspects	250
	10.4	A case study—short-cycle work	254
	10.5	A case study—long-cycle work	261

11 SYNTHESIS
- 11.1 Defining synthesis — 268
- 11.2 Collection and documentation of data — 270
- 11.3 Presentation of data — 276

12 ALLOWANCES
- 12.1 Allowances: introduction — 287
- 12.2 Relaxation allowances — 287
- 12.3 Process allowances — 294
- 12.4 The need for interference allowance — 300
- 12.5 Contingency allowance — 308

13 PRODUCTION STUDIES
- 13.1 Scope of the production study — 310
- 13.2 Production-study procedure — 311
- 13.3 Analysing the study — 314
- 13.4 An example of a production study — 317

14 ANALYTICAL AND COMPARATIVE ESTIMATING
- 14.1 Merits and demerits — 325
- 14.2 Analytical estimating — 326
- 14.3 Estimator training — 334
- 14.4 Comparative estimating — 335
- 14.5 Category estimating — 346

15 MEASUREMENT AND CONTROL OF INDIRECT WORK
- 15.1 Background — 353
- 15.2 The need for measurement — 354
- 15.3 Philosophy of measurement — 355
- 15.4 Systems — 355
- 15.5 Measurement and control — 356
- 15.6 Basic techniques of measurement — 363
- 15.7 Documentation — 370
- 15.8 Some important systems — 376

16 THE USES OF WORK MEASUREMENT
- 16.1 Introduction — 389
- 16.2 Use of work measurement in costing — 390
- 16.3 Use of work measurement in design — 392
- 16.4 Use of work measurement in costing a service — 400

16.5	Scheduling and planning	402
16.6	Application to remuneration	406

17 NEW DEVELOPMENTS
17.1	In which direction?	411
17.2	Computerized work measurement	414
17.3	Other uses	414
17.4	Computerized P.M.T.S.	

Bibliography
General and productivity (Chapters 1, 2, and 3)	417
Human relations (Chapter 4)	418
Basic concepts (Chapter 5)	419
Mathematics of measurement (Chapter 6)	420
Analysis stage, including activity sampling (Chapter 7)	422
Predetermined motion–time systems (Chapter 8)	423
Time study and rated-activity sampling (Chapters 9 and 10)	424
Synthesis (Chapter 11)	425
Allowances (Chapter 12)	425
Production studies (Chapter 13)	426
Estimating (Chapter 14)	427
Measurement and control of indirect work (Chapter 15)	427
The uses of work measurement (Chapter 16)	428
New developments (Chapter 17)	430

Index   431

1 Introduction to Work Measurement

1.1 INTRODUCTION

In industry, productivity measurement (of which the measurement of human work is an integral part) is a means to an end; it is not a discrete objective in itself. The need for measurement as a tool through which management is better able to exercise a real control of the business is unquestionable. What must never be lost sight of is that the individuals responsible for such measurements must have a clear and also a broad appreciation of the end-purposes for which the measurement is devised and applied.

The logic of the situation is clear: for better management control, firstly, measurement and quantification of the business situation and activity are essential; secondly, it is desirable that such measurements are coherent, pertinent, and as far as possible consistent not only between themselves, but also against a common philosophy and an easily reproducible and consistent basis. Thirdly, measurements must be so presented that management's ability to use them for useful purposes of control and improvement is supported as far as possible.

So that work measurement may be viewed in its proper context it would appear logical to deal first with the broader aspects of the measurement of productivity generally.

This has been generally defined as:

$$\text{productivity} = \frac{\text{output}}{\text{input}}$$

The form of this definition is a common one, constantly recurring, for example, in physics, with the study of levers, pulleys, and indeed other simple types of machine efficiency.

In principle this productivity ratio of measurement is a simple one, but in application, if an attempt is made to apply such a measurement over too large an area then considerable complexity can result. Fairly simple examples might be the number of key depressions per hour completed by a computer punch-operator, or the number of cubic yards dug by a labourer in a particular shift. Against this, attempts have been made to express productivity ratios for an entire company or indeed a group of companies. This method was first proposed by Smith and Beeching and expounded by them in 1948 in a paper entitled 'Measurement of the Effectiveness of the Productive Unit'. For this purpose it was necessary for them to redefine input as the input of manpower, materials, and capital equipment. Problems arose immediately in attempting the common quantification of multiple forms of output so that they could be added together.

Considerable complication obviously arises when a company is producing many varied products, although in the case of, for example, an airline or a company making one or two products only, the ratio can be valuable.

Equally the term 'input' needs to be considered in such a way that the constituent items of manpower, materials, and capital equipment can be properly summated. Here again problems can be met with when material prices vary considerably (for example copper), and also where much of the capital equipment involved will have been obtained in previous years when the value of money was quite different.

An ingenious approach was suggested by Smith and Beeching in their paper to express the materials and capital-equipment portion of input in terms of what they called 'Man-Year Equivalents'. These were obtained by dividing the actual sum of money spent on the resource by the average industrial income at that time. Capital equipment would of course require a further division by the years of amortization.

The thesis was that by adding together equivalent man-years of manpower, materials, and capital equipment an expression of input is derived which takes inflation into account. This sum was termed the 'resources-man-years'. (Presumably this stratagem does not take into account the situation where the basic value of a resource item in terms of man-years will vary with technological and other considerations.)

The concept as expounded by Smith and Beeching was further refined by Faraday in *The Management of Productivity* (1971) by taking the figure obtained from dividing output by the total resource man-years, i.e., the total productivity measure, and expressing the base year as 100, to provide an index called the total productivity index (T.P.I.).

Faraday argues that for the purpose of informing management that its policies are taking the organization in the right direction a decade would seem to be the most useful span of years through which the T.P.I. should operate. There is no doubt that the index could be a useful one, particularly in industries where output is not complex.

The total productivity index so derived acts, then, as a longer-term broad guide to senior management of the way in which the overall health of the concern is progressing; it does not assist in pointing to specific courses of corrective action that may be necessary.

It has been argued that this type of index could also be applied to a part of a business. While this is undoubtedly true, its value to management would only become apparent when a series of measurements along a time-scale showed a situation to be either deteriorating, improving, or remaining static. These indications might point to action by management, but the system as a whole does not offer an accepted yardstick or target against which such ratios can be considered, and appropriate action initiated.

These considerations raise a very basic issue, namely that productivity measurement of any kind will be of extremely limited usefulness to management if they cannot demonstrate effectively the need for improvement, or at least the appropriate corrective action to be taken. Unless such indicators motivate management to initiate improvement then they are of limited value.

It was consideration of these and other issues which led B.I.C.C. Ltd to develop Technical Cost Measurement. This approach was developed over several years within the B.I.C.C. group of companies by the Central Productivity Department of the group. A five-step approach, a tried and proved method of analysis, is used to establish the measurements involved.

Briefly the five steps may be outlined as:

Step 1. To determine the critical 'sensitive' area of the business that needs to be studied and measured. These areas, often referred to as key business areas, are defined as those parts of the business which, if they can be improved, will have the biggest overall impact upon the capital-employed performance. This is based upon the hypothesis that although there are many tasks to which a manager may set his hand in the course of his day-to-day work, only a few of these are likely to have significant effect upon the unit's profit performance.

Step 2. Where the areas are investigated in great detail and a theoretical proof is erected that there are substantial opportunities for potential improvement.

Step 3. When controlled experiments practically test out the improvement potential which was demonstrated in theory under Step 2.

Step 4. To develop with management detailed action programmes which specify precisely the actions and responsibilities involved with each manager, and the point in time by which sequential achievements must be made.

Step 5. Establishing performance measures and setting appropriate targets for achievement. The fixed points would be, firstly, the actual performance, and secondly, what is defined as the technically attainable performance, which is the highest productivity, at least cost, and so on, which can realistically be achieved by management within the key business area—derived through the use of appropriate studies and management-technique applications.

Each of the five steps could be elaborated a great deal further; for example, Step 1 may be accomplished initially by a penetrating overall company financial analysis, to obtain and analyse information from all sources—records, balance sheets, interviews, management views and opinions, and so on.

One method may be to examine financial accounts to evaluate the potential effect on profit of an improved use of specific resources—stocks, raw materials, scrap, debtors, and so on. For example, it might be possible to demonstrate that a 5 per cent reduction in scrap could lead to an improvement of 20 per cent in the profit returned on the capital employed. That is to say, the capital-employed performance of 10 per cent might be raised to 12 per cent by reduced scrap, resulting from a close and detailed study of those factors from which scrap resulted and an action programme of correction.

By this form of analysis a ranking model can be derived for all potentially sensitive areas, in the fields of profit, revenue, capital, and costs, and a further decision reached by gauging the practicality of improvement in each area, to allow an improvement programme to be planned against selected key areas.

Step 2 can be implemented from the investigation results, i.e. by using the scrap example again, from the comparison of scrap rates between like machines, or between different operating shifts, using different materials, and so on, or even by comparison with competitors' scrap rates, if these are available. The lowest scrap performance historically achieved might be used as one basis for target-setting.

By intelligent interpretation backed up by the result of detailed study an extending but acceptable target may finally be set, but before this, however, it is necessary to proceed to Step 3, when the above hypotheses are carefully tested out by rigorous practical experimentation under closely controlled conditions, so that the apparent theory of the situation is thoroughly tested and potential side-effects are well understood before final implementation takes place. Implementation of this step may place fairly heavy demands on local line-management or supervision, so that proper experimental control may be maintained; the lessons learned, however, lead inevitably to Step 4, where the accumulated experience crystallizes into specific action programmes, which can be monitored by more senior management.

After all this preparation it is possible to set real achievement targets, and it is suggested that a linear achievement-scale should be based on the following considerations. Firstly, from basic studies the actual performance in the given area must be quantified. For example, actual measured present scrap-rates might be postulated as 10 per cent of total production. Further, and from consideration of the many factors involved, it should be possible to suggest a maximum potential in the area in question, which is nevertheless believed to be realistically obtainable. This is the level called the technically attainable performance—it is a level of performance which is sincerely believed to be achievable over the course of time through competent management and first-class operating conditions. To continue our example, this level might be suggested as 2 per cent scrap.

It is therefore then possible to set an intermediate target, i.e., the budget performance which should be reached by the completion of the normal budget year.

Two important points must be made in conclusion. Firstly, no target should ever be set which is not acceptable and understandable to the manager whose eventual job it is to achieve it. Secondly, this form of investigational method is in general contrast to the normal way in which management techniques are commonly employed. It is customary for the manager to call in a management-technique specialist (Work Study, O. & M. Operational Research, etc.) to solve what the manager believes to be a particular problem; often to a preconceived solution. In a completely contrary fashion the investigation described above is one with only the broadest terms of reference, i.e., to follow through the five-step approach across the entire range of a company's interests, with the objective of maximizing business opportunity in terms of increased profit and/or reduced capital employed.

Selecting areas for study is the responsibility of the investigator, although his conclusions must be entirely supported by logic and be demonstrably sound to management. This approach is a most powerful tool for senior management, who should closely monitor progress towards targets which lead to real profit improvement in the business as a whole. It will be obvious that investigational work at this level and of this kind must be a most fruitful training ground for potential senior managers. It must inculcate problem-solving and general analytical ability, it brings the investigator in touch with all facets of a business, and above all his target-setting role must gain him invaluable experience in developing and maintaining relationships with senior management.

It is interesting to compare for a moment the relative potential impact on a business of the three approaches: the measurement of technical cost, total productivity indices, and conventional work study.

Technical-cost measurement initially studies the whole business, identifies key business areas, and makes management oriented studies into them with the objective of revealing potential improvement. It then ensures that tests practically validate possibilities of improvement; performance measurements and targets are set and defined, which are supported by detailed action-plans for management to work to.

The information required must obviously considerably assist the basic management–motivational processes of budgeting, management by objectives, and corporate planning. The total-productivity-index approach will, on the other hand, give an indication of the rate of improvement (or decline) of a company's performance against its own previous standards. It will not necessarily indicate areas where corrective action is required, nor does it particularly assist the three main managerial-motivational processes.

Conventional work study is used in general to solve a restricted range of industrial problems, i.e., basically those involving the shop floor and the man–machine relationship. These *may* often be of key importance, but without logical channelling and focusing of the use of this technique, the situation must often arise similar to that of the general practitioner treating a patient for an ingrowing toe nail when he was suffering from some terminal disease of the lungs. In the first place, the techniques of work study may not be being applied to key business areas, and therefore their potential impact must necessarily be limited. They will almost certainly be applied to the human work-situation, which is a most important field, but it must be borne in mind that there are many other business areas involving the use of working capital, marketing, costs, and so on, which could be of greater importance to a particular company at any specific time.

1.2 MANAGEMENT RATIOS

Another way of 'measuring' productivity is known as the management ratio. Management ratios are not measurements in the sense that they are related to defined basic yardsticks and can be reproduced independently at any place or time; they are rather financial and/or financial–physical ratios, of which examples might be:

$$\frac{\text{Net added value}}{\text{Man years}}$$

$$\frac{\text{Net sales value}}{\text{No. of personnel}}$$

By various simple arithmetical and algebraic principles management ratios can frequently be related to each other and can be formed into equations or 'pyramids'. It has been claimed that management ratios derived in this way can be used to compare the 'state of health' of different companies within the same group, or indeed between one company and another unconnected with it.

Management ratios suffer from two basic disadvantages. Firstly, both numerator or denominator frequently use money as a variable. Money unfortunately does not remain stable in value and the effect of devaluations, inflation, and so on, tend to make comparison difficult. It is sometimes possible to avoid distortion by, for example, using money in both the numerator and the denominator. This is shown by comparing two indices:

$$\frac{\text{Net added value}}{\text{Number of employees}}$$

$$\frac{\text{Net added value}}{\text{Total wages}}$$

In the second case changes in the value of money would effect both numerator and denominator equally, and would therefore cancel out.

Another major disadvantage in management ratios is that the quantities involved frequently depend upon accounting conventions and rules of thumb (which tend to be widely varied; there are over twenty different ways of depreciating assets recognized by the cost-accounting profession), and although this may not invalidate in-company comparisons, it must certainly make the use of comparisons between ratios across unconnected companies hazardous, to say the least.

1.3 MEASUREMENT OF WORK

Against this general background it is now possible to understand the correct place of work measurement in the tool-kit of the work-study engineer.
Work measurement is defined as:

> The application of techniques designed to establish the time for a qualified worker to carry out a task at a defined level of working (B.S. 3138: 1979, Term number 10004).

It must be stressed that work measurement should only be applied to the working situation after the observer is fully satisfied that the working method upon which measurement is to be based is the best that can sensibly be achieved.

Should the operator be able to institute a 'short cut' by way of an improved method, or even in some situations by not meeting proper standards of quality, then the true standard so derived will no longer be appropriate.

Performance targets based on work measurement are so derived that they can be met by a properly trained operator, suited and accustomed to his task, without excessive physical or mental strain. Properly applied and introduced work measurement contributes to:

> better use of the manpower resources through
>> balanced work-loading
>> improved manpower-planning
>> labour-cost control
>> improved manpower performance through rationally based incentive bonus schemes
>
> better use of capital assets through
>> soundly based production-planning
>> better materials-forecasting
>> standards for plant and machine operation

There are a number of measurement techniques in general use by work-study engineers. These range from simple timing of actual performances, through time study (where actual times are corrected by a performance-rating factor, and appropriate allowances incorporated for fatigue, personal needs, and contingencies), to various forms of activity-sampling synthesis from elemental times and estimating. A rather later, but more flexible set of measurement techniques stemmed from the concept of predetermined motion–time systems, where, broadly speaking, levelled times are ascribed to particular classified motions of the human body. There is unfortunately a plethora of such systems in use today, but in statistical terms there is no doubt that the system most widely used throughout the world is the one known as Methods–Time Measurement. The Work-Factor system probably comes second, whilst the rest are not widespread, except perhaps where localized to particular countries or industries. Predetermined motion–time systems (MTM especially) are dealt with in detail in Chapter 8.

1.4 MANPOWER-PRODUCTIVITY MEASUREMENT AND EVALUATION

As has been pointed out, productivity in its widest sense is a measure of efficiency with which the resources of labour, material and capital are used in producing the output of a manufacturing unit. In considering measurement in the particular field of manpower productivity it has been assumed that the main criteria will be to produce measurements which can evaluate:

> working performance, i.e., the speed and effectiveness with which manpower is applying itself to its given task, and
>
> the productive efficiency with which the manpower resource is being utilized.

The main purposes to which these forms of measurement would be put would give the following uses:

provide information which would assist management to take action to increase labour productivity as a whole, i.e., to increase performance and also effective utilization.
measure the results of such actions
set achievement targets
establish means whereby remuneration may be linked to improvement.

Accepting that these are the basic requirements of the situation, it follows that it is not the absolute value of performance that is so important as the change in performance from some reference point.

We may begin by defining labour productivity as the volume of output produced per man-hour of input, and we could express this as:

$$\text{Labour Productivity} = \frac{V}{H}$$

where V equals the volume of output, and H equals the man-hours required to produce this volume. We could also define the manpower performance factor as the number of standard hours' worth of work produced per man-hour input (*see* Chapter 3).

$$\text{Performance Factor} = P = \frac{W}{H}$$

where P equals the performance factor, W equals the work output in standard hours, H equals the man-hours required to produce the work.

In terms of the general effectiveness with which manpower is used on a particular task, i.e. the 'methods factor', a useful term to describe this could be 'Utilization Factor', which we would define as the volume of output per standard hour of input.

$$U = \text{Utilization Factor} = \frac{V}{W}$$

where U equals utilization, V equals the volume of output, and W equals the total standard hours required to produce the given volume.

By very simple algebra we can express labour productivity over all, i.e., incorporating both the factors of manpower performance and manpower utilization, as follows:

since $$P = \frac{W}{H}$$

therefore $$H = \frac{W}{P}$$

therefore $$\frac{V}{H} = P \times \frac{V}{W}$$

but
$$\frac{V}{W} = U$$

and therefore
$$\frac{V}{H} = P \times U$$

Thus, labour productivity is a product of the performance factor (or how hard manpower works) multiplied by the utilization factor (i.e., how effectively manpower is used).

The performance factor of an individual machine crew, work centre, or company, may be determined by abstracting the total work done in standard hours from bonus sheets, or other source documents, and dividing this by the total man-hours, usually obtained from clock cards. Since P is derived from standard hours and clock hours it has an absolute value which can be used to compare performance over different processes.

Its accuracy will depend upon the reliability and consistency of the work standards used (and also of course on clocking accuracy). A performance factor of 1·00 could be expected in a perfect work situation where there were no stoppages, breakdowns, etc., and where manpower were suitably motivated, trained and accustomed to its task. In fact, figures of less than 1·00 would almost always be met with, the precise level depending upon many environmental factors.

Conversely, it is impossible to obtain an absolute value for the utilization factor (U) since it is not possible to evaluate volume (V) in units common to all processes.

Since, however, it is the variation, i.e. improvement (or possibly decline) in productivity, that is significant as a pointer towards management action, an absolute value is not required; the change from some reference condition being far more important. As an example, to indicate changes in manpower utilization let the reference conditions be a volume equal to V_1, which is produced by an input of standard hours equal to W_1

i.e.
$$U_1 = \frac{V_1}{W_1}$$

Let the new improved conditions be the same volume of output, V_1, produced by a reduced number of standard hours input, W_2

i.e.
$$U_2 = \frac{V_1}{W_2}$$

Therefore
$$U_2 - U_1 = \frac{V_1}{W_2} - \frac{V_1}{W_1}$$

and
$$\frac{U_2 - U_1}{U_1} = \left(\frac{V_1}{W_2} - \frac{V_1}{W_1}\right) \div \frac{V_1}{W_1}$$
$$= \frac{W_1}{W_2} - 1 = \frac{W_1 - W_2}{W_2}$$

If we call this proportional change in the utilization factor the productivity index, denoted by the symbol D, we have:

$$D = \frac{W_1 - W_2}{W_2} = \text{work saved over work required,}$$

and since D is derived from the common unit of standard hours it is an absolute value in the sense that it can be used to compare the improvement in efficiency of different processes, although in itself it will not indicate the absolute level of their efficiency.

If remuneration increases are to be linked to the two factors of improved performance and improved utilization of manpower it will be obvious that improvements in performance from a given starting point will be most appropriately met by the introduction of a simple incentive bonus scheme, for example, straight proportional scheme or some form of geared scheme. (In particular, with the straight proportional incentive scheme the full saving deriving from an increase in performance above the norm is returned to manpower as a bonus, the company saving coming from a better utilization of overheads.)

Assuming, then, that effective incentive conditions are in operation, it follows that any increase in remuneration which is to be justified by manpower upon the grounds of increased productivity can only come from an increase in the utilization factor.

By further simple algebra the reduction in standard hours to complete a given task by virtue of an improved method or utilization can be simply factored by the known cost per standard hour to give the total manpower economy. This can be factored as appropriate to give shares to management and shop floor respectively.

To establish utilization factors certain reference conditions must be clearly understood:

1. It will be necessary to agree a reference mix of output which as nearly as possible represents the normal throughput of the operation.
2. It will be necessary to agree a volume of each constituent of the mix, which in total will represent a week's output from all shifts. This latter will be the reference volume V_1.
3. One must measure the standard hours required to produce this reference volume V_1; this will be the standard hours reference W_1.

Having established the various operational values, work references for the total process may then be calculated and the total value of W summated.

When a major improvement in methods is introduced, then a new work reference would need to be established. This situation could obviously involve conditions where new capital was involved, in which case some value for depreciation should be incorporated on the increase in capital. Depreciation could be related to manpower costs by simply calculating the ratio of depreciation over manpower costs per standard hour, so that the new work reference would be the sum of the actual standard hours plus the equivalent standard hours.

If no new capital is involved then the establishment of new work reference is much simpler.

1.5 SUMMARY

The first chapter of this book has been concerned with the assessment of productivity through the various productivity ratios and management ratios. The subject is investigated more specifically in Chapter 3.

Clearly, any attempt to assess the level of productivity relies on adequate, reliable, and appropriate means of measurement. Such means are available, and range from simple forms of estimating to the more advanced methods involving the use of mathematics.

Before attempting to understand the finer points of these work-measurement techniques it is instructive to see how the discipline of work measurement has evolved from its early primitive beginnings more than 200 years ago. This development is described in the next chapter.

2 Historical Development of Work Measurement

2.1 THE PHILANTHROPISTS

The measurement of work has no definite origins. Attempts have always been made to assess job times. Over countless centuries the early overseers, and latterly 'gaffers', foremen, and managers, have made their own assessments of effort and attempted their own interpretations of what constitutes 'a fair day's work'. Only comparatively recently have efforts been made to find ways of measuring work by more methodical means. The early concept of 'pace of working' was based on maximum effort with little regard for the well-being of the worker, the form of incentive being essentially negative. Unfortunately the amateur dabblings of these early managers, whose rough guesses and lack of regard for conditions, consistency, and human relations sowed the seeds of discontent, were to continue for many decades, and survive in some places to this day.

In the early days of the Industrial Revolution labour was still a cheap and abundant 'resource', and to work was regarded as a privilege rather than as a right. Consequently little heed was paid to the installation of sophisticated management controls.

At the end of the eighteenth century the Industrial Revolution and expansion of the factory system sharpened the division between employers and employed; the middle-class craftsmen and engineers finding themselves either with the minority who were swept up to the heights of employer of men, or sinking to the level of humble labourer from which few escaped. However, not all employers were blind to the needs and potentials of their workers. The philanthropic views of Robert Owen (1771–1858) are legendary. Owen was at once the founder of Socialism and a competent businessman. He was appalled by the waste of human potential in the factories of the nineteenth century. In addition to the improvements effected in working hours, wages, training of workers, and conditions of employment, Owen was responsible for the introduction of rudimentary method studies and work measurement. His concern for the humanitarian side of business inspired him to introduce rest and fatigue allowances for his workers, thus rejecting the idea that output was directly proportional to time worked.

Owen was far ahead of his time. Even fifty years after Owen, F. W. Taylor was being chastised for the alleged indifference to the feelings of employees of his 'scientific management'. Charles Bedaux was hated by many who accused him of exploiting labour. Nevertheless great advances were made by other pioneers in industrial psychology.

George Elton Mayo's contribution to the understanding of industrial behaviour through his research at the Hawthorne Plant in America opened a new era in industrial psychology. The development of theories of behavioural science continued through the work of Seebohm Rowntree, Mary Parker Follett and others, and lately through the efforts of Douglas McGregor, Abraham Maslow, Chris Argyris, Saul Gellerman, and Frederick Herzberg. Basically these researchers are saying the same thing, but with a slightly different bias and from varied viewpoints. The culmination of this work is apparent in the approach developed for the guidance of work-study practitioners, which is described in Chapter 4.

2.2 ORIGINS OF WORK MEASUREMENT

The origins of contemporary work measurement are to be found at the turn of the century with the first successful studies of Frederick Winslow Taylor, based on the principles of scientific management as propounded by him. As early as the 1890s he was advocating 'management based upon measurement plus control'. Taylor's scientific management was not well received by the trade-union movement of the day, and indeed the differences and suspicions engendered by the methods of Taylor and other pioneers still remain in current attitudes. In 1914 Professor R. F. Hoxie of Chicago University* undertook to investigate these differences on behalf of the American Commission on Industrial Relations. He found two major categories of objections. The objections to *measurement* were that scientific management regards workers as mere instruments of production, and that it drives them to their limits of endurance. Furthermore standards were based on extreme cases without due allowance for human differences. Other objections included the contribution of scientific management to overproduction and unemployment, and to the dependency of workers on the employers' concepts of fairness.

Over the next thirty years the methods of time study were improved and consolidated in the form in which they exist today. However, compared with the astounding technological advances made in other fields over the past half-century, developments in work measurement have been relatively negligible.

Probably the most significant advance was made in the 1930s in the area of predetermined motion–time systems. The introduction of the P.M.T. systems enabled more precise and consistent measurements to be made, which paradoxically became a serious disadvantage as the scope of measurement extended beyond the realms of the short-cycle, repetitive job. The original precise systems were soon to be augmented by second- and third-level systems which were more suitable for measuring long-cycle work.

About the same time the introduction of sampling theory to industrial applications facilitated the collection of greater volumes of data, while other

* R. F. Hoxie, *Scientific Management and Labour* (New York: Appleton, 1915).

techniques such as rated activity sampling were added later to the range of measuring tools.

The computer age of the 1960s, and more especially the advent of the desk-top computer and time-sharing systems, gave to work measurement the impetus needed to develop methods of more rapid application. The combination of computer technology and mathematics provided work measurement with some powerful new tools.

In recent years increasing numbers of companies turned to the measurement of indirect work in their search for new fields in which control and cost reductions could be effected. By the 1960s white-collar workers in the United States had outnumbered those in manual work, and the trend could be identified in many other countries. The clerical, maintenance, and other areas of indirect work demand a different approach. Accordingly, new techniques have been developed from the existing methods of measurement.

Work measurement has turned the full circle. It has gone through the stages of overall timing, elemental timing, and back to the overall measurement demanded by multiple-regression analysis and similar techniques. The systems of P.M.T. have passed from detailed, precise standard elements through second and third (broader) levels, and finally back to detailed systems, the use of which is again made feasible by the computer in methods such as Wocom and Sammie.

2.3 TIMING

For countless years jobs have been estimated and timed to provide crude, rough, job times which were found to be useful in assessing work-loads.

The work of Charles Babbage (1792–1871), published in 1835,* was quite extraordinary, anticipating as it did some of the contemporary methods and practices in manufacturing and work study.

Babbage was a mathematician and scientist, and Lucasian Professor of Mathematics at Cambridge. In addition to his many important contributions to science, which included the first computer (his 'Calculating Machine'), his work covered 'economy of human time', design of tools and contrivances for improving productivity, economy of materials, and much on the early theory of the division of labour. He also published what is probably one of the earliest examples of a time study and costing survey, for the manufacture of pins (Figure 2.1). However, time standards had been established for pin manufacture by M. Perronet in France some seventy years earlier. Perronet's times and costs, interpreted by Babbage are tabled in Figure 2.1. His assessment of output was 494 pins per hour.

Babbage studied the analysis of work into elements for the purpose of timing, as did another Frenchman, M. Coulomb, a contemporary of Babbage, who used

---

*Charles Babbage, *On the Economy of Machinery and Manufactures* (London: Charles Knight, 1835).

NAME OF THE PROCESS.	Time for making Twelve Thousand Pins.	Cost of making Twelve Thousand Pins.	Workman usually earns per Day.	Expense of Tools and Materials.
	Hours.	Pence.	Pence.	Pence.
1. Wire.....................	...	...	...	24·75
2. Straightening and Cutting .	1·2	·5	4·5	...
3. ⎧ Coarse Pointing	1·2	·625	10·0	...
Turning Wheel	1·2	·875	7·0	...
Fine Pointing	·8	·5	9·375	...
Turning Wheel	1·2	·5	4·75	...
⎩ Cutting off pointed Ends ..	·6	·375	7·5	...
4. ⎧ Turning Spiral	·5	·125	3·0	...
Cutting off Heads	·8	·375	5·625	...
⎩ Fuel to anneal ditto.......	...	...	...	·125
5. Heading..................	12·0	·333	4·25	...
6. ⎧ Tartar for Cleaning	...	...	...	·5
⎩ Tartar for Whitening......	...	...	...	·5
7. Papering	4·8	·5	2·0	...
Paper.....................	...	...	...	1·0
Wear of Tools	...	...	...	2·0
	24·3	4·708		

(a)

NAME OF THE PROCESS.	Workmen.	Time for making 1 lb. of Pins.	Cost of making 1 lb. of Pins.	Workman earns per Day.		Price of making each Part of a single Pin, in Millionths of a Penny.
		Hours.	Pence.	s.	d.	
1. Drawing Wire (§ 224.)	Man	·3636	1·2590	3	3	225
2. Straightening Wire ⎧	Woman .	·3000	·2840	1	0	51
(§ 225.) ⎩	Girl	·3000	·1420	0	6	26
3. Pointing.... (§ 226.)	Man	·3000	1·7750	5	3	319
4. Twisting and Cutting ⎧	Boy.....	·0400	·0147	0	4½	3
Heads.... (§ 227.) ⎩	Man	·0400	·2103	5	4½	38
5. Heading.... (§ 228.)	Woman .	4·0000	5·0000	1	3	901
6. Tinning, or Whiten- ⎧	Man	·1071	·6666	6	0	121
ing ...... (§ 229.) ⎩	Woman .	·1071	·3333	3	0	60
7. Papering ... (§ 230.)	Woman .	2·1314	3·1973	1	6	576
		7·6892	12·8732			2320

Number of Persons employed:—Men, 4; Women, 4; Children, 2, Total, 10.

(b)

Figure 2.1 Original analyses of standard times and costs prepared by (a) Perronet (translated by Babbage), and (b) Charles Babbage, about 1835

his times for planning purposes. Babbage was fully aware of the effects of direct observation on the worker's tempo. In his book he notes:

> For instance, if the observer stands with his watch in his hand before a person heading a pin, the workman will almost certainly increase his speed, and the estimate will be too large. A much better average will result from inquiring what quantity is considered a fair day's work. When this cannot be ascertained, the number of operations performed in a given time may frequently be counted when the workman is quite unconscious that any person is observing him.

Upon the ways of reducing experimental error he quotes M. Coulomb (in French) 'who has great experience in making such observations' as saying:

> I advise those who wish to repeat them [the experiments]; if they have no time to assess the results over several days of continued work to observe the output at different times during the day, without the workers being aware that they are being observed. One is not able to say how great is the risk of miscalculating either the speed of the worker or the effective time to carry out an operation on the strength of just one single observation of a few minutes' duration.*

Even today the advice, given in the last sentence, is sometimes ignored. Babbage also gave a description of a comprehensive job-specification form which included details of the operation, training, and skill required, processes, tools, reject analysis, and sketches of the workplace.

Modern management owes much to these first pioneers, and their contemporaries, Colbert, de la Hire, and Amontons, to mention only a few.

The history of time study is a troubled story, one of resistance, misunderstanding, misapplication, and often violent reaction. One of the many instances of labour troubles fomented by the application of time study occurred in the United States at Watertown Arsenal, and led directly to an investigation into the system in 1910 by the Interstate Commerce Commission. Such was the extent of the unrest that in 1913 the Government decreed that no part of the Senate Appropriation should be made available for the payment of anyone in the employ of the Government who was engaged in the practice of time study. Indeed this state of affairs survived until 1947, when a Bill passed by the House of Representatives finally recognized the use of time study.

In spite of the efforts of these workers, the meaning of 'a fair day's work' still defied precise definition. Frederick Winslow Taylor, a Quaker of Philadelphia, really earned his title 'Father of Scientific Management'. To Taylor must go the credit for laying the foundations of time study as it is practised today. His work is well documented in the literature of work study, and of course, in his own books.† Much of his work was concerned with the development of method study, but he also foresaw the need for research into the determination of a fair

---

* Taken from M. Coulomb's *Memoires de l'Institut*, Tome II, p. 247.
† F. W. Taylor, *The Principles of Scientific Management* (New York: Harper, 1929).

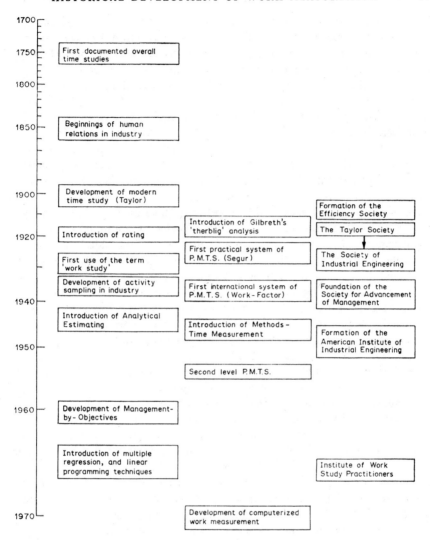

Figure 2.2 Some milestones in the history of work measurement (to a logarithmic time-scale)

day's work through the application of reliable work measurement. He presented his initial findings to the American Society of Mechanical Engineers in 1895, and again eight years later in the form of a paper entitled *Shop Management*, which was well received. It is interesting today to peruse Taylor's prescription for time study. Its modern approach to the use of standard tools, job description, layouts, costing applications, and detailed element breakdown is hardly distinguishable from present-day practice. Taylor's directions also advocated the use of the slide-rule and mechanical aids for calculating the results.

Premature curtailment of his schooling because of impaired eyesight at the age of eighteen forced Taylor to seek work in industry, where he had the opportunity of experiencing all sides of the business, rising from humble labourer to chief engineer at Midvale Steel Works when he was 31 years old. Most of his experimentation into the techniques of work study was carried out at Midvale. Taylor's methodical approach to the problems of general management, methods improvement, and work measurement proved beyond doubt that productivity could be greatly improved by the appropriate application of these procedures. However, although one of Taylor's objectives was to develop 'a spirit of hearty co-operation between management and the men', his persistent timing of their work and rest created a great deal of hostility, the resentment being reinforced by the accepted attitudes and views of managers of the day, who clung tenaciously to the ways of the Industrial Revolution. In fairness to Taylor he did state publicly before the Special Committee of the House of Representatives in his Testimony (25 January 1912): 'Never would I believe in applying scientific management unless it was thoroughly agreeable to both sides', and 'it is not scientific management until both sides are satisfied and happy'. Further, he protested that no one actually used his methods as originally intended. Taylor was in favour of letting men know that they are being timed. However, he does suggest secretive timing where open timing would 'only result in a row', and even gives details of a 'watch book' consisting of a framework, 'containing concealed in it one, two, or three watches' operated 'without the knowledge of the workman who is being observed'.

In Frederick Taylor's time the technique of *rating* was unknown, but he recognized that no two men work at exactly the same speed. His method for obtaining a representative standard was to observe first-class men only, and time them when working at their best. From these times the quickest times for each element were selected. It was then, to quote Taylor, 'a simple matter to determine the percentage which an average man will fall short of this maximum'. This is a crude form of rating assessment.

Taylor's death in Philadelphia in 1915 attracted as little interest in the United Kingdom as had his achievements during his lifetime. His impact in Europe came much later. The extent to which scientific management was, indeed, based on scientific principles has been the subject of controversy for many years. Unfortunately many practitioners and authors have attempted to surround work measurement with a pseudo-scientific aura which has done much to reduce the credibility of the concept. Carl Barth, for example, produced an empirical formula for handling time which includes a factor given to seven places of decimals.

Taylor's assistant at Midvale Steel was Henry Lawrence Gantt (1861–1919), who is best remembered for the chart which he perfected in 1917. His long associations with Taylor inevitably affected his working life. Gantt was concerned with the introduction of Taylor's differential piece-rates, subsequently developing his own Task and Bonus Scheme based on work measurement.

The man to whom the full credit for modern time study is due was born in France in 1886. Charles Eugène Bedaux transformed the relatively crude methods of timing into a relatively reliable technique of work measurement. Through his international management consultancies Bedaux, by now a naturalized American, installed his famous but controversial incentive plan based on time study.

One of Bedaux's main contributions to time study was the concept of the standard minute, which he named the 'B' (after his own name). His speed and effort rating system helped to achieve some measure of consistency in the resulting 'normalized' times, which were far superior to the selected times of Taylor.

Bedaux's consultants, known colloquially as Bedaux Men, were both feared and hated by workers and trade unions, who opposed his slave-driving methods at every opportunity. Nevertheless Bedaux's contribution included the idea of the Compensating Relaxation Allowance, and the first practical analytical tables for computing such allowances, in addition to the concept of rating. He is credited with the development of the Points System of job evaluation, still in current use in many industries.

Charles Bedaux enjoyed continued success until the outbreak of World War II, when he returned to France, where he worked until 1941. He died in America in 1944 in tragic circumstances.

Since Bedaux, the principles of time-study techniques have changed little, apart from the work done on improving allowances and in standardizing procedure. However, several attempts have been made by various researchers to improve upon Bedaux's speed and effort rating. It is a reflection of the reliability of the system that it still remains the most generally applied method.

A points system for rating was first described in 1927 by researchers in the United States of America, who were later to become famed as the originators of Methods–Time Measurement.* Known as the Westinghouse System, the method depended on the selection of the correct levels of skill, effort, conditions, and consistency for each job studied, points and hence ratings being allotted from tables. Two other notable attempts were made in the 1950s by R. L. Morrow† with his so-called *synthetic rating* and M. E. Mundel,‡ whose *objective rating* was first applied about 1958. All these systems are described later in this book.

The whole concept of time study has been analysed and criticized by many writers, one of the most prominent being William Gomberg in his work *A Trade Union Analysis of Time Study* (New York: Prentice-Hall, 1955).

By the 1920s sampling theory was well defined. In the cotton mills of Northern England the collection of data for the purposes of management control posed a difficult problem because of the nature of the layout of looms. An enterprising engineer named L. H. C. Tippett§ applied the theories of probability and sampling to the concurrent study of several machines and operatives. He called

---

* S. M. Lowry, H. B. Maynard, and G. J. Stegemerten, *Time and Motion Study* (New York: McGraw Hill, 1940).

† R. L. Morrow, *Motion Economy and Work Measurement* (New York: Ronald Press Company, 1957).

‡ Martin E. Mundel, *Motion and Time Study* (Englewood Cliffs: Prentice-Hall, 1960).

§ L. H. C. Tippett, 'Statistical Methods in Textile Research' *Journal of the Textile Institute Transactions*, Vol. 26, Feb. 1935.

the method 'A Snap-reading Method of making time studies'. The technique produced extremely reliable results, and, under the name 'Ratio Delay' was introduced into United States industry in 1940.

The technique of systematic sampling was applied to time study to create a new method, unfortunately called *rated activity sampling*. The application of advanced mathematics to work measurement continued with the emergence in the 1960s of measurement based on multiple-regression analysis and linear programming.*

This development heralded the completing of the cycle in the history of work measurement—a return to overall-cycle timing. Several other techniques have witnessed the turning of the wheel. *Estimating* can probably claim to be the oldest method of obtaining a job time. Considered to be too subjective in the period between the two world wars, it was largely displaced by time study and predetermined motion time systems. During World War II analytical estimating appeared as a refinement of basic overall estimating, and was based on synthetic data. Further attempts to improve the consistency of results culminated in the development of comparative estimating, and in about 1970 the extension to category estimating.

2.4 ELECTRONICS AND WORK MEASUREMENT

Until recently the stop-watch was the only piece of equipment available to the practitioner for the purpose of timing, although filming techniques had found limited application since the early part of the century.

Frank Bunker Gilbreth pioneered the use of both ciné and still photography, and although very little used by work-measurement practitioners for general timing purposes, the techniques found favour among the researchers of predetermined motion time systems. Photographic methods include micromotion (slow-motion) cinematography, developed by Gilbreth, who was also responsible for the cyclegraph and chronocyclegraph.† The latter were widely exploited by Anne Shaw.‡ Stroboscopic photography was well known to Work-Factor researchers,§ who used the method extensively in developing their system round about 1935.

The introduction of memomotion was a valuable way of augmenting the technique of activity sampling. A later development was the addition of mechanical devices for triggering the movie-camera at predetermined intervals. The mechanical equipment employed thus, and the mechanisms used in the chronocyclegraph, were later superseded by electronic circuitry.

Electronic equipment for measuring time was used very early in the history

* Dennis A. Whitmore, *Measurement and Control of Indirect Work* (London: William Heinemann Ltd, 1971).

† F. B. Gilbreth and L. M. Gilbreth, *Applied Motion Study* (New York: Sturgis & Walton, 1917).

‡ Anne G. Shaw, *The Purpose and Practice of Motion Study* (London: Columbine Press, 1960).

§ J. H. Quick, J. H. Duncan, and J. A. Malcolm, Jr, *Work-Factor Time Standards* (New York: McGraw-Hill, 1962).

of work measurement. Segur, working in the 1920s, made use of the Kymograph, an electrically controlled pen moving in sympathy with the input and tracing the results on a moving paper-tape. The device is capable of recording in thousandths of a second. In later years electronic machines with digital readout became available for measuring to ten-thousandths of a second.

Ralph M. Barnes* describes an electronic device called Wetarfac (Work Electronic Timer and Recorder for Automatic Computing), details of which were first published in 1956. All the usual data for a time study, including rating factors, may be fed into the machine via the portable keyboard.

Further research in the 1950s led to the application of tape-recorders to the measurement of work, the logical outcome of which was the development of Tape Data Analysis,† employing the tape-recorder as a means of rapid verbal recording of the elements of the job being studied.

From the very early days of work measurement, log tables, slide rules, and ready-reckoners have been used to perform the simple calculations of extension, and in the computation of standard times. The introduction of mechanical, and later electronic desk-calculators has speeded up the process. However, there still remained the problem of reducing the time for actual job-analysis, which in the case of predetermined motion time systems could amount to more than 90 per cent of the study time. In all but the shortest elements or cycles the use of first-level P.M.T. systems often proved prohibitive in time and labour. This was a major objection to the use of such systems, and a contributory factor to the reluctance of industry to replace obsolescent methods such as time study by first-level P.M.T.S. To some extent the second- and third-level systems alleviated the situation.

The availability of computer time, and the introduction of 'table-top' computers and time-sharing systems stimulated interest once again in predetermined motion–times and made feasible the measurement of long-cycle jobs by these methods. In 1964 research by workers in the United States produced a computer program in Fortran to analyse simple jobs. Of course there is no advantage in using the computer merely as a means of extracting element times from tables stored within its memory banks. It must be capable of performing detailed analyses after being fed with the minimum of input information. Results of such research have appeared in the form of computerized 'packages' such as Wocom, made available in the early 1970s. Such systems will produce detailed analyses of the work, complete with a basic time from the required input consisting of the co-ordinates of the various work-positions (of hands, materials, tote bins, etc.), shapes of pieces to be assembled, and other basic data.

2.5 PREDETERMINED MOTION–TIME SYSTEMS

Once again in the history of work measurement the name Frank Bunker Gilbreth is prominent, this time as contender for the title of originator of

---

* R. M. Barnes, *Motion and Time Study* (New York: Wiley, 1968), p. 418.
† F. Evans, *Tape Data Analysis* (Warrington: M.T.M. Association Ltd, 1973).

P.M.T.S. His 'therbligs', devised in 1924,* described the eighteen basic elements into which Gilbreth decided all work could be analysed. The list, shown in Figure 8.1 (Chapter 8), includes several manual elements but also the mental elements of 'inspect' and 'plan'. In modern systems it is not usual to include such factors as 'rest' within the basic system. It should be appreciated that therblig analysis was conceived originally as a means of recording the motions highlighted by micromotion studies.

It is interesting to note the striking similarity between the original therblig elements selected by the Gilbreths and those in use in contemporary P.M.T. systems over half a century later; yet another example of the foresight shown by this remarkable team.

Shortly after the appearance of therbligs a student of Frank Gilbreth, A. B. Segur, made his first application of a system derived from micromotion analysis which he called Motion–Time Analysis (M.T.A.).† He believed that the time required by experienced workers to perform defined fundamental manual motions remains constant.

More than a decade passed before the emergence of further systems, although with the appearance of Work-Factor new systems proliferated. Work-Factor was the first to gain international prominence, the initial application being in 1938 from research initiated in 1934. The system was published in 1945, and later in a book entitled *Work-Factor Time Standards*.‡ The originators were Joseph H. Quick, William J. Shea, and Robert E. Koehler. In later years contributors were the late James H. Duncan and James A. Malcolm Jr. Detailed Work-Factor sired the second- and third-level systems of Ready Work-Factor and Abbreviated Work-Factor and, in 1967, the technique for measuring mental work, Wofac Mento-Factor System, the culmination of nearly twenty years' research.

The period witnessed the appearance of many systems of relatively minor importance, including Holmes's Body Member Movements, the Engstrom-Geppinger method of Motion–Time Data,§ both from 1938, and, during World War II, a system of Elemental Time Standards from Western Electric Company, and yet another developed for use at the Springfield Armory by Capt. Olsen. None of the techniques achieved prominence outside the United States.

Ten years after the first application of the Work-Factor System, H. B. Maynard published, with G. J. Stegemerten and J. L. Schwab, another system for detailed analysis and measurement of work which was known as Methods–Time Measurement (described in Chapter 8). This was destined to become one of the most widely used methods of predetermined motion–times. The power of MTM has been increased through the recent introduction of higher-level data in MTM-2, MTM-3 and General Purpose Data.

* F. B. Gilbreth and L. M. Gilbreth, 'Classifying the Elements of Work' *Management and Administration*, Volume 8, August 1924.

† For a description *see* H. B. Maynard (ed.), *Industrial Engineering Handbook* (New York: McGraw-Hill, 1956).   ‡ Op. cit.

§ For a description of M.T.D. see Ralph M. Barnes, *Motion and Time Study*, 6th Edition (New York: Wiley, 1968).

The trend continued into the 1950s with Basic Motion Time study (B.M.T.), manuals for which were published in Canada from the work of Ralph Presgrave, G. B. Bailey, and J. A. Lowden,* followed by others from H. C. Geppinger (Dimension–Motion–Times), and from Irwin P. Lazarus. This decade was significant for the emergence of the first of the macro-P.M.T. systems, the most important being Universal Maintenance Standards (Maynard & Co.), and Master Clerical Data (Serge A. Birn Company),† both based on Methods–Time Measurement. A set of standard data tables for clerical work compiled by Paul B. Mulligan in the United States as early as 1940 formed the basis for the Clerical Work Improvement Programme sponsored by W. D. Scott and Company, an Australian company of consultants.

Most of the development in this field in the 1960s was concentrated on systems derived either officially or otherwise from Methods–Time Measurement. Such systems in the latter category include Master Standard Data, and the United Kingdom developments of Primary Standard Data, Simplified P.M.T.S., Basic Work Data, and Clerical Work Data. The MTM Associations have initiated research which culminated in the official systems of MTM-2 and MTM-3, and have been concerned with the development of MTM-2 Maintenance Data and Tape Data Analysis. Macro-P.M.T.S. was born of a need for predetermined data suitable for application to jobs other than those of very short cycle. The advent of computer time-sharing facilities and its part in reviving interest in P.M.T.S. as a viable means of measuring such work has been discussed elsewhere in this book.

* G. B. Bailey and R. Presgrave, *Basic Motion Timestudy* (New York: McGraw-Hill, 1958).
† S. A. Birn, R. M. Crossan, R. W. Eastwood, *Measurement and Control of Office Costs* (New York: McGraw-Hill, 1966).

3 Application of Work Measurement

3.1 THE RELATIONSHIP BETWEEN WORK AND TIME

The previous chapter has traced the development of work measurement from primitive, often stormy beginnings to the present day with its advanced methods of predetermined motion–time systems, and its computerized work-measurement 'package'.

The main objective of work measurement is the establishment of standard times for jobs, these standard times being set either at a defined, standard level of performance, or as 'target' times, for the purpose of planning, manning, costing, remuneration, and general control. A formal definition approved by the British Standards Institution is given in Chapter 5. The present chapter outlines the uses to which work measurement may be applied.

The term 'work measurement' is considered by some to be a misnomer on the grounds that it is the amount of *time* necessary for the completion of the job or any part of it which is sought by the work-measurement practitioner and not, in the strict sense of the word, the amount of *work*. It is significant that in the United States the term 'time study' is used generically to cover all techniques of work measurement. To the engineer and physicist 'work' is the amount of energy expended in performing a given task, and is entirely unconnected with time (unless one were considering the *rate* of working in horsepower or watts). The argument might appear somewhat pedantic, but it is essential that the trainee in work study be aware of the objectives and limitations of his trade right from the start. The measurement of work in terms of ergs is by no means impossible; indeed much research has been carried out on the human worker in his working environment, in the fields of ergonomics and kinetics, and in the study of fatigue and its compensating allowances, as well as in the actual measurement of work. However in the end it is the *time* for the job which is the most useful parameter. A man operating a lathe expends very little effort in turning the wheel to feed the tool slowly and carefully while the material is cut away. Clearly in such a case a knowledge of the number of ergs of work performed is of little help to the work-measurement practitioner. For the purposes of planning and manning it is necessary to know how many parts the machine is capable of producing, and this requires a knowledge of the machining time.

There are many ways of measuring work and these are described separately and in detail in Chapters 7–13 of this book. Briefly, these may be separated into three distinct categories, of timing, estimating, and predetermined motion time systems. This is shown in Figure 3.1.

APPLICATION OF WORK MEASUREMENT

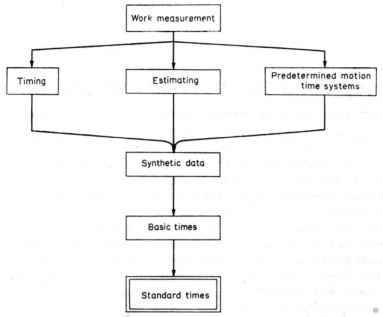

Figure 3.1 Use of the three basic methods of measuring work

The time standards established through the application of work measurement may be used by management in:

1. Assessing the correct level of manning for a department under specified conditions;
2. The loading of plant and equipment for economic utilization;
3. Planning for anticipated future demand;
4. Costing components and product manufacture and/or services supplied;
5. Supplying appropriate time-data useful when comparing different procedures, and assessing the merits of proposed method changes;
6. Establishing and maintaining sound incentive schemes.

The numerical calculations which are part of the above applications are facilitated by use of the various ratios known collectively as performance indices. Many forms of this measure exist, three of the most common being described in the following section.

3.2 UNITS OF WORK AND PERFORMANCE RATIOS

The aim of work measurement is to establish a reasonable and fair time for a job which an average qualified worker can achieve without undue exertion. The systems of P.M.T. have been built up from elemental times (i.e., times for each element of the jobs) which have been derived so as to satisfy the above requirement. Consequently the resulting basic times are also fair and reasonable, with an inherent consistency which enables jobs to be compared with one another.

The analyst using *estimating* as his method of measurement tries to assess the time for the job in accordance with the above criterion.

The facility of comparing on a common basis is essential if workers' performances against these times are to be used as measures of their diligence. Individual workers perform naturally at different speeds, even while engaged on identical work. Accordingly the basic time is established for the *job* and not for each individual. It is important that people paid on an incentive scheme and working at the same rate on different jobs should be paid a comparable bonus, which again requires basic times to be at a common level of pace.

The use of a timing technique demands some form of adjustment to be made to the 'raw' actual observed times so that they conform at a common level of working speed.

From the foregoing it is clear that neither the efficiency of workers nor their jobs can be judged or compared on numerical output alone. There can be no comparison between the tool-maker taking 60 minutes to complete his job, and a mass-production worker making 600 parts per hour because no knowledge of the effort put into the work is available. However, if it can be shown that the tool-maker performed 50 minutes' worth of work in his hour, and the mass-production operative should have completed his 600 parts in 50 minutes, there is a basis for comparison.

This basis is in the concept which measures the amount of work in each job in standard *work units*. Although work units may be of such a size that there are 100 to the hour, the most usual measure is the *standard minute*, and in some firms the *standard hour*. A man producing 300 units when each unit is worth one standard minute (a total of 300 s.m. of work) is working at the same rate and with the same effort as a colleague producing 150 units of a job worth two standard minutes per unit (again 300 s.m. of work).

The unit of work consists of work, together with the proportion of relaxation and other allowances allocated to that unit. If the unit of work selected is the standard minute, a person working at *standard performance* will complete 60 units of work in one hour. A rate of working equal to half standard performance yields 30 work-units in an hour.

Any job consists of a certain number of (and fractions of) work units; one for instance could be worth 8·4 s.m. per cycle. In the standard hours system this is equivalent to 0·14 s.h. The total number of units of work in the job is known as the *standard time*, defined by the B.S.I. (Term number 43032) as:

> The total time in which a job should be completed at standard performance (i.e. basic time plus contingency allowance plus relaxation allowance).

In the above example a worker engaged for seven hours and producing a standard performance for the day would complete each cycle in an *average* time of 8·4 minutes, but would not necessarily complete individual cycles in that time because the units of work include relaxation-allowance time, which may not be taken during each cycle. Furthermore he could achieve a higher performance during the morning, reducing his pace as the day passed.

3.3 APPLICATION OF WORK UNITS AND PERFORMANCE INDICES

The methods of deriving measures of utilization have been described in the previous section. The various ratios are useful for the purposes of forward planning as well as for assessing performance retrospectively. All projections and forecasts are based on past records and trend lines, and for manning a section or department it is necessary to know the *effective rate of production* in addition to the actual quantity produced.

3.31 USE OF WORK UNITS IN LOADING AND MANNING

When attempting to establish the number of people required to produce a certain level of output, be it goods or services, the planner is faced with the problem of assessing the probable rates of production likely to be achieved over the next planning period. Individual differences cause people to work at levels which are natural and particular to them, and not uniformly at a fixed 'standardized' level selected by management, nor do they work consistently day after day at any one level. The planner, in trying to resolve the difficulty, may decide to use actual performance figures attained over a previous reference-period, or alternatively to use the theoretical output based upon the so-called *standard performance*. The former offers a more realistic choice; theoretical outputs may be adjusted by the application of one of the measures of performance defined in the previous section.

The theoretical output from a group of workers may be calculated in the following way. Suppose six workers are engaged in work whose standard-time value is 1·2 standard minutes. By definition (*see* p. 26) the standard time for a job is the average time a qualified worker would take to perform the task (including time for relaxation) were he to *work on average at standard performance.**

Thus a person working at this standard performance could produce $\frac{60 \text{ min}}{1\cdot2 \text{ s.m.}}$

or 50 cycles per hour, which is equivalent to 375 over a day of $7\frac{1}{2}$ hours. Six people working for $7\frac{1}{2}$ hours each day produce 375 × 6, or 2,250 per day.

By this means a standard output for the group has been established. The standard may be used to estimate the number of working days required to complete an order. If the company received an order for a quarter of a million of these parts the time schedule for its completion is:

$$250{,}000 \div 2{,}250 = 111 \text{ days approximately.}$$

Should it be necessary to fulfil the order in, say, 40 days, the number of operatives required would be calculated thus:

Theoretical daily output of one worker	= 375 parts
Number of parts produced in 40 days	= 15,000 parts
Therefore, number of workers to complete the order in 40 days	= 250,000 ÷ 15,000
	= 17 people approximately

* The concept of standard performance is explained in the following chapter.

Alternatively all the above calculations may be performed using the concept of *work units* by finding the work content of the job in terms of work units, and the number of work units the operatives are capable of producing.

In the above example each person is able to produce 450 standard minutes of work each working day of $7\frac{1}{2}$ hours, remembering that work units (in this case, standard minutes) include appropriate allowance for relaxation and other conditions.

The number of work units in 250,000 cycles of work = 250,000 × 1·2
= 300,000 s.m.

To accomplish the required task in 40 days, work must be performed at the rate of

$$\frac{300,000}{40} \text{ s.m. in a day}$$

Therefore the number of people to satisfy the requirement is:

$$\frac{300,000}{40} \div 450$$

$$= \frac{7,500}{450}$$

$$= 17 \text{ people as before}$$

Conveniently these calculations have made use of the idea of standard performance, but planning based on this level of working could be unrealistic and lead to inaccurate forecasting, because this theoretical working pace might not reflect the true position. Where the variance between the standard and actual performances is significant the theoretical values may be adjusted. It is necessary to collect information about the past achievements of the workers concerned over a representative period, using the data in one of the performance indices. Since the output will be subject to the delays normally experienced, the department performance is used in the calculation. It could well be that the average department performance over a representative period stands as low as, say, 85, in which case the daily output of one average worker may be adjusted to $375 \times \frac{85}{100}$ parts, or 319 approximately. For six workers this represents 1,914 parts daily.

A better estimate of the number of days required to produce a quarter of a million components is given by:

250,000 ÷ 1,914, which is about 131 working days

While it may be said that the average performance of six workers engaged on this work is 85, there may be no evidence to suggest that this average will be appropriate for all workers subsequently employed in this task. However, failing further information it must be assumed that this is so. Thus:

The number of workers required to produce 250,000 parts in 40 days, working at standard performance, is equal to 17

APPLICATION OF WORK MEASUREMENT

The number of workers required to produce this output, but working at a performance of 85, is equal to $17 \times \frac{100}{85} = 20$ people

Once again the alternative method using work-units may be employed in the foregoing calculations. The potential number of work-units capable of being produced by a worker in a day of $7\frac{1}{2}$ hours is 450 s.m. Working at an 85 performance he is able to produce 382·5 s.m. of work. The subsequent calculations proceed as before, using the new value of standard minutes. The values obtained in the preceding description for manning may be used in assessing the loading of machines and processes, and this is described in the following section.

3.32 WORK UNITS AND MACHINE LOADING

The application of work units to the planning of machine utilization will be demonstrated by way of an example. Consider the following data for a lathe situated in a machine shop engaged on specialized work.

Average take-down and set-up time between jobs = 36 s.m.
Standard time for job C 60 (per component) = 2·8 s.m. (30 needed)
C 61 (per component) = 5·2 s.m. (50 needed)
C 62 (per component) = 4·5 s.m. (6 needed)
B 86 (per component) = 3·2 s.m. (350 needed)

The efficiency of the man/machine unit has been estimated previously at 82 per cent, a figure based on past experience.

The total time required from the end of the job which preceded job C 60 to the end of job B 86 is found from:

Time for job C 60 (2·8 × 30) = 84 s.m.
C 61 (5·2 × 50) = 260 s.m.
C 62 (4·5 × 6) = 27 s.m.
B 86 (3·2 × 350) = 1,120 s.m.
Total = 1,491 s.m.

At an efficiency of 82 per cent the time necessary for completion must be extended to $1{,}491 \times \frac{100}{82}$ or 1,818 minutes. To this must be added the total time for 'take-down and set-up' of 144 s.m. (36 × 4 occasions). Thus the total time allowed is

$$1{,}818 + 144 = 1{,}962 \text{ minutes}$$
$$= 33 \text{ hours}$$

A form of visual presentation facilitates the loading of machines and allocation of work to people, especially where several are working concurrently on different jobs. A type of Gantt chart for this purpose is illustrated in Figure 15.4 (p. 361). This example shows the application to clerical work.

3.33 WORK UNITS AND STANDARD COSTING

In the more advanced costing systems the value of goods or services produced is calculated at the more important stages throughout the processing. The system of evaluation is known generally as *standard costing*. In essence it is the comparison of *actual costs* incurred in progressing the item or service up to a given stage against the predetermined *standard costs* which should have been incurred. The variation of the actual costs from the standard cost is known as the *variance*, and this may be positive or negative.

This particular branch of cost accounting is very wide-ranging, and it is not the purpose of a book such as this to describe the many techniques of standard costing. However, the role of work measurement in the field of cost accounting is an important one, and for the purpose of describing the application of work-units it is sufficient to take a simplified view of cost analysis.

A detailed description of this application is given in Chapter 16, but basically the total cost comprises the elements of labour cost, materials, and overhead cost. The costs may be further sub-divided, as described in Chapter 16. Material and overhead costs may be obtained from appropriate sources, but the cost of providing labour for the manufacture of a particular product, or for the provision of a certain service, may be calculated from the two components of (i) labour cost per hour, and (ii) standard time for the job.

Thus if labour in a particular section costs £6·92 per hour, and the job is measured, establishing a standard time of 3·0 standard hours, then the job standard-labour cost is 3 × £6·92, or £20·76.

Of course, the *actual* time taken may differ from the standard time, and this will produce a *variance*. Suppose the worker took 3·5 hours to perform the work. Then the actual cost will be 3·5 × £6·92, or £24·22, producing a variance of £20·76 − £24·22, or −£3·46. Lost time may be costed too. In the above department each minute of waiting time is a loss of 11·5 pence to the department.

3.34 APPLICATION OF WORK MEASUREMENT TO REMUNERATION

During the early part of the century work measurement was widely employed to establish standard times on which more reliable incentive-payment schemes could be based. So powerful was the impact that even today in many quarters work measurement and incentives are regarded as inseparable quantities, and even as synonymous terms.

In order that incentive schemes pay wages which are fair to all those concerned regardless of the type of work they are doing, it is essential that output be measured, not in terms of items produced, but in work units, because these are independent of the nature of the work. Wages earned are related to the performance, which in turn is calculated from standard times set by work measurement.

The subject of incentives based on work measurement is discussed in Chapter 16.

4 The Practitioner and his Approach

4.1 THE ROLE OF PSYCHOLOGY

The twin disciplines of psychology and sociology are essential to the implementation of work-measurement systems. For centuries it has been accepted that the doctors and lawyers, engineers and artisans should be well versed in the art of their profession and use of their equipment.

The work-measurement practitioner, too, is trained in the use of the tools of his trade. Like the line manager he is directly concerned with people, and especially with the jobs which provide their means of livelihood. People have inherent basic needs which must be safeguarded; they will fight to protect their security and to maintain status and prestige. Any move against their jobs will be regarded with suspicion. They are naturally curious, wanting to appreciate and understand what is going on around them. Much has been learned about the way people act under given circumstances, but although 'general principles' of behaviour have been derived people are still individuals and act in individual ways. However, although individually they are different, in general people are gregarious, tending to adhere to formal groups, and to form informal groups inside and outside the working environment.

One of the main threats (both real and imagined) to this jealously guarded security and status is seen to be the practitioner of work study, who seeks to raise the productivity of the company and whose aims may often appear or be represented to be in conflict with those of the work-force.

It is clear that an acute human-relations problem exists when attempts are made to introduce time standards. Through years of experience in applying the techniques of industrial management, researchers of industrial psychology have defined certain general principles which have stood the test of time, and should be incorporated into any work-study procedure. The practitioner of work measurement must be competent in the use of his techniques and equipment, but he should be able, additionally, to understand and apply the basic psychological and sociological concepts which affect the behaviour patterns of both his subjects and himself.

The results of many years of experience show that people have needs which must be satisfied if frustration is to be avoided. Frustrated needs will stimulate aggressive or apathetic attitudes in employees, or tendencies to non-co-operation.

Industrial strife may result from genuine grievances harboured by employees, or through the actions of agitators whose aims are to stir up trouble in the company or industry. In both cases the source of the trouble may be traced back to the frustration of some action, or a need which was not satisfied. In certain instances the company or persons against whom the reaction is directed may not be the source of the original frustration; they could be acting merely as a substitute for the real cause. Sometimes the work-measurement practitioner is seen to be the epitome of much that is bad in the management, and he may suffer unwarranted antagonism.

It is because of such apparently illogical actions that it is necessary for the practitioner to recognize these situations and appreciate that there may be more below the surface than is apparent to the more casual observer. Realizing this, he may be in a better position to deal with the situation adequately.

4.2 NEEDS AND DRIVES

The needs to which reference has just been made include the physiological needs, which are of little concern in the present context, and the psychological needs.

The so-called 'exploratory drive' explains much of the desire for participation and for consultation possessed by most people. The manifestation of *curiosity* is obvious in very small children, and this gives rise to the theory that curiosity is innate, and inherited from our ancestors who needed to understand their world in order to survive.

Whatever its origins, it is clear that most people have a need to know and to understand what is going on around them. In normal circumstances workers do not blindly accept or obey orders; they perform their duties more readily if they understand that there is some reasonable objective in their work. When asked to do work which apparently serves no useful purpose often people will discharge this duty but with reluctance, and grumble about it to others within earshot. Curiosity stimulates the building up of knowledge, which in turn increases the efficiency of the individual to cope with the problems of his life and work.

From this it is apparent that the satisfaction of curiosity helps to promote productive working. Work study is one of the prime instigators of change and hence movement into the unknown, which makes understanding and clarification of the situation by managements extremely important.

Psychologists such as Frederick Herzberg* and Abraham Maslow† have described how people have a need for *security* which, again, is inborn. This need appears to remain with us throughout life. The strength of this drive varies, and depends on the degree to which the security is threatened. The need is stronger in some workers than in others; some taking risks, going 'out on a limb' while others are plodders, preferring to stay in the same rut doing the same safe job. Once again, work study may be instrumental in conceiving changes which appear to threaten jobs and security.

The so-called 'resistance to change' is born of fear, and presents the greatest

---

* F. Herzberg, *Work and the Nature of Man* (New York: World Publishing Company, 1966).

† A. H. Maslow, *Motivation and Personality* (New York: Harper & Row, 1954).

barrier to improvements and the acceptance of work-measurement standard times. This resistance may be reduced by open discussion to dispel suspicion and mistrust.

Work measurement may be seen as a threat to security, a legacy from former days when the techniques were widely abused. The need for security is coupled with another, the need of a individual to *belong* to a group, be it a working group within his occupation, or a social group or family group outside his work. People feel safer in the company of others (hence the popularity of trade unions), and any move to disrupt the harmony of the group or against one of its members is seen as a threat to security. In the work situation an operative may be singled out as a subject whose job is to be measured. Later the results of work measurement may be used for the purposes of manning and reallocation of jobs. Both may be interpreted as threats to the security of groups or individuals. Once again, diplomatic handling of the situation is required of the practitioner and of management.

Everyone has a status need, assessed by Maslow as one of the more potent in his 'hierarchy of needs'. A worker must maintain his status or self-esteem and even makes great effort to enhance it. If he is not acquainted with the facts of a study an operative may feel that he is not considered important enough to warrant being informed. This will have the effect of reducing his apparent status, yet another reason for the incorporation of some form of consultation and communication of information to staff.

Where it is possible to do so, the practitioner of work measurement should strive to avoid deflating a worker's ego, and in cases of deadlock must allow him a means of escape which preserves his dignity. Failure to accommodate the other person in this respect will cause him to resort to forms of ego adjustment, some of which are innocuous enough, but others may be damaging to the situation.

Work study is a prime instigator of change, examples of which have been given above. Frustration of needs coupled with unfortunate past experience of work study have helped to foster resentment and stimulate active opposition to the techniques and managements that allow it. There will always be conflict between the aims of the workers and the objectives of managements.

Most industrial psychologists agree that people work more willingly and hence more effectively if they are allowed the freedom to complete the work without close supervision or control, using initiative and demonstrating that they are responsible workers. Usually this is incompatible with traditional work-study policy, which precisely specifies work-patterns, and exerts control through fairly rigid standard times. The inhibiting of individual initiative occasioned by the inflexible application of control procedures can be a disadvantage of work measurement. One of man's greatest attributes is his ability to make decisions and to solve problems which may beset him. Such qualities should be encouraged to maintain a purposeful drive. The conflict between this and the aim of work measurement to impose standard times and 'defined methods' is fortunately not formidable at operative level, because most people employed on repetitive work accept the inherent boredom and absence of opportunity which are the concomitants of such work. Implementation of the recommendations of work study and use of the time standards are the responsibility of management, in whom is vested the power to apply a fair interpretation of the situation.

Those theories of motivation which affect work measurement may be identified as in two forms, both having a common objective. One form, that of direct motivation, is concerned with the stimulation of people to work because they *wish* to do a good job, rather than to perform a task because they *must* to avoid punishment or gain reward. Indirect motivation, the second form, stimulates co-operation and team-spirit by providing an atmosphere conducive to harmonious working through worker participation and full consultation on all relevant matters. Current motivation theories as outlined above highlight the importance of maintaining good relations and co-operation through the medium of consultation.

4.3 CONSULTATION

The case for adequate consultation has been stated in the preceding section. Adequate consultation implies communication between all parties concerned, and this includes management, for often it is more difficult to convince the hierarchy than the shop floor that change is necessary. Consultation may include a degree of participation, affording the operative the opportunity of adding to the pool of information about his job.

The 'interested parties' may include members of line and functional management, supervision, representatives of trade unions and of groups of workers, and individual operatives. Those actually consulted will vary from company to company. In some firms, especially those with weak worker-representation, it is considered inadvisable to consult with workers' representatives: 'What has it to do with those people?' Past experience has shown the value of gaining the co-operation and tacit or expressed approval of staff, and the eradication of difficulties and objections to new moves at an early stage. This obviates the risk of a confrontation and possible loss of face at a later time.

There is no specific phase in a study during which consultation should take place; it is a continuing process. For convenience, the contact between practitioner and workplace will be described in three parts: consultation prior to the study, during the study, and at the installation.

4.31 CONSULTATION PRIOR TO THE STUDY

Resistance to change is considered by some to be an attitude of mind generated by management. While this may not be wholly true, managers can contribute to the reduction of these barriers by removing some of the secrecy and hence the suspicion which often surrounds work measurement. Uncertainty about the future may lead to frustration, which is manifest in aggressive or regressive traits. Regression increases the worker's susceptibility to rumour, causing him to believe the worst.

Understanding of this situation has prompted managements to develop procedure for communicating with the parties concerned in a work-measurement study. While conforming to a set pattern the procedure must be flexible enough to accommodate the peculiar circumstances of individual applications.

The nature and amount of information to be disseminated will be governed by company policy, tradition and 'custom and practice' within the firm, and the relevance of such information to the current application. It will usually include an outline of the scope of the study and of the techniques to be employed, the objectives and likely outcomes in broad terms, and redeployment and redundancy policy.

Exactly to whom such intelligence is conveyed again will depend upon similar criteria, and on the agreed policy of the company. At corporate level the generally accepted procedure is to consult the representatives of the appropriate trade unions or groups of workers, agreeing in principle the application of work measurement. It is usual for management to reserve the right to use whatever technique of work measurement may be appropriate to individual studies. However, this has not deterred local shop-stewards from raising objections to certain methods.

Following corporate-level consultation, each application may be preceded by local discussions prior to the actual study. This involves meeting the section supervisor, through whom liaison with the section should be made. If nothing else this shows courtesy, and affords the supervisor his due respect as head of the section. The first study at least will be preceded by a short meeting between the manager, supervisor, and a representative from the area to be studied.

The initial contact between the practitioner and the section should be made through the supervisor, who introduces him to the operatives who are to be the subjects of the study. This establishes the authority of the supervisor and demonstrates departmental approval of the study. The supervisor will offer a few words of explanation to the operatives, who should already be aware of the possibility of being studied.

The onus is now upon the practitioner to proceed with the study.

4.32 CONSULTATION DURING THE STUDY

The technical aspects of taking a work-measurement study will be considered in succeeding chapters of this book. The present chapter is concerned solely with the 'human side' of such a project.

It is impracticable to formulate a set of rigid rules or code of behaviour to bind the practitioner during a work-measurement application. Nevertheless experience has highlighted certain proprieties which should be observed if conflict is to be avoided.

1. The practitioner should adopt a friendly attitude toward the subject of the study, conversing as the occasion demands. He should not be unduly talkative and should restrict his conversation to the early and latter parts of the study, allowing the operative to work without unnecessary distraction. This also enables the practitioner himself to concentrate on collecting accurate data

during the actual recording phase. Equally he should not force the conversation even on the introduction or conclusion of the study where it is apparent that this is not encouraged.
2. The practitioner should behave in a manner befitting a representative of management, being friendly without being too familiar. It is usually accepted that an observer should stand while making recordings, but exceptions to this rule are permissible, especially when the data are being collected through a protracted interview. But a slouching, hands-in-pockets attitude does not impress, nor does it inspire confidence.
3. Details of alterations and implementation of new methods should not be discussed with operatives, although this does not preclude sympathetic attention to ideas put forward by staff and due acknowledgment of these.
4. Similarly the practitioner should avoid being drawn into discussion about the data he has recorded such as observed times, ratings, and relaxation allowances.
5. One of the most potentially dangerous subjects to be raised will be that of wages and monetary payments. The wise practitioner will have little to do with such questions, politely but firmly advising the questioner to consult his supervisor on such points. Even vague mention of the possibility that increased earnings 'might result' from the study may be construed as a promise of extra money.
6. The golden rule on matters affecting policy should be to refer all questioners to management.
7. The occasional disgruntled operative in his search for allies will tempt the observer into the trap of criticizing management or company policy. At other times a practitioner, in his eagerness to gain the co-operation of the staff, may appear over-sympathetic. The unwitting practitioner, by his unguarded comments, will be creating a situation from which it may be difficult to withdraw subsequently and which could severely embarrass management in the event of a disagreement.
8. It is not uncommon for the practitioner to be made the butt of the jokes which are almost invariably part of the ritual. Alternatively, there will be occasions when sarcasm or even bitterness is shown toward him. The experienced observer will have heard it all before, and merely shrug off the comments, proceeding with his study. However, in extreme cases, which if the original consultations have gone well should be rare, he must never retaliate nor become involved in arguments, but should disengage himself and report back to the supervisor or his own superior. Should the practitioner meet with an unco-operative subject he should take similar steps and arrange for a different operative to demonstrate the job.

4.33 CONSULTATION DURING INSTALLATION

On concluding the study the practitioner politely acknowledges the forbearance of the operative and his co-operation during the study. Finally he will thank the supervisor.

It is usual for supervisors and their immediate superiors to discuss the results of the study with the practitioner, to agree the standards before their application in the area. It is not unusual for managements to invite shop-floor representation at such meetings. Indeed some organizations extend this invitation to cover the whole application, and include simultaneous study by practitioner and workers' representative.

The practitioner must be prepared to defend his findings, which, presumably, he has attempted to verify before attending the meeting. Results will almost certainly be challenged—by either side. In some instances of deadlock a compromise may be reached by the addition of policy allowance, but this is the prerogative of management. Work study must guard against attempts to adulterate the time values for the purposes of appeasement. Within accepted limits of error, ratings and observed times obtained by experienced practitioners cannot become subjects for bargaining, and cannot be adjusted at the whim of managers or to fit the circumstances. Genuine errors are occasionally inevitable and these, of course, must be rectified. Studies which are in doubt for good reason may be restudied or verified through the medium of the production study, but it should be emphasized that this is exceptional. Workers should not have recourse to this outlet as a general rule.

In most cases challenging the standard times is part of the ritual acted out by workers' representatives, who are assuming the role of 'devil's advocate', demonstrating that they are for ever vigilant. Regardless of the validity of standard times, the innate defence-mechanism often engenders a reluctance to accept without some resistance.

5 Basic Concepts of Measurement

5.1 OBJECTIVES

The main objective of work measurement is to establish a fair and acceptable time for a specified job which may be achieved by trained, qualified workers who are applying themselves with reasonable effort. Unfortunately many of the words used in this description are subject to individual interpretation and are without precise definition. One of the deficiencies apparent in the written word is its inability to describe qualities such as effort and speed in absolute terms, and such a description is fundamental to any definition of work measurement, implicit in the following B.S.I. account:

> Work measurement is the application of techniques designed to establish the time for a qualified worker to carry out a job at a defined level of working. (Term number 10004)

There is an unfortunate inconsistency in the above definition and the classification of activity sampling as a technique of work measurement in the same Glossary, because activity sampling does not satisfy the criterion of 'at a defined level of performance'. Moreover other systems of measurement, principally some of those described in Chapter 15, use *actual* times achieved for the work.

In general it is true to say that most techniques of work measurement specify the times for completion of jobs at a common and clearly defined level of performance. The obvious advantage in standardizing times in this way is that dissimilar jobs may be compared for work content. Once jobs have been reduced to a common level in terms of time and pace their standard times afford bases for planning and allocation of work-load, manning jobs, measuring operator and department effectiveness, implementation of incentive schemes, costing the labour content of jobs, and many other managerial controls. These are dealt with in Chapter 16.

5.2 STAGES OF MEASUREMENT

The measurement of work may be effected basically in three ways. The time required to carry out a task may be *estimated* using past experience of the work as a guide, or by comparison with specimen bench-mark jobs. Alternatively a chronometric device may be employed to time the work directly. The third fundamental method is through the application of predetermined motion–time systems, which provide tabulated times for the basic human movements.

Although standard times may be established by such diverse methods it is possible to identify common aspects in the procedure for arriving at the standard. The stages are known respectively as analysis, measurement, and compilation.

5.21 OUTLINE OF PROCEDURE

The first stage in a work-measurement application is the analysis of the work into its constituent *elements*. The size of elements varies with the type of measurement employed. For the purpose of estimating times, elements may be of the order of a minute, while those for time study usually range between 0·1 and 0·3 minute. At the extreme, a P.M.T.S. element such as 'grasp' occupies as little as one-thousandth part of a minute. The elements of micro-P.M.T.S. are often measured in terms of ten-thousandths of a minute (Work-Factor) or hundred-thousandths of an hour (Methods–Time Measurement). As a consequence of this wide variation in element size, analyses are individual to the mode of measurement employed.

In the second stage, measurement is effected in one of the three ways mentioned above. Elements are assigned basic times appropriate to the amount of work involved and the difficulty inherent in the work. The use of a timing system may require the application of rating to extend the observed or actual times to basic times, as described on p. 42.

Compilation* is the term for the reassembly of the elements previously analysed for purposes of measurement. The stage involves the computation of basic times for elements; the sum of average elemental basic times producing the basic time for the job. The standard time is computed, and the whole study 'worked up' and defined on some form of study sheet. The standard time is presented to management for use in planning and controlling work.

A fourth stage of selecting suitable elements for inclusion in the synthetic-data bank may be the logical conclusion of the study. Most departments of a firm are concerned with the execution of work of a similar nature, with the inevitable result that some elements are common to many of the jobs. If certain elements recur in other jobs then their basic times will be known and may be extracted from the data-bank, obviating the need for further timing, except for those elements for which synthetic data are not available.

5.22 EXTENSION OF OBSERVED TIMES

Systems of work measurement which are based on timing usually† require some form of adjustment to be made to the observed times in order to represent these times at standard rating. Observed times reflect the application of the worker to his task rather than define the time the average qualified worker needs to complete the task at a defined level of performance. Consequently, observed times do not present satisfactory measures for comparing different jobs in their 'raw' state, and must first be adjusted to represent the time the worker *would*

---

* This is sometimes known as 'synthesis', but this should not be confused with the application of synthetic (or standard) data. In this present book the term 'synthesis' is used in the context of the B.S.I. definition (see p. 268).

† There are exceptions to this rule, notably the system of VeFAC Programming described in Chapter 15.

have taken had he worked at standard rating. This adjustment is known as *extension*. Standard rating is defined by B.S.I. Term number 41025 as:

> The rating corresponding to the average rate at which qualified workers will naturally work, provided that they adhere to the specified method and that they are motivated to apply themselves to their work. If the standard rating is consistently maintained and the appropriate relaxation is taken, a qualified worker will achieve standard performance over the working day or shift.

An operator may take 0·4 min (observed time) to complete a cycle of the job, when working at half the standard rating. A fast worker may take 0·13 min while working at $1\frac{1}{2}$ times the standard rate of working. The first employee would have taken half the observed time had he performed at the standard rating, completing the cycle in 0·2 min. The second worker would have taken $1\frac{1}{2}$ times as long had *he* performed at standard rating, i.e. $1\frac{1}{2} \times 0·13$, or 0·195 min—almost the same extended time as before.*

From the above it is apparent that the extended time for the job is independent of both the observed pace of working and the person doing the work, that is, the extended time is established for the *job* and not for the person doing the job.

The result of this extension process is the *basic time*, and is calculated by adjusting the observed time by the fraction by which the worker's pace varies from the standard pace:

$$\text{basic time} = \text{observed time} \times \frac{\text{observed rating}}{\text{standard rating}}$$

The two components of basic time are:

Observed time, which is the actual time taken by the observed worker to perform an element or cycle of his task, measured on some form of chronometric device, and

Observed rating, which is the speed with which he performs the work. The methods employed to assess this speed are discussed later in this chapter.

5.3 ALLOWANCES

The basic time for a job yielded by the application of work measurement is the average time required for carrying out each cycle of the work. Over the shift or working day the worker will tire, and the rate at which fatigue overtakes him depends on the energy expended in doing the work. It is unreasonable to expect a worker engaged on heavy manual work to maintain a given performance as readily as another whose job is less demanding. The former will require relatively longer periods of rest to recover from the effects of his exertions. It is generally accepted that all workers need time to relax their efforts during the working period, if only to satisfy their so-called 'personal needs'. The tangible

---

* The discrepancy between 0·2 and 0·195 being due to error in assessing the ratings of $\frac{1}{2}$ and $1\frac{1}{2}$, and to subtle changes in motion pattern which occur at different speeds of working.

recognition of this exhibits itself in the form of the *relaxation allowance*, an addition to the basic time which effectively extends the basic time by a predetermined percentage.

The addition of a relaxation allowance of 10 per cent to a basic time of 0·5 b.m. increases this to 0·55, enabling the worker to relax his pace of working and reducing the rate at which he becomes fatigued. Alternatively he may maintain a high pace, using the accumulated allowance in periodic rest-periods. Although the basic time for a cycle is 0·5 b.m. the operator need only complete the cycles at an average of 0·55 min to achieve standard performance.

Some aspects of the job occur so infrequently or unpredictably that measurement would be uneconomic or impossible. Such elements are termed contingencies, and allowed for in a blanket *contingency allowance.*

While working in a team, or operating a group of automatic machines, it is almost inevitable that either operator or machine will be subjected to enforced delays owing to differences between machine-running time and the operator's working time. If the machine-minder has so many machines to attend that he cannot get around to them before they become idle, allowance is made for this loss of production in the form of a machine-interference allowance (described in Section 12.4). With fewer machines the worker will have insufficient work and must be compensated by way of an *unoccupied time allowance* (described in Section 12.4).

The allowed time is completed by the addition, where applicable, of policy allowance, or a bonus increment. The standard time should stand inviolate in the face of all pressure from outside to increase its value, provided the practitioner is genuinely convinced that the study and its computation are correct. On occasions, especially in conditions of incentive payment, management may feel that the standard time is such that the worker will not earn sufficient bonus to motivate him effectively. At its discretion management may decide to inflate the time by adding the further allowance. The allowed time is that time which, if achieved, will pay the worker a wage equivalent to what he would normally have earned had he completed the work in the standard time.

Other allowances which may be pertinent from time to time are *learner allowance* for compensating trainees for their inability to attain a reasonable performance during their learning period, and *changeover allowance*, which covers the waiting time occasioned by the changing over of equipment or material between jobs or batches of work. These allowances do not constitute part of the standard or allowed times, but are separately applied.

Allowances are described in more detail in Chapter 12.

5.4 OUTLINE OF WORK MEASUREMENT

The foregoing sections of this chapter have described the fundamental parts of measurement techniques. This section deals with the major components of all measurement systems, culminating in the individual descriptions of the three main divisions.

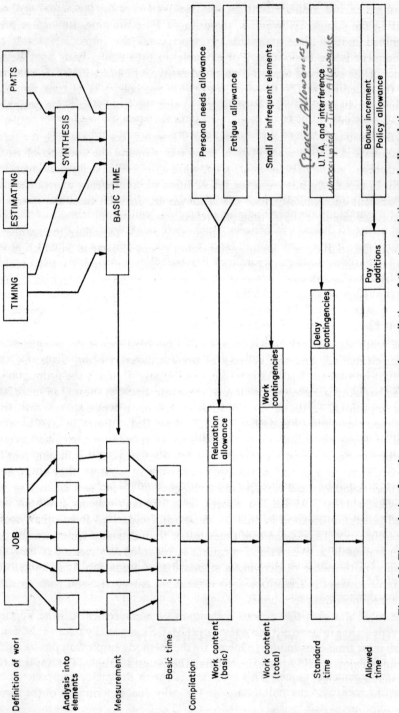

Figure 5.1 The analysis of work, measurement, and compilation of the standard time and allowed time

Most systems of measurement are very similar except for the actual method of setting the standards. Whether these be set by estimating, timing, or predetermined motion time standards, in most cases the immediate result of measurement is the basic time. For example, in time study, basic time is the product of the observed time and observed rating, whereas a P.M.T.S. analysis (with the exception of CWIP—*see* Chapter 15) produces a basic time directly, although the basic time may be adjusted to suit the local norm (for example, the conversion of the T.M.U. values of MTM to 100 B.S.I. values).

To the basic time is added the relaxation allowance (*see* Figure 5.1), the total being known as *work content: basic*. Some jobs demand the addition of work contingencies, giving *work content: total*, while others need delay contingencies as well. In those jobs not requiring the addition of contingency allowances the work content and standard time are synonymous. In such cases the standard time consists merely of basic time plus relaxation allowance.

Finally, for payment purposes the firm may decide to inflate the standard time with the addition of a bonus increment or policy allowance, in which event the inflated time is known as *allowed time*. Additionally, either of the two process allowances for interference or unoccupied time may be appropriate, and supplement the standard time.

5.41 METHODS OF WORK MEASUREMENT

The methods of work measurement fall into the three main categories of timing, estimating, and the systems of predetermined motion–times. Of the multitude of systems, both basic and derived (many of which are proprietary), all may be traced back to one or more of these categories, as shown in Figure 3.1.

The traditional textbook classification of the main techniques is into five categories: time study, synthesis, P.M.T.S., analytical estimating, and activity sampling. However, this division is rather restrictive because it excludes certain important techniques and proprietary systems. Furthermore it is inconsistent in including some generic and some specific terms. For example, time study and analytical estimating are two specific techniques of timing and estimating respectively, whereas P.M.T.S. is a generic term for many systems. Synthesis is a second-generation method in that its standards are derived from other, more basic techniques. Unless *observed* times rather than extended times are acceptable as standards, activity sampling cannot be regarded as a method of measuring work in the sense of deriving a standard time for a job. It is, therefore, preferable to use the separation into three main categories, and then to subdivide into individual techniques.

The rates at which the various techniques of measurement can be applied vary widely, but in general the systems of P.M.T.S. (especially micro-P.M.T.S.) are the most time-consuming, followed by the methods employing timing, with estimating enjoying the advantage of being the fastest to apply. The application ratio may be defined as:

the ratio of *time required to set the standard time*, to the *standard time itself*

Results of research carried out by the present author reveal the ratios given in Figure 5.2.

System	Approximate order of magnitude of application ratios	Approximate study time required to measure analyse, compile, and issue a job basic time for a cycle of 15 min
1st level P.M.T.S.	1 : 150 to 1 : 350	60 hr
2nd level P.M.T.S.	1 : 70 to 1 : 150	20 hr
3rd level P.M.T.S.	1 : 30 to 1 : 50	10 hr
Macro-P.M.T.S.	1 : 15 to 1 : 40	8 hr
time study	1 : 10 to 1 : 120[1]	5 hr
rated activity sampling	1 : 8 to 1 : 80[1]	4 hr
analytical estimating	1 : 0·4 to 1 : 3	0·2 hr
comparative estimating	1 : 0·005 to 1 : 0·05[2]	0·03 hr
category estimating	1 : 0·001 to 1 : 0·015[3]	0·015 hr

Notes: (1) Depends on the variability of the work, and hence the number of cycles studied.
(2) Does not include the time to set up the bench-marks and associated standard times.
(3) Does not include subsequent statistical checks on the quality of estimating.

The study times in the last column are for jobs of 15 min cycles in which all elements are dissimilar, and do not repeat within the cycle.

Figure 5.2 A comparison of speeds of application for various methods of work measurement

Of course to achieve increases in speed of application sacrifices in other directions are inevitable; the victim in most cases being precision of measurement. In many work areas such as maintenance engineering, and work which is essentially long-cycle or variable in nature, this inexactitude may cause little concern, especially over a long period of application. While the standard time may be hopelessly inadequate to describe individual cycles of the work, over many such cycles it may afford a representative yardstick for purposes of comparing *average* times taken.

5.42 PREDETERMINED MOTION TIME SYSTEMS

Originating in the 1930s, the systems of predetermined motion–times (P.M.T.S.) have been slow to gain recognition as a major technique of measurement in this country. Time study still retains its place as the premier system in spite of its subjective rating-factor. However, the reluctance of British managements to accept P.M.T.S. may be due in part to the high application ratio of the detailed systems. To some extent this objection has been nullified with the advent of second- and third-level systems.

On the assumption that a job will be performed in exactly the same basic time on each occasion it is carried out (provided that the conditions and motion-pattern remain unaltered), it is possible to pre-record basic times for fundamental human motions in tabulated form on data-sheets. These elemental times are derived from extremely large samples of different jobs, operators, and observers, thereby covering widely differing conditions and largely cancelling effects of bias.

Subsequent analyses of jobs enable the analyst to extract the appropriate elemental times from the relevant data-sheets. When summed the elemental times produce a basic time for the complete job.

Most predetermined time-data are considered to have universal application. The second P.M.T.S. division is that known as *synthesis*, the data for which is being prepared for use within the firm from which it originated. Synthetic data (sometimes known as standard data) do not exhibit the same degree of precision as micro-P.M.T.S. They are of different magnitude, being of the order of minutes rather than the milli-minutes of micro-P.M.T.S. elements. Banks of synthetic data are compiled from previous work-measurement studies taken in the firm.

5.43 TIMING

Jobs may be timed *in situ* using some form of chronometer such as electronic timer, stop-watch, or wall clock. The most widely used system in this category is time study, a system of stop-watch timing carried out while observing the worker actually performing the work.

The job is first analysed into smaller 'elements', the start and finish of which are defined by break-points. Elements are then timed separately to produce element 'observed times'. Time study is distinguished by the process of applying *rating* (or pace assessment) to convert the observed times into basic times at a defined level of performance.

Rating is a feature also of a somewhat controversial system called rated activity sampling. This method is an extension of time study; a hybrid from time study and activity sampling. Unlike time study, activity sampling does not recognize elements as natural periods for timing, but sets its own, either at random or at regularly spaced intervals.

5.44 ESTIMATING

By far the fastest, and equally the most coarse, group of techniques for setting standard times are based on estimating. Of the several variations of estimating two are prominent.

Analytical estimating is a more refined method in this category, in which the job is broken down into elements which are separately estimated to produce a more appropriate standard. The second important method, *comparative* estimating, relies on the establishment of specimen 'bench-mark' jobs acting as reference levels against which jobs being studied may be compared for work content.

5.5 RATING

None of the methods of timing produces anything but the times actually being taken by operators to perform their tasks. In some systems this is acceptable (*see* Chapter 15), but in techniques such as time study and rated activity sampling the basic time required is not that *actually* being taken, but that which would be taken had the worker been applying himself with the required degree of effort. This necessitates the modification of the observed times by a factor determined by the pace at which the worker was performing when observed. The process of assessing the value of this factor is known as *rating*.

> To rate is 'to assess the worker's rate of working relative to the observer's concept of the rate corresponding to standard rating. The observer may take into account, separately or in combination, one or more factors necessary to the carrying out of the task, such as: speed of movement, effort, dexterity, consistency.' (B.S.I. Term number 41027)

According to the stop-watch readings one operator may be performing successive cycles of his task at an average time of 0·50 min while his colleague may be taking 0·40 min over the same work. A third worker may succeed in completing each cycle in 0·36 min. The task of the work-study observer is to select a time which is most representative of the work.

One method of realizing this objective is to take an average of the three observed average times. The average, 0·42 min, may be unrealistic if the three workers were extending themselves beyond that which one would consider to be a reasonable pace for the work and the prevailing conditions. On the other hand they may be working extremely slowly, very slowly, and slowly, respectively. In the first case the time of 0·42 min is the average for very fast-working, and in the second example for very slow workers; neither of which could be regarded as acceptable standards.

If the concept of 'standard rate of working' is accepted then it must be possible to express the times produced by the above (and all) workers engaged on a common task and working under identical conditions and to the same method, as a unique basic time by the modification of rating.

Rating is a purely subjective assessment of rate of working (even objective rating demands a subjective comparison), and the observer can expect no assistance from any form of instrument when making his judgment. In order to perform his rating assessment the observer must compare the performance of the worker in carrying out each element of the job with a preconceived mental standard. Thus the basic time determined through the application of this factor is dependent upon the competence of the observer to perform this comparison, and on his ability to retain the mental image of the standard pace. Once this mental image of the standard rate of working has been appreciated by the observer he is able to judge the rate he is observing as a percentage of the standard, with an accuracy and consistency according to his competence. Not all rating is assessed in terms of percentages, and it is instructive to study some of the existing rating scales.

5.51 RATING SCALES

During the early part of this century when rating was in its infancy little attempt was made to standardize procedure, and consequently several methods of rating and of recording these ratings were evolved. Most of these were based on some sort of rating scale defined by two theoretically 'fixed' points.

One of the first scales was based by Charles E. Bedaux on a points system where 60 signifies the 'normal' rate of working (*see* Figure 5.3). Bedaux considered that workers paid on a daywork system (i.e. no financial incentive) will perform 60 minutes' worth of work in an hour. Hence a worker performing at the 'incentive' rate assumed to be one-third faster than this normal rate, would do an extra 20 minutes' worth of work in an hour, that is, 80 minutes of work. The rating which achieves this output is referred to as 80 B's, and the performance of this worker as 80 B hour. This scale is otherwise known as 60/80 scale. In the nineteen-twenties and 'thirties it was adopted by many firms, some of which still retain the scale in preference to the other, later scales.

Other scales have their bases in 100, which is considered to be a more manageable and logical standard, facilitating calculation and allowing rating to be

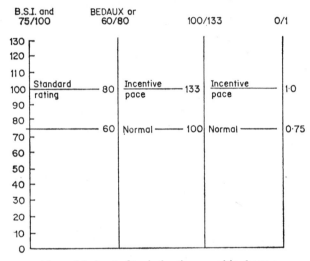

Figure 5.3 A set of typical ratings, used in the text

made as percentages of the standard. On a scale widely used in the United States and to some extent in the United Kingdom the normal level is defined by 100 rating. Unfortunately the equivalent to Bedaux's 80 rating is 133⅓ (100 + 33⅓). Consequently the scale is known as the 100/133. Other researchers put the incentive level at a slightly lower level of 125. There has been a tendency in recent years to extend the basic times at the higher rating level which necessitates the use of 133, 80, or 125 in the denominator of the calculation. Clearly the computation is facilitated if the higher level is denoted by 100. As the scales are linear the 'normal' level becomes 75 in this, the 75/100 rating scale. An adaptation of the scale was the one adopted by the British Standards Institution.

On the B.S.I. (0/100) scale the 100 rating is referred to as standard rating and is defined by B.S.I. Specification 3138 (Term number 34016). The definition is given on p. 42. Thus rating is a tempo of working at any instant; consequently standard rating is a defined level of instantaneous rating. On the other hand, if a worker completes over a defined period an amount of work which corresponds to that which he would have done had he worked at an average rating of 100 B.S. and taken his approved allowances, he is said to have made standard performance:

> The rate of output which qualified workers can achieve without over-exertion as an average over the working day or shift provided they adhere to the specified method and provided they are motivated to apply themselves to their work. (B.S.I. Term number 51004)

5.52 B.S.I. PREFERRED SCALE

The Mechanical Engineering Standards Committee were responsible for defining and standardizing work-study terms in their B.S. Glossary 3138, revised in 1969. To achieve consistency it was desirable to rationalize the range of rating scales available. Faced with the problem of selecting the most suitable scale which could be recommended for use nationally the Committee decided to adopt the existing 0/100 scale, advancing the following reasons in support of their choice.

Earlier it was shown how the arbitrary levels of 'normal' and 'incentive' rates originated. The M.E.I.S. Committee rightly question the validity of the assumptions that time-workers perform at three-quarters of the pace of workers on incentive payment. People do not work at a universally specified pace but at their own individual levels, dictated by their motivation at that particular time, by the conditions obtaining, or by their ability and other factors. Therefore the value of the actual pace which is designated as *standard* is not important, as it is to be used merely as a reference level against which individual ratings may be assessed. However, the situation will be more realistically described if the standard is set at the point around which 'average qualified workers' who are adequately motivated will work naturally. Since the scale is linear it is necessary only to define one point (the standard rating) in addition to zero. The Committee felt that the 75 rating had no special significance.*

5.53 FACTORS TO BE RATED

Rating is necessary to counteract the attributable or voluntary changes in the operator's pace of working. Several researchers in the field have attempted to establish satisfactory methods of assessing this variable. Among the best known are Maynard's Westinghouse levelling procedure, Mundel's objective rating, and Nadler's pace comparison. The method most widely used in the United Kingdom was derived from the work of Charles Bedaux, and is known as 'speed and effort rating'. This latter technique recognizes four factors which must be rated simultaneously. Latest information theory has cast some doubt on the ability of the human observer to rate four factors on many levels of pace. The other systems have the advantage of rating only two factors or even one factor at a time.

* From B.S.I. 3138/1959

Opinions vary as to what is to be rated, but with speed and effort rating some measure of agreement has emerged.

Speed of movement. The most obvious factor to rate is the pace or tempo of the worker's manual movements. Pace is the primary factor, and the one which is predominant in manual work. In most systems of rating the speed of movement is modified by the degrees of difficulty inherent in the work, such as the following:

Effort. In some techniques of rating, notably speed and effort, and Westinghouse levelling, it is appreciated that the pace of working may be affected by other factors, including the physical or mental exertion expended by the worker in carrying out the work. While a man may walk at about six feet per second without unduly exerting himself, the same person carrying a hundredweight bag of cement will be hindered, taking longer to cover the same ground. When rating for speed and effort the rating depends on the amount of work done or energy expended. Therefore the observer must compensate for lack of pace by taking into consideration the effect of the hindrance.

Consistency. One of the reasons for between-element variations in the observed times is non-adherence to the defined motion pattern for the job. Thus it is necessary for the observer to adjust his rating to compensate for the deviations from the motion pattern specified. Similarly, inconsistency in speed of movements can affect the observed times and must be duly taken into account.

Dexterity. A worker with good dexterity and anticipatory perception will perform more productive work than one working with the same effort but with poor dexterity; much of his effort being expended on non-productive fumbling movements.

Job difficulty. In some systems the above factors and others are combined into a comprehensive factor of job difficulty. Mundel's objective rating is such a method.

5.54 RATING LEVELS

Rating scales in theory are continuous variables, but in practice are discrete because of the inability of observers to rate only on certain levels. Usually an observer is expected to assess performance to the nearest five points, but rating in steps of ten may be more realistic if consistency is to be maintained. Some researchers, notably Kanon, argue that the average human channel-capacity (i.e. the amount of information he can appreciate) is seven levels, and the average human being is incapable of differentiating between information on more than seven levels.

The observer is expected to rate only when he sees workers performing at ratings between 70 and 160 (on the B.S.I. scale). The reason for this is that most people have difficulty in mentally assimilating and recognizing performances which fall outside limits of working commonly encountered in day-to-day life, i.e. very high or very low paces of working.

Rating to within five points gives nineteen increments, whereas rating to the nearest ten points produces ten increments, which is more readily attainable.

However, most operatives perform close to the 100 rating, thereby reducing the range of ratings. Furthermore, the most important question is whether the observer is able to distinguish between performance levels differing by a mere 5 per cent, especially when the reference point is but a mental image.

It is difficult to describe the process of rating, that is, whether the observer rates absolutely, or compares the tempo he sees against his concept of standard rating. Whatever the procedure, the worker requires a comprehension of the standard rating level. Some authorities insist that the standard should be particular to each individual firm, and based on the prevailing tempo. But the most widely favoured is the concept of a universal standard rating label defined by bench-mark rating films. Advocates of individual standards support this concept because they doubt the abilities of employees of low-tempo firms to achieve the standard. However, if the universal standard be accepted and basic times set at this level the inclusion of *target times* which may be lower or above the standard is not precluded. Thus the standard rating may be 100, but the target rating (i.e. that which the firm *expect* of the employee) may be 80.

The target time is applicable to learning allowances, where the trainee cannot be expected to reach the vicinity of standard rating until experience has been gained. A learning curve based on his projected progression may be compiled, and from this, target times which the trainee should achieve daily or weekly may be found by interpolation.

5.55 RATING CONSISTENCY

Inconsistency in rating, which is the cause of much concern among work-measurement practitioners, has been the subject of research projects over the past forty years. Some of the earlier researchers include A. J. Keim and R. H. Lehrer.

A study in 1954 by W. Rogers and J. M. Hammersley of groups of observers rating particular jobs established a range of about plus and minus 40 per cent of the mean value of a particular rating (with a 95 per cent confidence level). This indicates a spread between 60 and 140 rating on a job which the group on average rated as 100.

The first really comprehensive study of rating was made in 1950 by researchers of the College of Engineering, New York University, for the Society for Advancement of Management (S.A.M.). Notable figures in the rating committee were Marvin E. Mundel and Ralph Barnes, both pioneers in work study, and H. B. Maynard and G. J. Stegemerten—originators of Methods–Time Measurement.

The objective of the survey, which was carried out on a nation-wide basis, was to establish a universal definition of a fair day's work. The medium for comparisons was the ciné-film. Entitled 'Rating of Time Studies', the film embodied twenty-four operations of widely varying, but short-cycle jobs. The film was to become recognized as the authoritative standard and is still widely used in training work-measurement practitioners to this day. As each operation was depicted at five different rating levels it was possible to obtain 120 assessments from each of the 1,800 observers working throughout the United States. The results of the survey showed that the concept of a universal standard was a valid one, and that there was a significant degree of consistency among observers in America.

Professor N. A. Dudley,* writing in 1968, describes a similar rating survey which was carried out in the United Kingdom at about the same time as the American survey. In October 1949 Thomas U. Matthew established the Work Measurement Research Unit, which made its pilot study in 1950 with the objective of collecting the relevant data.

The actual survey, which again was nationwide, was carried out in 1950–51. The objective was to investigate the standards of consistency and accuracy obtaining in time-study rating practice achieved by individual engineers, by firms, and by industries. Other objectives were to determine the allowances required as compensation for fatigue or other factors in different occupations and under different working conditions.

Most of the 750 observers drawn from 286 firms were from manufacturing industries. Fortunately the Birmingham researchers were able to secure the loan of films used by the American surveyors. Nine of the twenty-four operations were selected for the U.K. survey, the others being regarded as unsuitable for various reasons.

5.56 WESTINGHOUSE SYSTEM

A system developed for Westinghouse by Lowry, Maynard and Stegemerten, (also known as the L.M.S. rating system) adjusts the average observed times in retrospect after the readings have been taken. The observed times are levelled according to four factors, of skill, effort, conditions, and consistency. Rating factors are assigned according to the degree of each factor exhibited by the worker, and conditions of work pertaining. The factors are tabulated in Figure 5.4. Thus an operator working with good skill ($C1 = +0.06$), good effort ($C2 = +0.02$), with excellent consistency ($B = +0.03$) under average conditions ($D = 0.00$), is assessed as working at:

$$0.06 + 0.02 + 0.03 + 0 = +0.11, \text{ or } 111 \text{ per cent rating}$$
(i.e. 11 per cent above the normal 100 rating)

This value is applied in the usual way for extending observed times.

Although at first sight the system appears to inject into rating an element of objectivity through the use of tabulated data, the application of the data still remains largely a matter of individual judgment. Furthermore, subtle variation in the motion patterns used may not be appreciated when assessing skill and effort. This last disadvantage, of course, is not unique to the Westinghouse system.

5.57 OBJECTIVE RATING

The process of rating is notorious for the subjectiveness inherent in its performance. Many attempts have been made to institute some form of objective

* N. A. Dudley, *Work Measurement: Some Research Studies* (London: Macmillan, 1968).

basis on which rating can be founded. Little success has been attained in creating a 'time study machine', but one attempt at reducing the subjective element is Mundel's *objective rating*.* Whether in fact the method satisfies the criterion of 'objectiveness' completely is very doubtful.

Mundel, while accepting that the operator's pace must be a function of factors such as skill, aptitude, and others, nevertheless refutes the contention that these can be assessed separately because of their interdependence. Consequently, he argues, all those factors which affect the pace of working are really contributing to *job difficulty*.

SKILL			EFFORT		
+0·15	A1	Superskill	+0·13	A1	Excessive
+0·13	A2		+0·12	A2	
+0·11	B1	Excellent	+0·10	B1	Excellent
+0·08	B2		+0·08	B2	
+0·06	C1	Good	+0·05	C1	Good
+0·03	C2		+0·02	C2	
0·00	D	Average	0·00	D	Average
−0·05	E1	Fair	−0·04	E1	Fair
−0·10	E2		−0·08	E2	
−0·16	F1	Poor	−0·12	F1	Poor
−0·22	F2		−0·17	F2	

CONDITIONS			CONSISTENCY		
+0·06	A	Ideal	+0·04	A	Perfect
+0·04	B	Excellent	+0·03	B	Excellent
+0·02	C	Good	+0·01	C	Good
0·00	D	Average	0·00	D	Average
−0·03	E	Fair	−0·02	E	Fair
−0·07	F	Poor	−0·04	F	Poor

Figure 5.4 Performance-rating table ('Westinghouse') for rating (reproduced by kind permission of McGraw-Hill Book Company, New York, from S. M. Lowry, H. B. Maynard, and G. J. Stegemerten, *Time and Motion Study, and Formulas for Wage Incentives*, 1940)

Objective rating attempts to simplify the task of the observer by reducing the factors to be rated to just those of pace and job difficulty. Furthermore it attempts to replace the subjectiveness of comparing job skill, effort, consistency, and other factors with mental standards, with a method based on objective measurements. Two distinct steps are identified in objective rating as a consequence of this:

Step 1. The observer makes his assessment of the speed or pace of working as he sees it, without concern for any other limiting conditions which may retard the worker. The pace is assessed against what is alleged to be an objective pace standard. The rating is applied in the usual way to modify the observed time.

* Marvin E. Mundel, *Motion and Time Study* (New Jersey: Prentice-Hall Inc., 1960).

Step 2. A second adjustment is applied as a percentage increment from Tables to allow for difficulties and restrictive factors which may be present in the work.

The objective pace-standard required by Step 1 is defined by just one 'benchmark' job which uniquely illustrates the speed of movement necessary to achieve the standard rating. This bench-mark may be selected in various ways, but usually does not purport to represent a *universal* standard. Instead it is peculiar to the firm in which the standard is to pertain, being mutually agreed by the interested parties (management, observers, unions, and others). However, the choice of a national or universal standard is not entirely excluded, and some firms prefer to adopt this concept.

The media for demonstrating the standard bench-mark pace are the actual demonstration, and the use of films which depict the job at various paces of working. One such movie is the 'step film' loop, which shows step-by-step variations in speeds for carrying out the bench-mark job. The *multi-image* film projects several paces simultaneously in the same frame by dividing the frame into say four, six, or twelve areas, one for each pace. One of the scenes is selected as the standard pace (say 75 B.S.), and the other scenes then assume their proportionate ratings based on this standard.

The concept of difficulty in performing tasks was of course appreciated in the early days of P.M.T.S.; the Work–Factor system, developed in the 1930s, and others emerging later, introduced adjustments to the basic elemental times in the form of degrees of difficulty to which the operator was subjected in the performance of his task. In a similar way difficulty is assessed in objective rating. Some factors suggested by Mundel are (*a*) amount of body used, (*b*) foot pedals, (*c*) degree of bimanualness (i.e. use of two hands assisting or working simultaneously), (*d*) eye–hand co-ordination, (*e*) handling or sensory requirements, (*f*) weight or resistance effects.

The above factors must be assessed separately for each element, and the percentage adjustments read from the appropriate tables. Finally these are summed to give the total percentage difficulty-adjustment.

5.571 Performing the rating. The foregoing outline has described the procedure for setting up the conditions for performing objective rating. In practising the method the observer studies the elements of the job and assesses the observed pace against his concept of the standard rating as defined by the film loop. Alternatively it is suggested that the observer compares a film of the observed pace and compares this simultaneously with a multi-image film loop. While excellent in intention, this may be somewhat expensive in time and money for some firms, and may only be used, of course, for block or overall rating because each element will be performed at a different rating.

It must be appreciated that the observer is concerned solely with pace of movement; degree of difficulty is assessed separately.

After adjusting or extending the observed time by the pace rating, this second adjustment for difficulty is effected. The observer selects the appropriate percentages from the tables of difficulty and sums them to obtain the total adjustment.

This is calculated and applied subsequently to the pace adjustment and cannot be added to it. Thus the pace may be assessed as 120 per cent and the difficulty as 15 per cent. The observed time is first multiplied by 120 per cent and the result multiplied by 115 per cent. The required basic time cannot be derived in one calculation of multiplying by 135 per cent.

Advocates of objective rating claim greater accuracy and less variability for the system over comparable methods of rating. While it is indisputable that direct comparison of filmed observation against a standard film must facilitate the rating assessment, there still remains a certain degree of subjectiveness in the method of comparison, however much exposure to the standard the observer may have experienced.

5.58 SYNTHETIC RATING

It is well known that the 'average, trained, experienced' operator works with good consistency when the conditions are conducive to this. An interesting approach to synthetic rating was suggested by the American industrial engineer R. L. Morrow,* who used the ability of people to work consistently to advantage in his *synthetic rating* system. Morrow's method bases its assessment not on subjective assessment, but on time values obtained from P.M.T.S. analyses of selected elements.

On the assumption that the operator works at a consistent pace through all elements a rating factor is obtained from comparing the observed time for the element with the P.M.T.S. derived basic time. Thus:

$$\text{rating factor} = \frac{\text{P.M.T.S. derived basic time for the element}}{\text{average observed time for this element}} \times 100$$

This rating factor is applied to other-element average observed times for the job. The system is applicable only if the original assumption of consistency is valid, and some researchers, notably J. L. Schwab, have cast doubt on the validity of the method.

Example

A job is analysed into five elements for the purposes of work measurement. A time study produced the following average observed times: Element A = 14 cmin, Element B = 8 cmin, Element C = 15 cmin, Element D = 20 cmin, Element E = 12 cmin. An MTM analysis of Element D was made, suggesting a basic time of 360 t.m.u. (equivalent to 0·18 b.m. when converted to B.S.I. rating).

$$\text{Therefore the rating factor} = \frac{18}{20} \times 100$$
$$= 90 \text{ rating}$$

Applying this to all elements:

Basic time for element A = 0·14 × 90 = 0·126 b.m.
element B = 0·08 × 90 = 0·072 b.m.

* R. L. Morrow, *Time Study and Motion Economy* (New York: Ronald Press, 1946).

element C = 0·15 × 90 = 0·135 b.m.
element D = 0·20 × 90 = 0·180 b.m.
element E = 0·12 × 90 = 0·108 b.m.

5.6 RATING ERRORS

Rating checks may be performed by comparing the observer's rating of elements demonstrated by actual performance, or from films projected at a carefully controlled speed, and for which the actual ratings of the scenes depicted are known. The comparison may be performed using any of the recognized mathematical or statistical analyses.

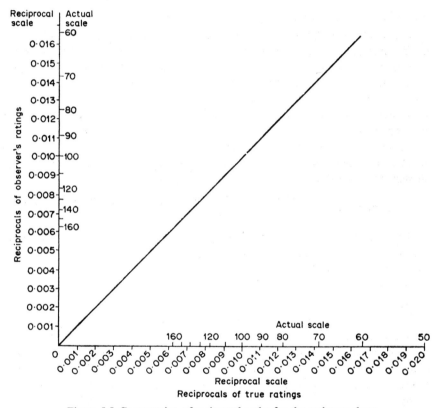

Figure 5.5 Construction of reciprocal scales for the reciprate chart

By far the most obvious methods of demonstrating errors are the graphical analyses, although rating parameters afford easier means of objective comparison.

Graphical representation is achieved by plotting the *reciprocal* of the observer's rating on the y or vertical axis, and the reciprocal of the true rating on the x (horizontal) axis (Figure 5.5). The reason for using reciprocals is given later in this section. It is from the graph compiled in this way that the rating errors have received their names. Instead of converting each rating into its reciprocal and plotting it on a linear graph, a reciprocal conversion-scale may be added, as shown in Figure 5.4.

There are three errors to which rating may be subject, and these classifications are applied to groups of readings but are inapplicable to individual values.

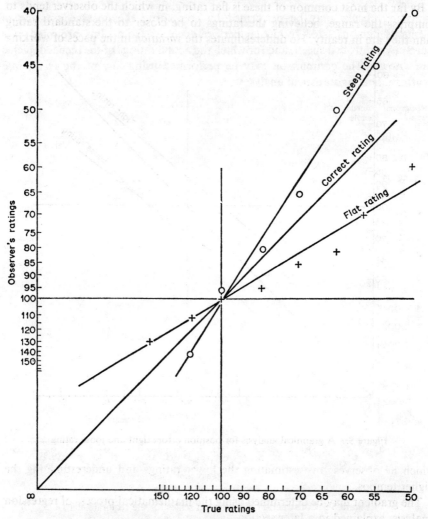

Figure 5.6 A graphical analysis for gradient error: flat and steep rating

Gradient error. In Chapter 6 it was stated that any straight line on a graph may be defined by the equation

$$y = mx + c,$$

where m is the gradient of the line, and

c is the constant which fixes the position of the line.

BASIC CONCEPTS OF MEASUREMENT 59

When the line describes the observer's rating ability, the gradient (m) indicates the uniformity of rating. Assuming equal scales on both axes, correct rating will produce a line with a gradient of unity. Gradients of more than 1 are indicative of *steep rating*, while a shallow line with a gradient of less than 1 describes *flat rating* (Figure 5.6).

By far the most common of these is flat rating, in which the observer tends to compress the range, believing the ratings to be closer to the standard rating than they are in reality. He underestimates the *variation* in the paces of working

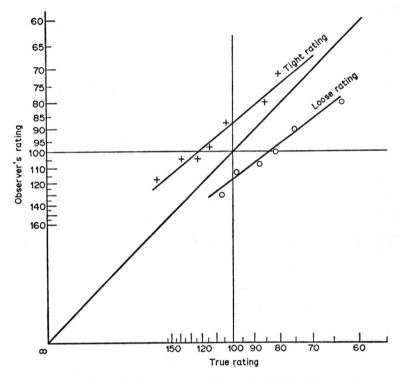

Figure 5.7 A graphical analysis for position error: tight and loose rating

which he observes; overestimating the lower ratings and underestimating the higher tempos.

The gradient may be determined using the mathematical process of regression analysis, explained in a later section.

Position error. In addition to a gradient error, rating is subject to a further error of position. All ratings may be over- or under-estimated; the errors being known as *loose* and *tight* respectively (Figure 5.7).

Consistency. Rating may be consistent or erratic, that is, an observer may rate consistently tight, consistently loose, consistently flat or steep. Errors which occur with consistency may be compensated for by mental adjustment by the observer as long as he is aware of his defects. On the other hand erratic rating shows inability to rate effectively and is difficult to correct. Consistency is improved by experience and practice.

Rating consistency may be assessed in several ways, including the use of co-efficients of correlation and regression (*see* Chapter 6).

One method of analysis, based on the reciprate method, was devised by D. J. Desmond of the University of Birmingham, England, and published by

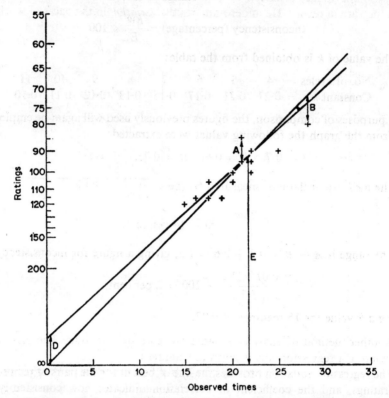

Figure 5.8 A graphical method for measuring rating-consistency and degree of flatness (by permission from N. A. Dudley, *Studies in Management—Work Measurement, Some Research Studies*, London: Macmillan, 1968)

Professor Dudley.* The method is a graphical interpretation, measuring flatness and inconsistency. A normal reciprocal rating chart is used (see Chapter 9, Section 9.25) on which are plotted the ratings and times to be analysed. A line is plotted through the origin so that it passes through the middle of the plotted points (Figure 5.8). This is called the Estimated Operation Line. A second line—the Study Line—is now plotted as a line of best fit for the points. The values of *A*, *B*, *D*, and *E* are measured on the graph and used in the equations given below. The measurements are in cm. *A* is the point which lies at the maximum

* N. A. Dudley, *Studies in Management—Work Measurement: Some Research Studies* (Macmillan: London, 1968).

distance above the Operation Line, and B is the point at the maximum distance below the Operation Line.

The measures of quality of rating are:

$$\text{Flatness (percentage)} = \frac{D}{E} + 100$$

$$\text{Range} = A + B = w$$

$$\text{Inconsistency (percentage)} = \frac{kw}{E} \times 100$$

The value of k is obtained from the table:

No. of cycles =	4	5	6	7	8	9	10	11
Constant k =	0·27	0·21	0·17	0·15	0·13	0·12	0·11	0·10

For purposes of comparison, the figures previously used will again be employed. From the graph the following values were extracted:

$$A = 0\cdot7, \quad B = 0\cdot6, \quad D = 0\cdot75, \quad E = 5\cdot5.$$

The measure of flatness, measured by the ratio $\frac{D}{E} \times 100$ is

$$\frac{0\cdot75}{5\cdot5} \times 100 = 14 \text{ per cent.}$$

The range is $A + B = 0\cdot7 + 0\cdot6 = 1\cdot3$, giving a figure for inconsistency of:

$$\frac{0\cdot07 \times 1\cdot3}{5\cdot5} \times 100 = 2 \text{ per cent,}$$

using a k value for 15 readings of 0·07.

Another method of analysis is based on regression analysis. The method of calculating a regression line is given in Chapter 6.

The regression equation provides the line of best fit for the pairs of reciprocals of ratings, and the coefficient of correlation indicates how consistently the ratings have been assessed.

An indication of gradient error may be obtained from the gradient (m) of the regression line.

Perfect rating-ability, in terms of consistency, is indicated when the coefficient of correlation is unity. This coefficient may be used to compare the abilities of different observers, or to chart the progress of individual observers. As a general rule a beginner at his first attempts will range between 0·82 and 0·90. The experienced practitioner usually achieves between 0·96 and 0·98. These results are applicable to sample size of 25 rating pairs, such as are found in the standard rating-film sequences.

The method of regression analysis is particularly suitable if a computer link or 'table top' computer (such as the Olivetti Programmer 101) is available in the work-study office.

5.61 WHY A RECIPROCAL-RATING CHART?

Unfortunately the analyses which use the observer's and corresponding control ('true') ratings plotted on linear scales are not valid if the observer is rating either flat or steep.

Several writers* have discussed rating analyses and have pointed out the fallacy of using linear graphs for this purpose. The argument is based on the theory of Desmond's Reciprate Analysis. As shown in Chapter 9, if the observer plots his ratings against observed times his readings should describe part of a hyperbola (Figure 9.2), because:

$$\text{basic time} = \frac{\text{observed time} \times \text{observed rating}}{\text{standard rating}}$$

i.e.,
$$b = \frac{r.t}{s}$$

If the rating is correctly carried out, b and s are constants, thus $r.t = k$ (a constant), and this is the equation of a rectangular hyperbola.

Let R_o = the observed rating, and R_t = the true rating, which corresponds to R_o.

then,
$$R_t.t = k$$

and,
$$\frac{1}{R_t} = \frac{t}{k} \tag{A}$$

This is the straight line for accurate rating if t be plotted against $(1/R_t)$.

On a reciprocal-rating chart this straight line passes through the origin with a slope of $1/k$ (Figure 5.6).

If rating is flat with a slope of $1/k'$ then the line will not pass through the origin but will have an intercept on the y (R) axis. If c be that intercept (Figure 5.5) then:

$$\frac{1}{R_o} = \left(\frac{1}{k'}\right)t + c. \tag{B}$$

If it is required to study the relationship between observer's rating and true rating rather than observer's rating and observed time, then an equation which excludes time is preferred. Substituting equation (A) in equation (B)

$$\frac{1}{R_o} = \frac{k}{k'R_t} + c$$

$$= \frac{k + k'R_t.c}{k'R_t}$$

therefore
$$R_o = \frac{k'R_t}{k + k'R_tc}$$

* D. J. Desmond, 'The Statistical Approach to Time Study *Proc. R. Statistical Society*, 1950, p. 36; D. J. Desmond, 'The 1950 National Survey of Work Measurement', *Work Study and Industrial Engineering*, vol. 6, 1962; B. Moores, *The Assessment of Work Load* (M.Sc. Thesis, Manchester University, 1964).

BASIC CONCEPTS OF MEASUREMENT

$$= \frac{R_t}{\left(\dfrac{k}{k'} + R_t c\right)}$$

This would result in a curve if plotted on linear graph-paper, so it is necessary to use a chart with *both* axes calibrated in terms of reciprocals.

In order to illustrate the shape of this curve some actual results will be considered. Suppose a job has a basic time of 50 centiminutes at a rating of 100

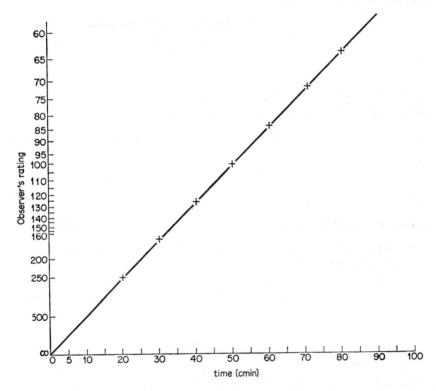

Figure 5.9 The line of perfect rating

B.S. Then this job, performed at 50 B.S. rating (assuming there are no changes in method, etc.), will take 100 cmin, and so on. A table may be constructed for a range of times and corresponding ratings for this job:

Rating	50	55·5	62·5	72	83	100	125	167	250
time (cmin)	100	90	80	70	60	50	40	30	20

These values, when plotted on a reciprocal-rating chart (Figure 5.9), produce a straight line. The time scale may be replaced by another reciprocal-rating scale for the reciprocals of 'true' ratings.

Suppose that the observer is rating flat on the cycle times given above:

true ratings	50	55·5	62·5	72	83	100	125	167
observed ratings	60	70	80	85	95	100	110	130
reciprocal t.r.	0·02	0·018	0·016	0·014	0·012	0·01	0·008	0·006
reciprocal o.r.	0·0165	0·014	0·0125	0·012	0·0105	0·01	0·009	0·0075

When plotted on a reciprocal-rating chart, Figure 5.5, again a straight line results, but on a linear-rating graph (Figure 5.10) the form is a curve.

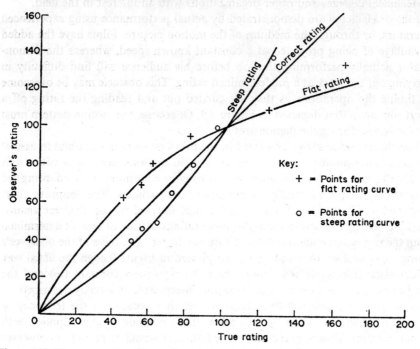

Figure 5.10 The appearance of lines of best fit for steep rating and flat rating respectively when plotted to linear scales

Similarly with steep rating:

true ratings	50	55·5	62·5	72	83	100	125
observed ratings	40	45	50	65	80	95	140
reciprocal o.r.	0·025	0·022	0·02	0·0155	0·0125	0·0105	0·0072

In this case a reverse curve results on the linear graph.

From Figure 5.10 it will be appreciated that the gradient varies according to where the ratings occur. Flat rating can appear to be steep at the lower end of the rating scale.

The existence of the curves has been demonstrated by experimenters using actual clinical data and the results appear to bear out the theoretical conclusions.*

* B. Moores, op. cit.

5.62 MAINTAINING RATING CONSISTENCY

Whether one accepts the universal concept of rating, or adheres to the idea of a unique standard for each firm, it is essential that within a particular organization there should be a minimal degree of variation between individual observers' concepts of standard rating. Consistency among practitioners is maintained in most firms by regular rating sessions held by the firm, with a feedback of results analysed by one of the recognized methods. Rating clinics are held periodically by technical colleges and other organizations with an interest in the field.

Jobs or elements are demonstrated by actual performance using experienced operators, or through the medium of the motion picture. Films have the added advantage of being projected at a constant known speed, whereas the demonstrator actually performing the job before his audience will find difficulty in carrying out the work at a predetermined rating. This obstacle may be overcome by timing the operations as they are carried out and reading the rating off a chart similar to that depicted in Figure 5.9. Of course, the motion pattern must not be varied during the demonstration.

The charts and analyses described in the previous section are useful in assessing individual abilities. Where regular formal sessions are not feasible it is possible to monitor the practitioners' consistency continuously by adapting the charts for use in time studies rather than rating studies. The reciprate chart described in Chapter 9 is used to plot ratings assessed by the observer against the observed times corresponding to these ratings. A line of best fit determined from the regression equation will indicate the degree of flatness of the observer's rating. A typical set of results is shown plotted in Figure 9.3. In the usual way a regression line is plotted, from which the regression constant (m) and the coefficient of correlation are extracted and interpreted, as already described.

It has been pointed out by several writers that because of the tendency of raters to flatness in rating the so-called 'true' or guide ratings supplied with standard rating-films may themselves be flat. This would mean that an observer who was rating such films would appear to be correct, or even *steep*, even though in reality his ratings were flat. This certainly is a point to be borne in mind by both users and producers of rating films.

6 The Mathematics of Measurement

6.1 FUNDAMENTAL CONCEPTS

The role of mathematics in work measurement is a very important one. Not only does the discipline figure in the simple calculations which are an integral part of the analyses, but more advanced methods such as regression analysis are being used increasingly in determining time standards. An understanding of the theory of sampling and probability is essential to the practitioner who desires the

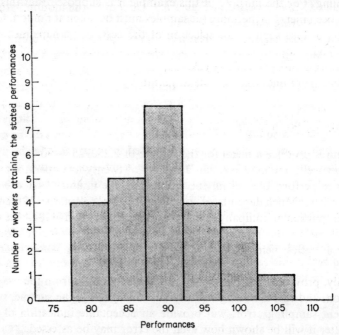

Figure 6.1 A histogram showing the number of workers attaining the stated performances

ability to apply work measurement in a comprehensive way. Knowledge of the underlying theory enables him to appreciate the suitability and limitations of the respective techniques, and to interpret correctly the results he obtains.

Much of work measurement is carried out on a sampling basis, and indeed, with relatively few samples. The time-study practitioner observes but 20 or 30 work cycles to derive his basic times, as does the P.M.T.S. applicator.* Sampling

---

* A statistical means of determining the number of cycles to observe for a predetermined level of accuracy is given in Chapter 7.

can never tell the whole story; at best it indicates what is *probably* the case. This being so, the practitioner must be aware of *how* probable is the result he obtains. Hence the importance of sampling theory and probability theory.

One of the best ways of exhibiting 'static' data is the frequency distribution. This results from accumulating data in the form of a tally, and joining up the tops of the groups. If an extremely large amount of data is plotted the distribution curve will be continuous, but where data are arranged in a small number of discrete groupings (or classes) the *histogram* is more appropriate. A typical histogram is depicted in Figure 6.1.

6.11 SAMPLING

Suppose it is required to estimate the average performance for workers in Britain by taking samples from various firms. A prerequisite to the success of this survey is the application of a common measure of performance throughout the sampling. For the purpose of this example it is supposed that this requirement has been met. Furthermore the samples must be taken at random to obviate any effects of bias such as the selection of the best, or perhaps mediocre performers. The random samples are to consist of performances of twenty-five workers during one particular week.

A sample from one firm may yield the ratios:

80, 100, 90, 90, 85, 90, 85, 95, 85, 95, 85, 80, 80, 90, 95, 90, 85, 100, 90, 95, 105, 90, 80, 90, and 100

These figures produce a mean for the sample of $2{,}250/25 = 90{\cdot}0$.

But how well, one may ask, can a sample of twenty-five operators represent the average performance of *all* the workers in the country? Or even in that particular firm? Much depends, of course, on how representative of the 'average' firm is this particular company. Often a sample of twenty-five is all the information available to the researcher; time study being a particular case of this number of cycles being taken as representative of the whole work of an operator.

Actually, provided that the sample is taken at random from the *whole* of the population, and this is homogeneously mixed originally, the sample, subject to an inherent sampling error, will provide an acceptable indication of the true value. Later it will be shown how even the error may be assessed.

To obtain a better assessment of performance a sample of twenty-five from a second firm may be taken which yields a mean performance of $107{\cdot}5$. Pooling the results improves the estimate and indicates a mean performance of $197{\cdot}5 \div 2 = 98{\cdot}8$. A really representative figure demands the collection of data from as many firms as possible, and in this particular case an imaginary nationwide survey of 180 companies produced a mean performance ratio for that week of $92{\cdot}82$. This is a much more representative sample, as it covers 4,500 workers. Again there will be an inherent statistical error, but it will be very much reduced, and again one which may be assessed.

6.12 PARAMETERS OF A DISTRIBUTION

The mean is one of several *measures of central tendency*; other averages being the *median*, or middle value in the distribution, and the *mode* or highest occurring value. In the above example the following parameters apply:

The mean value was 90·0.

The median, the middle figure when the values have been placed in ascending order, is the 13th figure and has a value of 90.

The mode is also 90 because this performance ratio occurs eight times. The fact that it is the highest value may be seen from Figure 6.1.

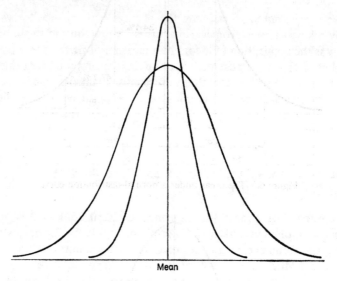

Figure 6.2 Two normal-distribution curves with different standard deviations or dispersions, but having the same mean

In this particular case the mean, median and mode have identical values. This is because the frequency distribution is reasonably symmetrical. Indeed, differences in the mean, median, and mode are used to measure the skewness of other distributions which may be asymmetrical.

One of three parameters which distinguish individual frequency-distributions from one another is the shape of the curve (it may be rectangular, skew, bell-shape, or any one of many forms). Another parameter is the value of the mean, which fixes the curve in space. Figure 6.2 illustrates one particular frequency-distribution shape to which any data could be attributed on the horizontal scale. However, once the mean is assigned to the curve its position is fixed in space (Figure 6.4).

The distribution is not uniquely described by its shape and the value of its mean alone. This is clear from Figure 6.2, which depicts two similar distribution shapes with the same mean value. A third parameter which measures the *dis-*

persion or spread of the data is clearly needed; the two measures of note being the simple measure of range, and the more useful one of *standard deviation*.

The shape of the frequency distribution will depend on the character of the data plotted. Where the values have equal chances of occurring a rectangular

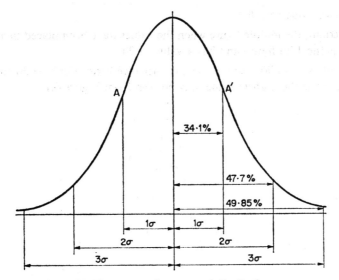

Figure 6.3 The areas under a normal-distribution curve

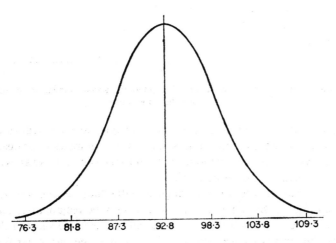

Figure 6.4 Actual data from the example in the text, with a normal distribution superimposed

distribution results. Although the curve may take virtually any form, one particular shape of great importance in the study of work measurement is the 'normal' or Gaussian distribution.

Some 'raw' data such as heights, weights, or I.Q.s of people may assume the form of the normal distribution. Similarly dimensions of some manufactured items will conform to this curve when plotted. In many cases the shape of the distribution may be other than this. However, one very important occurrence of the curve is in the field of sampling. *In all cases of sampling, the mean values of the samples may be regarded as originating from a normal distribution, whatever the shape of the original distribution of raw data from which the samples were taken.* This phenomenon is of fundamental importance to the work-study practitioner, and, indeed, to any investigator concerned with obtaining data through the medium of sampling.

Before discussing the application of the normal distribution it will be instructive to consider the attributes of the curve.

6.13 THE NORMAL DISTRIBUTION

From Figure 6.2 the characteristic shape of the normal distribution will be apparent. In theory the extremities of its bell-shape tail off to infinity, but in application, of course, it has more practical limits. Because the theoretical range is infinite a method of artificially dividing up the range into meaningful and more useful parts must be devised. As there are no definite discontinuities or corners to the curve some other way of defining the unit size of the divisions must be sought. At point A in Figure 6.3 the curve will be seen to change direction. From the left of the curve, moving toward the right, it is curving upward in an anti-clockwise direction, until it reaches point A, when it starts to curve clockwise. The reverse occurs at point A'. Points A and A' are known to mathematicians as 'points of inflexion'.

One of the most useful and important measures of spread which utilizes the point of inflexion as its basic measure is the standard deviation. The distance from the mean value to the point A (or A'), measured along the horizontal axis, is assigned the value of one standard deviation. If a scale be placed on this axis, as has been done in Figure 6.4, the actual value of this standard deviation may be ascertained: the value may be determined also from a formula used later in this chapter. Once this dimension has been determined the scale may be marked off in standard deviations and fractions of standard deviations. The standard deviation will always be the same size in relation to the size of the normal distribution, and only the actual *value* will differ from problem to problem. This fact provides a fundamental basis for all problems concerned with normal distributions. In fact for all practical purposes *all normal distributions have a width of six standard deviations* approximately; all that remains when approaching individual problems is to calculate the actual value of the standard deviation.

In solving problems of the above nature, the first step is to sketch a normal distribution with the standard deviations already marked in. In some organizations pre-printed charts are available for this purpose. Secondly, the calculations of the mean and standard-deviation values is undertaken. Using these values the horizontal scale then is marked off. The data in Table 6.1 illustrate values of performance ratios previously used which happen to be approximately normal. The calculation of standard deviation (usually denoted by the Greek letter sigma σ) utilizes the formula:

THE MATHEMATICS OF MEASUREMENT 71

$$\sigma = \sqrt{\frac{\Sigma f d^2}{n-1}}$$

where f is the frequency of occurrence of each of the values
d is the deviation of each of the considered values from the mean of all values
n is the number of values in the calculation

The numerical calculation is performed in Table 6.1, but described here. It will

TABLE 6.1

CALCULATION OF STANDARD DEVIATION

Performance	Frequency	Deviations	(Deviations)²	Frequency × (Deviation)²
	A	B	C	D
80	4	−10	100	400
85	5	−5	25	125
90	8	0	0	0
95	4	5	25	100
100	3	10	100	300
105	1	15	225	225
				1,150

$$\sigma = \sqrt{\frac{\Sigma f d^2}{n-1}}$$

$$= \sqrt{\frac{1,150}{24}}$$

$$= 6 \cdot 92$$

be seen that the formula demands the addition of the squared deviations from the mean value ($\Sigma f d^2$). This is readily accomplished by first calculating the deviations (column B in Table 6.1) and squaring each deviation in turn (column C). Next the total squared deviations for each performance (column D) are calculated by multiplying each value in column C by its corresponding frequency of occurrence ($f d^2$). The sum of column D is $\Sigma f d^2$. This total is divided by 24 (i.e. 25 − 1), and when the square root is taken the standard deviation of 6·92 is the result.

It is important that the work-measurement practitioner appreciates the magnitude of the errors to which the techniques he uses are subject. Mathematical procedures such as averaging or standard-deviation calculations tend to produce answers which, mathematically, are very precise, but which are unrealistic in practice because of the impracticability of measuring to such a degree of precision. However, for the purpose of illustrating the statistical concepts dealt with in this chapter the two decimal places resulting from the calculations will be retained.

The value of the standard deviation calculated above (6·92) may be used as an estimate of the population standard deviation because it has been corrected to obtain a closer fit to the true standard deviation by the use of the denominator $n - 1$.*

Now it is a fundamental fact that with all normal distributions the areas under the curve and between certain fixed limits are always the same. This is illustrated by Figure 6.3, from which it will be seen that 68·2 per cent of the values lie within one standard deviation of the mean. Similarly 99·7 per cent of the values (virtually all of them) are within three standard deviations each side of the mean. Notice that the given areas pertain to measurements of standard deviations from the *mean* in each case. Hence the area bounded by the curve and one standard deviation from the mean is 34·1 per cent, but the area between this limit and one more standard deviation is not equal to 34·1 but to the difference between the areas bounded by the mean and two standard deviations, and the mean and one standard deviation, i.e. 47·7 − 34·1, or 13·6 per cent.

The importance of the foregoing theory cannot be stressed too strongly and its applications will become apparent later in this chapter.

6.14 SAMPLE MEANS

Another important phenomenon upon which work study capitalizes is concerned with the distribution of sample means. An interesting circumstance which enables the practitioner to use sampling and from this make certain predictions is the fact that when samples of any data are taken and the *means* of the sample values plotted, provided a large number of sample means are available a normal distribution will be formed. This will happen even though the original distribution may be anything but normal. This fortunate situation makes it unnecessary to know the shape of the distribution of the original raw data when dealing with sample means. Any sample means can be assumed to come from a normal distribution of means.

The theory given above for individual values which are normally distributed is applicable equally to sample means. It is usual for statisticians to refer to the standard deviation of sample means as the standard error (or more fully, the standard error of the mean), otherwise the application of the theory to which reference was made earlier in this Chapter is valid in the same way. Figure 6.5 illustrates the distribution of raw data. Superimposed on this are the distributions of sample means for samples taken four at a time, and twenty-five at a time respectively. Two things are apparent from this Figure: the distributions are quite 'normal' in shape, and also the distribution for samples taken twenty-five at a time is more constricted and has a narrower base than that for means of samples of size four.

The standard error is calculated from the formula:

$$\text{standard error} = \frac{\text{standard deviation of individual readings}}{\sqrt{\text{sample size}}}$$

* For an explanation of the reasoning behind this the reader is referred to M. J. Moroney, *Facts From Figures* (London: Penguin Books, 1964).

or in symbolic form

$$\sigma_s = \frac{\sigma_p}{\sqrt{n}}$$

The standard deviation of all individual readings (or population) will probably never be known exactly as it is impossible to obtain all readings which exist. For example, it is impossible to measure the millions of cycles of a job executed by the mass-production worker in a factory: this is why sampling techniques such as time study are performed in the first place. This problem is overcome by estimating the standard deviation from a sample taken from the population. An illustrative example is given below.

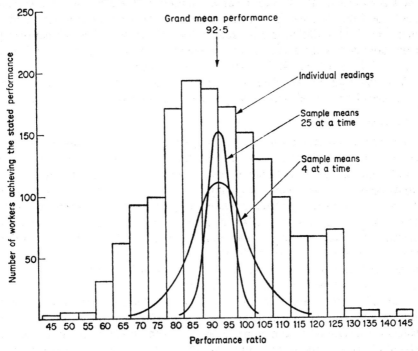

Figure 6.5 Normal distributions for sample means of sample sizes (a) four and (b) twenty-five, superimposed on the histogram for individual values

6.15 SUMMARY

At this stage, before proceeding and in order to consolidate the foregoing principles, the most important points will be summarized.

1. When plotted in the form of a frequency distribution, data may assume any form, but if it happens to take the characteristic bell shape of the normal distribution, certain predictions may be made about the data.
2. The spread of the normal distribution may be measured in standard deviations.
3. The areas under the normal curve and between limits measured in standard deviations are constant whatever the manner of the data.

4. For practical purposes it may be assumed that the range from one end to the other of a normal distribution is approximately equal to six standard deviations.
5. Whatever the shape of the distribution of raw data, *means* of samples taken at random from the data will form their own normal distribution.
6. The standard deviation of sample means (or standard error) is given by the formula $\sigma_p/\sqrt{n}$.

6.2 APPLICATIONS

In the previous example the standard deviation was equal to 6·92. This may be taken as an estimate of the population standard deviation σ_p.* The mean of 90 may also be used as the population mean. But how accurate is this mean as an estimator of the true population mean? Or more precisely, how far out from the true mean will this estimated mean be?

To answer this question recourse must be had to the normal-distribution theory, use of which is quite legitimate as we are considering means, and means are normally distributed. Furthermore, because we are considering means it is the standard error and not standard deviation which must be used.

$$\text{Now the standard error for the sample of } 25 = \frac{6 \cdot 92}{\sqrt{25}}$$

$$= 1 \cdot 38$$

Using the areas of Figure 6.3 it will be seen that 68 per cent of the area lies between plus and minus one standard error, i.e. ±1·38, and 95 per cent of the area lies between plus and minus two standard errors from the mean, i.e. ±2·76. From this it can be deduced that there is a 95 per cent chance that the true mean will lie within ±2·76 of the estimated mean of 90 performance.

Thus the true mean lies somewhere between the limits 87·24 and 92·76, with 95 per cent confidence.

The concept of confidence is very important to the statistician, and to the statistically minded work-study practitioner, as much of the theory of work measurement uses time standards subject to statistical errors.

Logically, if he wishes to be more certain of his facts the practitioner must inverview a larger number of workers, and the time-study officer must extend his study over a larger number of cycles. Statistically this has the following effect. In the complete survey mentioned above, 180 samples of 25 workers, 4,500 in all, produced a mean ratio of 92·82. The sceptic will view this result with some apprehension, especially as it is *outside* the 92·76 limit calculated previously. There must be some explanation for this apparent incompatability.

Suppose the standard deviation of the 4,500 performance ratios in the survey was calculated as 5·5. Then the standard error inherent in the mean of 92·82 is 5·5 ÷ $\sqrt{4{,}500}$, or 0·082, which is an exceedingly small error. Two standard errors equal 0·164, so with 95 per cent confidence we can say that the true value

* It was to improve this estimate that $n - 1$ and not n was used in the standard deviation formula on p. 71.

THE MATHEMATICS OF MEASUREMENT 75

of the mean for the whole working population lies between 92·82 − 0·164 and 92·82 + 0·164, or between 92·66 and 92·98 approximately. It will be appreciated that the true mean has been pinpointed very closely to 92·82.

One question may yet linger in the mind of the doubter. Why was the previous estimate of 87·24 to 92·76 so far out? Logically the reply is that it was because the sample size of 25 was so very small. Statistically we say that we were only 95 per cent sure that the limits were pertinent anyway, and as it happened the 5 per cent chance that the true value was outside the limits happened to turn up!

Example 2

Extracting further information from the survey, it may be necessary to know the probability that any individual's performance will be less than, say, 82, or the probability that the mean of any group of twenty-five workers at random will be less than 89·5.

To answer the first part of the problem it is necessary to determine the distance of the individual worker's performance from the mean *not* in actual performance, but in standard deviations, and then to relate this to the area under the curve (from Figure 6.3). In dealing with such problems of probability it is usual to equate the total area under the normal curve to total probability, that is, to unity. Thus the probability of any value occurring by chance between the mean and one standard deviation is 0·341 (the 34·1 per cent area). In the present problem the survey disclosed a mean of 92·8 and a standard deviation of 5·5 performance units, so one standard deviation below the mean gives a value of 92·8 − 5·5, or 87·3 performance, while two standard deviations below the mean gives 92·8 − 11·0, or 81·8 performance (Figure 6.4). As this is almost the particular value stated in the problem, the question is resolved into one of finding the probability that any individual performance will be less than two standard deviations below the mean. The whole area below the mean is clearly one-half, and the area between the mean and two standard deviations from Figure 6.3 is 0·477. Therefore the area outside the two-standard deviation limit is 0·500 − 0·477, or 0·023, which means there are 23 chances in 1,000 that any one worker chosen at random will have a performance ratio below 82 per cent. Alternatively, out of every 1,000 workers, about 23 will not achieve the 82 performance during that week.

The second part of the problem is tackled in the same way, this time using the standard error which is compatible with the distribution of means.

$$\text{standard error} = \frac{5 \cdot 5}{\sqrt{25}} = 1 \cdot 1 \text{ performance units}$$

$$\text{number of standard errors from the mean} = \frac{\text{distance from mean}}{\text{standard error}}$$

$$= \frac{92 \cdot 8 - 89 \cdot 5}{1 \cdot 1}$$

$$= 3$$

(Note that this formula could have been used in the problem just solved above, using standard deviation in place of standard error.)

The given value of 89·5 is three standard errors from the mean
The area below three standard errors is 0·5000 − 0·4985 = 0·0015

Therefore the probability that the mean value of a group of twenty-five people at random being less than 89·5 by chance, is just better than a one-in-a-thousand chance.

If the chance is so slim, how could it happen? The answer is that most probably it is not due to chance, but to something amiss, such as the workers feeling off colour, wrong information returned by the observer, or some other

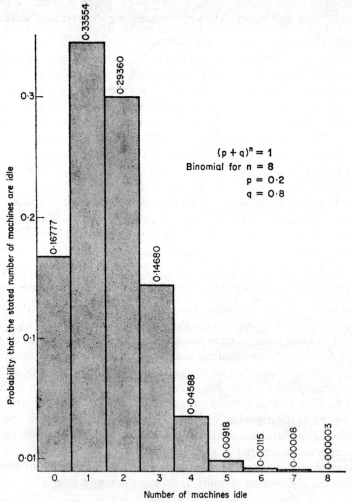

Figure 6.6 A binomial distribution for a sample of eight machines which have a probability of being idle of 20 per cent

disturbing contributory factor. Incidentally, this type of analysis could be used by management as a warning indicator of trouble.

6.3 THE BINOMIAL DISTRIBUTION

In previous sections the data have been in the form of variables, the values even being continuous variables. In practice operators do not conveniently work at 85, 90, or any other rounded figure, but are more likely to perform at speeds between these figures. The grouping to the nearest five units is an artificial convenience to make the figures more manageable.

Some areas of work measurement, principally one technique known as activity sampling, are concerned with *attributes* data rather than *variables* data. The mass-production checker using a 'go not-go' gauge is sampling using the attributes of 'pass' and 'reject'. The activity-sampling observer uses 'working' and

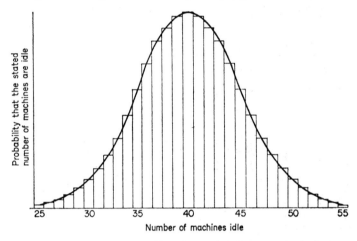

Figure 6.7 Another binomial probability distribution, similar to that of Figure 6.6, but for 200 machines, and with a normal distribution superimposed for comparison

'unoccupied' as his two attributes. The frequency distribution which covers the situation involving only two cases is the *binomial* distribution.

In a coil-winding room of eight machines which automatically stop when the wire runs out, the activity-sampling observer may calculate the probability that any number of machines will be idle, provided he knows the probability of a machine running out of wire. The resulting binomial distribution is shown in Figure 6.6. With a large number of machines—say 200—the frequency distribution will be as illustrated in Figure 6.7, which is reminiscent of the normal distribution. This fortuitous similarity provides a valuable link with the theory outlined in previous sections of this chapter.

The formula for the standard deviation (and standard error) of proportion for a binomial distribution, happily, is no more complex than:

$$\sigma = \sqrt{\frac{pq}{n}}$$

In the formula p represents the probability that one event will occur, while q is the probability that the complementary event will happen. As usual, n is the sample size. This formula is for the standard error of proportion, that is, the standard error of the distribution of p. The formula for standard error for a binomial distribution also takes the form $\sqrt{npq}$, which is the standard deviation of individual values. The difference is illustrated in Example 3, which follows. For reasons stated in Chapter 7 (p. 124), for the purpose of activity sampling in work measurement the proportional form given above is more convenient.

Example 3

A work-measurement practitioner is observing a clerk in an office to determine the extent of his useful work.

He visited the office twenty times at random intervals during the week and noticed that the clerk was working on fourteen occasions. He therefore concluded that the clerk was working for $\frac{14}{20} \times 100$, or 70 per cent of the time, and, consequently, idle for 30 per cent. But as he had based his results on only twenty observations he was not sure that this result was very reliable. Apart from any bias or human error, there is a statistical error present whenever samples, especially small ones, are taken. This error may be calculated exactly, as shown in earlier examples.

The standard error, using the formula just given, is:

$$\sqrt{\frac{70 \times 30}{20}}, \text{ which equals 10 per cent approximately.}$$

To be 95 per cent confident, two standard errors must be taken, i.e. 20 per cent is the error to which p is subject. From these observations all the practitioner can state, with 95 per cent confidence, is that the clerk was working between 50 and 90 per cent (i.e. 70 ± 20 per cent) of the time. This is somewhat vague, but to be more precise he would need to observe more than twenty times.

On the other hand, the standard error of the *distribution* is given by the formula quoted above, that is, $\sqrt{npq}$, which in this case is equal to

$$\sqrt{20 \times 0.7 \times 0.3} = \sqrt{4.2}$$
$$= \pm 2 \text{ observations (approximately)}$$
or for two standard errors $= \pm 4$ observations

Thus the result this time is in *actual observations*, and not in percentages as it was before. From this it is concluded that the clerk was working 14 times ± 4 times, which compares with 70 ± 20 per cent of the previous result. The actual number of times the clerk was working means very little, but the *percentage* of time he worked is more meaningful, and this is why the standard error of proportion is preferred.

In general the limits of error (L) for 95 per cent confidence are given by:

$$2\sigma = L = 2\sqrt{\frac{pq}{n}}$$

where p is the percentage of the time the observed event is occurring
q is the percentage of the time it is not happening
n is the number of observations

This is usually written as:

$$L = 2\sqrt{\frac{p(100-p)}{n}}$$

because $p + q = 100$ per cent, thus $q = 100 - p$.

To find the number of observations necessary to conform to certain predetermined limits, first square both sides of the equation.

$$L^2 = \frac{4p(100-p)}{n}$$

and rearranging:

$$n = \frac{4p(100-p)}{L^2}$$

6.4 THE STRAIGHT LINE

A study of the mathematical basis of work measurement would be incomplete without the inclusion of the straight line. So much of the data processed by the practitioner is linear, or can be transformed into a linear state. The importance of this member of the group known as *conic sections* is due to the applicability of the basic formula to such diverse areas as financial incentive schemes, synthetic data, determination of basic times by graphical means, checking of rating ability, and the derivation of time standards through regression analysis. The conic sections referred to are so named because they are formed by dissecting cones in certain planes, and include the ellipse, circle, parabola, and hyperbola. The hyperbola is important in work measurement, being the form of the graph which results when observed ratings are plotted against observed times in time study.

The basic form of one straight line (which is one of a pair produced from the conic-section equation) is $y = mx + c$. Whenever two variables (y and x) are present in an equation, and they are first-order, i.e. not squared, cubed, etc., the resulting graph is linear: the value m being a measure of the slope or gradient of the line, and c the constant intercept, or point on the y axis where the line cuts it.

Any graph of two variables which is linear has this basic form, the values of m and c defining the particular line uniquely.

Referring to Figure 6.8 and to the line AB in particular, the general form $y = mx + c$ may be adapted to describe the line AB by determining the value of its constants. The line is seen to cut the y (vertical) axis at the point where $y = 5$. As this is the definition of c, then it follows that $c = 5$.

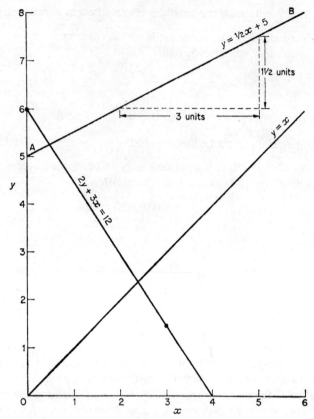

Figure 6.8 Examples of some straight lines, and how the gradient is measured

The gradient m is the inclination of this line to the horizontal, defined by the ratio:

$$\frac{\text{distance of the vertical}}{\text{length of the base}}$$

of any right-angled triangle constructed with its base parallel to the x axis and using the line AB as the hypotenuse, as shown in Figure 6.8. In this case the vertical distance is $1\frac{1}{2}$ units and the base is equal to 3 units; m is therefore $\frac{1}{2}$. The required equation for this line is: $y = \frac{1}{2}x + 5$.

Next consider the reverse procedure, of plotting a line from a given equation, and the equation to be plotted is $2y + 3x = 12$. This is not in the standard form $y = mx + c$, and must first be transformed:

$$2y + 3x = 12$$
$$2y = -3x + 12$$
$$y = -\tfrac{3}{2}x + 6$$

It is now in the standard form, and on comparing the coefficients of x and y with the standard equation it is apparent that $m = -\frac{3}{2}$ and $c = 6$. To plot this line only two points are required, and one point to use is clearly the point of intersection of the line and the y axis, i.e. 6. The other point may be determined by assigning any arbitrary value to x and solving the equation, thus:

let $\quad x = 3,$
then $\quad y = -(\frac{3}{2} \times 3) + 6$
$\quad\quad\quad = -4\frac{1}{2} + 6$
$\quad\quad\quad = 1\frac{1}{2}$

At the point where $x = 3$ and $y = 1\frac{1}{2}$ a second point is marked, whence joining the two points gives the required line (Figure 6.8).

A third line is plotted on Figure 6.8 and marked $y = x$. This may be rewritten in the standard form as $y = 1x + 0$, where $m = 1$ and $c = 0$. This line passes through the zero (or origin) with a gradient of 1. On a graph of an incentive scheme this would be the direct-proportion line.

6.41 LINE OF BEST FIT

On occasions the practitioner may be faced with plotting data on a graph to which a mean line, or line of best fit, is appropriate. Such a case is that considered in Chapter 5, the comparison of an observer's ability to rate certain jobs. Throughout the field of work measurement the practitioner is required to deal with graphical analyses, so an ability to handle numbers and interpret the results is of extreme importance.

By way of example the reader is asked to consider the two sets of data listed in Table 6.2. These are pairs of values (y and x) which have been plotted on Figure

TABLE 6.2

CALCULATION OF A REGRESSION EQUATION

	A Y data	B X data	C A × B	D A^2	E B^2
	95	90	8,550	9,025	8,100
	105	100	10,500	11,025	10,000
	95	95	9,025	9,025	9,025
	80	70	5,600	6,400	4,900
	115	120	13,800	13,225	14,400
	120	130	15,600	14,400	16,900
	115	105	12,075	13,225	11,025
	90	90	8,100	8,100	8,100
	85	95	8,075	7,225	9,025
	100	100	10,000	10,000	10,000
	115	110	12,650	13,225	12,100
	90	80	7,200	8,100	6,400
	75	75	5,625	5,625	5,625
	105	110	11,550	11,025	12,100
	100	90	9,000	10,000	8,100
Totals	1,485	1,460	147,350	149,625	145,800
Mean	99	97			

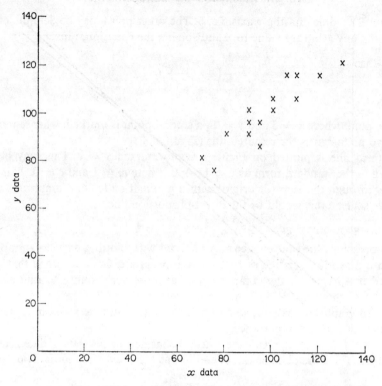

Figure 6.9 A scatter diagram for data from Table 6.2

6.9, and for which a line of best fit is required. This may be accomplished either by sketching in the line by inspection, or more precisely by the mathematical process of *regression analysis*.

Regression analysis has many applications in work measurement, most of which are given in this book. The usual method of analysis is the method of least squares. The calculations are given in Table 6.2 and described below.

1. The objective is to define the line of best fit in the standard form $y = mx + c$, and then to construct the line on the graph field when it should pass through the middle of the plotted points (Figure 6.10).

2. The regression-equation constants m and c are isolated by solving simultaneously the two equations:

$$\Sigma y = m\Sigma x + nc$$
$$\Sigma xy = m\Sigma x^2 + c\Sigma x$$

This involves summing the values of y, x, xy, and x^2, so appropriate columns are prepared (Table 6.2).

3. The constants are inserted in the regression equation $y = mx + c$, giving in this case $y = 0 \cdot 76x + 25$, after they have been determined using the solution by simultaneous equations.

THE MATHEMATICS OF MEASUREMENT

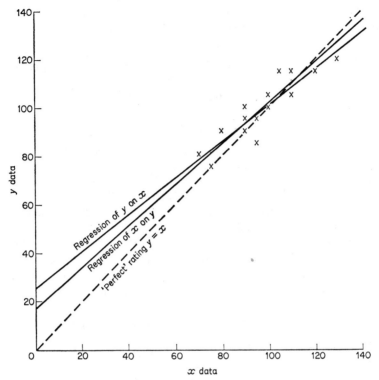

Figure 6.10 Regression lines plotted through the points of Figure 6.9

Calculation

$$\Sigma y = m\Sigma x + nc \qquad 1{,}485 = 1{,}460m + 15c \qquad \text{(A)}$$
$$\Sigma xy = m\Sigma x^2 + c\Sigma x \qquad 147{,}350 = 145{,}800m + 1{,}460 \qquad \text{(B)}$$

Multiply (A) by 292 $433{,}620 = 426{,}320m + 4{,}380c$ (C)
Multiply (B) by 3 $442{,}050 = 437{,}400m + 4{,}380c$ (D)
Subtract (C) from (D) $8{,}430 = 11{,}080m$
  $m = 0 \cdot 761$ (E)

Substitute m in (A) $1{,}485 = 1111 + 15c$
  $374 = 15c$
  $c = 25$

Regression equation is $y = 0 \cdot 761x + 25$

A much more elegant method of calculating the coefficients, and one which facilitates the use of electronic calculators, is through the following formulae:

$$m = [n\Sigma xy - \Sigma x \, \Sigma y] \div [n\Sigma x^2 - (\Sigma x)^2]$$

and

$$c = [\Sigma x \, \Sigma xy - \Sigma x^2 \, \Sigma y] \div [(\Sigma x)^2 - n\Sigma x^2]$$

These formulae are derived in the following way. Rearranging the regression equations

$$\Sigma y = m\Sigma x + nc \quad \text{and} \quad \Sigma xy = m\Sigma x^2 + c\Sigma x$$

produces
$$c = \frac{\Sigma y - m\Sigma x}{n} \quad (F)$$

and
$$c = \frac{\Sigma xy - m\Sigma x^2}{\Sigma x} \quad (G) \text{ respectively}$$

Equating (F) and (G),
$$\frac{\Sigma y - m\Sigma x}{n} = \frac{\Sigma xy - m\Sigma x^2}{\Sigma x}$$

From this, $\quad \Sigma x \Sigma y - m(\Sigma x)^2 = n\Sigma xy - nm\Sigma x^2$

or $\quad m[n\Sigma x^2 - (\Sigma x)^2] = n\Sigma xy - \Sigma x \Sigma y$

and $\quad m = [n\Sigma xy - \Sigma x \Sigma y] \div [n\Sigma x^2 - (\Sigma x)^2]$

In a similar way the formula for c may be derived.

The calculation just performed is the regression of y on x. This form may be used when it is required to read the values of y corresponding to chosen values of x. In the rating example just discussed, the expected value of the observer's rating for any value of the true rating may be read off the graph using the regression line of y on x. However, if it is necessary to read the graph from the y values to find out the resulting value of x a second regression line, the regression of x on y, must be used.

The regression of x on y is calculated in exactly the same way as that for y on x, but using the following equations in place of those just given:

$$\Sigma x = m\Sigma y + nc$$
$$\Sigma xy = m\Sigma y^2 + c\Sigma y$$

It will be seen that the x and y values in the two sets of equations have been transposed.

The equation for the regression of x on y is $x = 1 \cdot 077y - 9 \cdot 3$, which may be converted to $y = 0 \cdot 928x + 8 \cdot 6$ and plotted in the usual way. The two regression lines are shown in Figure 6.10.

6.42 CORRELATION

When plotting graphs of work-measurement data it is very tempting to sketch or calculate a line of best fit through the points on the assumption that the data are related in a linear fashion. Sometimes this is a dangerous exercise as the assumption may not be a valid one. For example, if observed ratings be plotted on the y axis and observed times on the x axis of a graph the points which result appear to be in a reasonably straight line within the limits of human error. In fact they will be part of a parabola, and this is discussed further in Chapter 11, when dealing with graphical interpretation of time studies.

Furthermore it is only assumed that the data are related in the first place, and from this certain questions arise: is the relation good enough to justify the representation of the values by a straight line? Just how good is this relationship? The measure of the degree of relationship between two sets of variables is known as *correlation*.

The most usual ways of calculating the coefficient of linear correlation are

THE MATHEMATICS OF MEASUREMENT

(i) use of the correlation formula, and (ii) use of the regression-line coefficients. The coefficient of correlation measures the amount of deviation of the plotted points from the regression line, therefore it is not surprising that there is a close connection between regression and correlation. The formula for this deviation is:

$$\frac{\Sigma uv}{\sqrt{\Sigma u^2 \, \Sigma v^2}}$$

In this formula u and v are the deviations of the x and y values from their respective means. The data from Table 6.3 will be used to illustrate the calculation:

$$r = \frac{2{,}810}{\sqrt{(2{,}610 \times 3{,}695)}}$$

$$= \frac{2{,}810}{3{,}105}$$

$$= 0{\cdot}905$$

TABLE 6.3
CALCULATION OF A COEFFICIENT OF CORRELATION

	Y data	X data	Devs. of Y (u)	Devs. of X (v)	$u \times v$	u^2	v^2
	95	90	−4	−7	28	16	49
	105	100	+6	+3	18	36	9
	95	95	−4	−2	8	16	4
	80	70	−19	−27	513	361	729
	115	120	+16	+23	368	256	529
	120	130	+21	+33	693	441	1,089
	115	105	+16	+8	128	256	64
	90	90	−9	−7	63	81	49
	85	95	−14	−2	28	196	4
	100	100	+1	+3	3	1	9
	115	110	+16	+13	208	256	169
	90	80	−9	−17	153	81	289
	75	75	−24	−22	528	576	484
	105	110	+6	+13	78	36	169
	100	90	+1	−7	−7	1	49
Totals	1,485	1,460	0	+5	2,810	2,610	3,695
Means	99	97					

The second method of calculating the coefficient of correlation utilizes the regression coefficients of the two regression equations. The respective regression equations previously determined were:

$$y = 0{\cdot}76x + 25 \quad \text{B / A}$$

and

$$x = 1{\cdot}077y - 9{\cdot}3$$

From these equations, the values of m respectively are 0·76 and 1·077. The coefficient of correlation is simply the geometric mean of these two values, i.e. $\sqrt{(m_1 . m_2)}$. Thus in this case the coefficient is 0·905.

In general, the interpretation of the value of correlation depends on the size of the sample, or number of pairs of readings considered. However certain values have definite meanings. For instance:

a coefficient of correlation of $+1$ indicates perfect correlation between the data;
a coefficient of -1 means perfect *inverse* correlation, i.e. as one set of data increases the other variable decreases;
a coefficient of zero shows that there is no connection between the data;
values between zero and unity indicate increasingly better fit as the value of 1 is approached.

It is possible to test the coefficient of correlation in order to find out if the connection is significantly better than just a chance relationship, but this topic is outside the scope of the present book. Several examples of the application of the concept will be seen in the following chapters.

6.5 INTRODUCTION TO MULTIPLE-REGRESSION ANALYSIS

The mathematical technique of regression analysis has many applications in work measurement. In addition to its use in rating checks just described, regression analysis may be applied to the analyses of synthetic data (*see* Chapter 10), in the extension of basic times from observed times by graphical means (Chapter 9), and in fact, whenever data are presented in graphical form. The more complex multiple-regression analysis may be used for establishing basic times for indirect and variable-element work from *over-all* times.

Multiple-regression analysis (M.R.A) is a very flexible and powerful device, and has been applied in many areas of business management. Originally devised by Sir Francis Galton (1822–1911) at the turn of the century regression analysis has only recently found favour in work-measurement circles. Until the advent of accessible computer facilities, applications in all fields were confined to relatively simple problems because of the sheer volume of data and calculation required. However, the simplicity of the calculations, as demonstrated in Table 6.2, make the technique an ideal subject for the computer program.

The example of the previous section was concerned with just one independent variable, the resulting regression equation being a 'line of best fit' through the scattered points of the observations (Figure 6.10). M.R.A. is used where more than one independent variable must be considered. The general form of the equation is:

$$Y = a_o + a_1 x_1 + a_2 x_2 + a_3 x_3 + \ldots a_j x_j$$

where there are j independent variables. Clearly this is not representative of a straight line as we know it, because the equation exists in more than two dimensions. While it is not possible to visualize more than three dimensions, fortunately it is possible to perform the necessary calculations.

In order to produce a regression equation many of these equations (each known as an 'observation') are required, the number of observations being in excess of the number of independent variables (at least thirty to forty for three or four variables).

The final regression-equation is to be used to predict the value of the dependent variable Y in a given set of circumstances. For example, Y could represent the standard time to complete a job while the variables x_1, x_2, x_3, etc. are the elemental times for the job. The constants a_0, a_1, a_2, a_3 ... represent the quantities of the respective independent variables, or the number of times the various elements occur in a cycle.

The equation shown above is typical of a linear regression-equation, but of course not all data are in linear form. However, it is possible to transform non-linear data into linear form; examples of this are given in Chapter 9.

The major steps in making a study using M.R.A. are summarized below:

1. The data are collected from observations made, or from past records of the situation.
2. The dependent variable is selected. (In work measurement almost invariably this is basic or standard time.)
3. The independent variables are selected as those which most probably have the greatest effect on the dependent variable. It does not matter if, inadvertently, inappropriate variables are chosen, because these may be eliminated from the final equation when the proper statistical tests have been made (*see* item 7).
4. Data are formed into equations incorporating those independent variables selected in (3).
5. The equations are solved using one of the recognized techniques of solution. Often this is so complex that recourse must be had to the facilities of a computer.
6. From the values assigned to the independent variables as a result of these calculations the regression-equation is formed.
7. Several statistical tests may now be performed on the regression-equation and its predictors to validate, and to eliminate, any independent variables which do not correlate very highly with the dependent variable, or which correlate highly with other variables.
8. The amended regression-equation may be used for the intended purpose.

The applications of M.R.A. in business are many, but in this present book it is the use of M.R.A. in work measurement which is of more interest. The technique will now be described in more detail, using a work-measurement application to illustrate the procedure.

6.6 M.R.A. IN WORK MEASUREMENT

Among the many possible applications in work measurement a particular case of collecting books to complete orders in the warehouse of a large publishing company will be considered.

It was decided to use the technique of M.R.A. to establish a regression equation which could be used to predict standard times for repetitive cycles of the complete task of perusing the order, collecting the book(s) from various locations, completing the delivery note, and packing the order. The regression-equation would, in fact, be a formula for synthetic data, although later it might be possible to eliminate certain variables which would then invalidate such values as synthetic data.

Consequently a team studied the work and decided to adopt the following factors as their independent variables (x), because these appeared to have the greatest direct effect on the total basic time (Y):

$x_1 =$ the number of books collected
$x_2 =$ the number of different locations visited
$x_3 =$ the size of the parcel in cubic metres
$x_4 =$ the number of straps used to tie the parcel on the automatic binding-machine
$x_5 =$ the number of entries to complete on the delivery note (including unavailable books if any).

It was essential that good representative times for the 'average' conditions were obtained, so the work-measurement team decided to make at least thirty observations. Effort rating (see Chapter 5) was not performed by the team, although this could have been done by using average ratings to extend the observed times (Y). The results were collected and tabulated ready for analysis (see Table 6.4). Because of the volume of data and number of variables, the analysis was performed on a computer using a standard 'package' program.

The computer program was designed to provide the following data:
1. The regression coefficients, which give the values of a_o, a_1, a_2, a_3, and so on, in the general equation:
$$Y = a_o + a_1 x_1 + a_2 x_2 + \ldots + a_j x_j$$
where there are j independent variables.
2. The index of determination, which indicates the percentage of the dependent variable explained by the input variables.
3. A correlation matrix, from which the intercorrelations between the variables and between the variables and the dependent variable (Y) may be ascertained.
4. The various statistical measures and other parameters which may be used to assess the error limits, 'F'-test and 't'-test values, and other checks.
5. The differences between the observed values of Y and those calculated from the regression-equation.
6. Additional facilities may be called upon to fit various curves to the data to obtain the curve which best represents the data in case it should be other than linear.

TABLE 6.4

DATA COLLECTED AND TABULATED FOR A MULTIPLE-REGRESSION ANALYSIS

Collection of books to make up orders

Observation No.	Observation time Y	No. of books collected x_1	No. of locations x_2	Size of parcel (cu. m.) x_3	No. of straps around parcel x_4	No. of entries x_5
1	7·38	10	3	0·008	3	12
2	13·90	20	5	0·027	4	3
3	10·10	16	4	0·012	3	3
4	10·32	12	8	0·012	3	6
5	19·70	25	12	0·027	6	20
6	10·72	12	6	0·009	3	2
7	5·80	5	5	0·006	1	1
8	3·32	2	2	0·002	1	2
9	7·00	12	1	0·014	3	2
10	31·64	60	10	0·058	10	14
11	19·70	30	8	0·030	6	6
12	6·80	12	1	0·012	3	2
13	8·00	6	6	0·005	1	7
14	17·08	24	12	0·030	5	4
15	13·00	10	8	0·008	2	12
16	4·20	4	2	0·005	1	2
17	9·90	15	4	0·020	3	4
18	14·50	15	10	0·014	3	16
19	9·00	12	2	0·013	2	10
20	17·40	30	8	0·029	6	4
21	9·20	15	3	0·017	3	2
22	18·20	28	10	0·032	6	6
23	15·20	26	8	0·025	5	5
24	11·70	16	5	0·014	3	8
25	2·50	1	1	0·002	1	4
26	6·66	10	2	0·008	2	2
27	3·80	4	1	0·003	1	5
28	14·96	20	8	0·022	4	7
29	3·05	2	1	0·003	1	4
30	14·31	22	5	0·025	4	2

Facility is available also for removing those variables which do not contribute to the improvement of the index of determination, or which contribute little to it.

The data of Table 6.4 were input, and the computer programmed to analyse them to produce the required output. The various parts of the output will now be described and explained separately.

The regression cofficients given in Table 6.5, reproduced from the computer print-out, may be inserted in the general multiple-regression equation:

Y (basic time) = (constant) + a_1 (number of books) + a_2 (number of locations) + a_3 (parcel size) + a_4 (number of straps) + a_5 (number of entries)

Table 6.5

A COMPUTER PRINT-OUT FOR PARAMETERS IN M.R.A.

MULFT$

VERSION OF 09/03/1968

VARIABLE	REGR COEFF	MEAN VALUE	STD DEV
0	1.7506	11.3013	6.37288
1	.423244	15.8667	11.8924
2	.547196	5.36667	3.47884
3	50.3177	.0164	1.23389E-02
4	−.499019	3.3	2.07032
5	.122066	5.9	4.72229

After insertion of the values, the equation becomes:

$$Y = 1{\cdot}7506 + 0{\cdot}423x_1 + 0{\cdot}547x_2 + 50{\cdot}32x_3 - 0{\cdot}499x_4 + 0{\cdot}122x_5$$

Often it is instructive to compare the actual times observed (Y) with the theoretical times calculated after inserting the variables from Table 6.4 into the equation (*see* Table 6.6).

Table 6.6

A COMPUTER PRINT-OUT FOR OTHER PARAMETERS IN M.R.A.

SOURCE OF VARIATION	DEGREES OF FREEDOM	SUM OF SQUARES	MEAN SQUARE
TOTAL	29	1177.8	40.6136
REGRESSION	5	1161.93	232.387
ERROR	24	15.861	.660873

INDEX OF DETERMINATION: .986533
F-RATIO TEST STATISTIC: 351.636

ACTUAL	CALCULATED	DIFFERENCE	PCT DIFFER
7.38	7.99491	.614906	7.6
13.9	12.6802	−1.21983	−9.6
10.1	10.1842	.084249	.8
10.32	11.0463	.726253	6.5
19.7	19.7038	3.84015E-03	0
10.72	9.31265	−1.40735	−15.1
5.8	6.52776	.727758	11.1
3.32	3.53723	.217231	6.1
7	6.82825	−.171748	−2.5
31.64	32.2544	.614379	1.9
19.7	18.0733	−1.62669	−9

THE MATHEMATICS OF MEASUREMENT

ACTUAL	CALCULATED	DIFFERENCE	PCT DIFFER
6.8	6.72762	−7.23828E-02	−1
8	8.18027	.180274	2.2
17.08	17.9775	.897519	4.9
13	11.2299	−1.77009	−15.7
4.2	4.53467	.334673	7.3
9.9	10.2856	.385611	3.7
14.5	14.7317	.231669	1.5
9	8.80067	−.199326	−2.2
17.4	17.7789	.378864	2.1
9.2	9.34333	.143331	1.5
18.2	18.4219	.221852	1.2
15.2	16.5057	1.3057	7.9
11.7	11.4424	−.257592	−2.2
2.5	2.81092	.310921	11.
6.66	6.72607	.066073	.9
3.8	4.25304	.453038	10.6
14.96	14.5584	−.401569	−2.7
3.05	3.28448	.234484	7.1
14.31	13.304	−1.00604	−7.5

The equation reproduced above may be used as synthetic data formula as it stands, but it does require a knowledge of five different variables in order to calculate one value of the basic time. Thus it is in the interests of economy to reduce the number of variables if possible by rejecting those which contribute little to the predictor. Clearly in order to do this there must be available some method of measuring this effect. The measure used is the index of determination. The procedure is one of process of elimination, and is carried out as follows:*

TABLE 6.7

A COMPUTER PRINT-OUT FOR ELIMINATION OF LESS EFFECTIVE VARIABLES

INDEX OF DETERMINATION	F-RATIO	VARIABLES IN ANALYSIS				
.985901	335.661	1	2	3	4	5
.975909	253.182	2	3	4	5	
.956722	138.165	1	3	4	5	
.985779	433.241	1	2	4	5	
.985542	426.035	1	2	3	5	
.981713	335.529	1	2	3	4	

WHICH VARIABLE IS TO BE ELIMINATED ? 3

* Alternatively the reverse procedure may be used in a 'stepwise' program which begins with one independent variable and *adds* successively the different variables until an optimum situation is reached. Computer programs are available for this purpose.

(b)

INDEX OF DETERMINATION	F-RATIO	VARIABLES IN ANALYSIS			
.985779	433.241	1	2	4	5
.959393	204.762	2	4	5	
.954603	182.241	1	4	5	
.985387	584.418	1	2	5	
.981617	462.786	1	2	4	

WHICH VARIABLE IS TO BE ELIMINATED ? 4

(c)

INDEX OF DETERMINATION	F-RATIO	VARIABLES IN ANALYSIS		
.985387	584.418	1	2	5
.684147	29.2415	2	5	
.946948	240.967	1	5	
.980653	684.287	1	2	

WHICH VARIABLE IS TO BE ELIMINATED ? 5

(d)

INDEX OF DETERMINATION	F-RATIO	VARIABLES IN ANALYSIS	
.980653	684.287	1	2
.679555	59.3786	2	
.917026	309.457	1	

The table of indices of determination (Table 6.7a) is inspected to find the combination of variables which has the least effect on the dependent variable when removed, i.e. look for the highest index of determination. The variable which is *not* included in the combination producing the highest index is the one to eliminate, because without this variable the index of determination is increased. In the example it will be seen that the overall index is 0·985901, but the highest individual index is 0·985779, for which the missing variable is x_3.

The program is re-run without this variable, and once again the resulting table (Table 6.7b) is scanned. This time the highest index is 0·985387 and the missing variable is x_4, the next choice for elimination. The new overall index of determination with these two variables deleted is 0·985387, which although lower than the first index, with all variables included is very little below this original value.

The next elimination is x_5 (Table 6.7c), leaving just x_1 and x_2, which together produce an index of determination of 0·980653. For convenience the results are summarized below:

THE MATHEMATICS OF MEASUREMENT

Variables included	Index of determination (overall)	Percentage of Y explained by variables
1, 2, 3, 4, 5	0·985901	98·6
1, 2, 4, 5	0·985779	98·6
1, 2, 5	0·985387	98·5
1, 2	0·980653	98·0
1	0·917026	91·7

It will be seen that very little difference is found in the percentage of Y explained by the regression-equation by dropping variables x_3, x_4, and x_5. Without these variables the regression equation still explains as much as 98 per cent of the Y variable. Even when just one variable (x_1) is included (the number of books collected) as much as 91·7 per cent is covered.

If it is decided to discard all variables except x_1 then the resulting regression-equation becomes:

time to make up order = 3·159 + 0·513 (number of books collected)

For the little extra work involved in counting the number of different locations the equation containing x_1 and x_2 is superior:

basic time = 1·747 + 0·393 (number of books) + 0·618 (number of locations)

TABLE 6.8

A COMPUTER PRINT-OUT OF A CORRELATION MATRIX

CORRELATION MATRIX (INPUT VARIABLES)

	Y	X 1	X 2	X 3	X 4	X 5
Y	1.000000	.957615	.824351	.939734	.947404	.508470
X 1	.957615	1.000000	.663843	.977478	.977856	.362637
X 2	.824351	.663843	1.000000	.665636	.697575	.548053
X 3	.939734	.977478	.665636	1.000000	.965691	.309629
X 4	.947404	.977856	.697575	.965691	1.000000	.384097
X 5	.508470	.362637	.548053	.309629	.384097	1.000000

The table of intercorrelations (Table 6.8) is useful to compare the degree of relationship between each independent variable (x_i) and the dependent variable (Y), and also between each of the independent variables. A high correlation between any independent variable and the dependent variable indicates a good predictor, but intercorrelations between any two variables is detrimental because it means that both variables are saying the same thing. Inspection of the table shows a high degree of correlation between $x_1 - x_3$, $x_1 - x_4$, and $x_3 - x_4$, indicating that it is unnecessary to include all these variables.

Other statistical measures are considered before the final equation is formed; the reader may like to pursue the subject in greater depth with a statistician (*see also* Bibliography).

The use of the formula is illustrated in the following example:

number of books	8	4	2	3	1	1	3	= 22
number of location	C2	F3	A1	H8	K7	P11	G2	= 7

Basic time for making up and packing an order of 22 books, collected from 7 locations is:

$$\text{basic time} = 1 \cdot 747 + (0 \cdot 393 \times 22) + (0 \cdot 618 \times 7)$$
$$= 14 \cdot 72 \text{ b.m.}$$

Clearly, because other variables have been eliminated the time for collecting one book is not 0·393 b.m. This value is merely a constant for variable x_1.

The work-load for the department may be found by using the mean time for collecting the average order, which from Table 6.5 is seen to be 11·3013. With the addition of a 15 per cent relaxation allowance this becomes 12·9965 s.m., or 13 s.m. approximately.

Thus in a day of 460 min the output per man at standard performance is $460/13 = 35$ orders (to the nearest whole number).

The combination of M.R.A. and the computer provides the work-measurement practitioner with a very powerful tool for measuring jobs with a highly variable work content. In the foregoing description certain statistical checks have been omitted, not because they are unimportant but for reasons of space. It is recommended that the analyses, and investigation of variables, is treated seriously, and by one trained in the use of statistical method. The practitioner then has at his disposal another interesting and valuable technique.

6.7 LINEAR PROGRAMMING (L.P.)

As an alternative to M.R.A., analysis of the equations may be performed using linear programming. The approaches are similar in the initial stages, but the methods of solution are based on different mathematical concepts. M.R.A. is based on the method of least squares, whereas L.P. is based on least differences.

One of the most common methods of solution for this type of problem is one known as 'simplex'. The procedure, while simple in essence, is extremely time-consuming, and requires much calculation. The method of simplex is described in many textbooks devoted to operational research or advanced mathematics (*see* Bibliography). As in the case of M.R.A., once again computer 'package' programs are available for this procedure, which is ideally suited for solution by computer.

Basically the problem is to solve a set of equations (or often inequalities) compiled from the observations. For example, using the data from Table 6.4, the inequalities would be:

1. $10a_1 + 3a_2 + 0 \cdot 008a_3 + 3a_4 + 12a_5 \leqslant 7 \cdot 38$ min.
2. $20a_1 + 5a_2 + 0 \cdot 027a_3 + 4a_4 + 3a_5 \leqslant 13 \cdot 90$

....

30. $22a_1 + 5a_2 + 0 \cdot 025a_3 + 4a_4 + 2a_5 \leqslant 14 \cdot 31$

THE MATHEMATICS OF MEASUREMENT

These may be converted into equations by the addition of 'slack' coefficients to each model:

1. $10a_1 + 3a_2 + 0.008a_3 + 3a_4 + 12a_5 + k_1 = 7.38$
2. $20a_1 + 5a_2 + 0.027a_3 + 4a_4 + 3a_5 + k_2 = 13.90$

. . . .

30. $22a_1 + 5a_2 + 0.025a_3 + 4a_4 + 2a_5 + k_{30} = 14.36$

Column totals 476   161   0·497   99   177

The expression to be maximized is formed from the totals of the column counts, thus:

$$476a_1 + 161a_2 + 0.497a_3 + 99a_4 + 177a_5$$

It is not possible to demonstrate the process graphically because of the impracticability of displaying the data in six dimensions. However, the principle will be illustrated by considering a much simplified example of just one independent variable (collecting books, x_1), with a second (number of locations, x_2). Suppose the study were made for the following:

x_1 = walk to specified location, select the required book(s), if available, and return to the work station with those books collected.

x_2 = number of different locations visited.

If six observations are made on six different orders, the results could appear thus:

Observation	No. of books	No. of locations	Time
1.	20	1	20·4
2.	13	11	14·6
3.	22	26	27·0
4.	10	20	20·0
5.	15	4	13·2
6.	1	1	1·4

Inequalities may be derived from the general form: $a_1 x_1 + a_2 x_2 + \leqslant Y$, thus:

$$20a_1 + a_2 \leqslant 20.4$$
$$13a_1 + 11a_2 \leqslant 14.6$$
$$22a_1 + 26a_2 \leqslant 27.0$$
$$10a_1 + 20a_2 \leqslant 20.0$$
$$15a_1 + 4a_2 \leqslant 13.2$$
$$a_1 + a_2 \leqslant 1.4$$

Using maximum values, these may be treated as equations for the purposes of calculation:

$$20a_1 + a_2 = 20.4$$
$$13a_1 + 11a_2 = 14.6$$
$$22a_1 + 26a_2 = 27.0$$
$$10a_1 + 20a_2 = 20.0$$
$$15a_1 + 4a_2 = 13.2$$
$$a_1 + a_2 = 1.4$$

A graph of a_1 against a_2 may now be plotted in the usual way, by putting a_2 equal to zero, and plotting a_1 on the Y axis, and then a_1 equal to zero and plotting a_2 on the X axis, and joining the two points. This is repeated for all six lines. The complete graph is shown in Figure 6.11.

The 'feasible' area is shaded in Figure 6.11. This area contains the feasible solutions to the problem. The six observations were originally listed as 'less than' inequalities, and any solution within the shaded 'feasible area' is a solution which, in fact, *is* less than the given values. The idea is to find the maximum point within the area which also satisfies the *objective function* which, as explained above, is formed from the column totals. In this particular example the

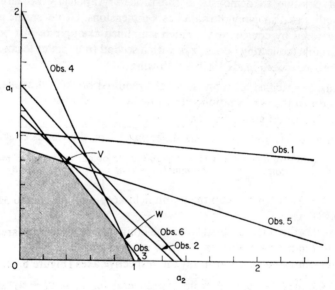

Figure 6.11 Models for six observations of the task of collecting books: the feasible area is shaded

column totals, respectively, are 81 (for the values of the x_1s) and 63 (for the x_2s). Maximization is achieved as follows:

1. The maximum values of a_1 and a_2 obviously lie as far away from the zero (origin) as possible, but as the solution must lie within the feasible area, the furthest point from zero must be one of the vertices on the boundary of the shaded area.
2. The objective function formed from the column totals is: $81a_1 + 63a_2 = K$, where K is any constant.
3. This may be rearranged into standard form for a straight line, $y = mx + c$, and becomes:

$$a_1 = \frac{63}{-81}a_2 + \frac{K}{81}$$

THE MATHEMATICS OF MEASUREMENT

4. The equation may be plotted on the graph as a line with a negative slope (m) equal to $\frac{63}{81}$ by joining the point 0·63 on the a_1 axis to the point 0·81 on the a_2 axis (line MN). Because K is an unknown quantity the line may exist *anywhere* on the graph as long as it has the constant slope of $-\frac{63}{81}$.
5. To maximize the objective function within the shaded area a ruler is placed

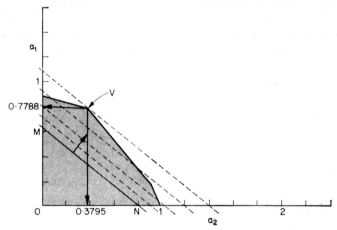

Figure 6.12 The objective function added to Figure 6.11 to obtain the vertex of maximum value

on the line MN, and moved up the graph field keeping it parallel to MN (see Figure 6.12).
6. The *last* vertex the ruler reaches (V) on the boundary of the shaded area is the solution to the problem. (The first vertex reached was W.)
7. The values of a_1 and a_2 are read off the respective axes (Figure 6.12). These are $a_1 = 0\cdot7788$, and $a_2 = 0\cdot3795$.

Linear programming is another very useful technique to augment the methods of work measurement available to the practitioner who wishes to explore the wider fields of the discipline of measurement.

7 Analysis Stage

7.1 BASIC PROCEDURE

Previously, in Section 3.1, it has been shown that all work measurement falls conveniently into three distinct stages, which may be called analysis, measurement, and compilation. Additionally there is a preceding stage of preparation, and a succeeding stage of synthesis. Although differing in detail, the principles of preparation, analysis, compilation, and synthesis are common to all techniques of measurement: it is the form taken by the measurement which distinguishes one technique from another. Therefore it is possible to present a basic-procedure pattern which should be followed in any work-measurement study. The basic procedure may be varied to suit the method of measurement selected.

Although the main objective is to derive a standard time for a defined operation, clearly it is necessary to make adequate preparation to ensure that the study is conducted in a methodical way, with minimum disruption of the job being studied. The five stages will be considered in detail in the following chapters.

The schematic diagram of Figure 7.1 shows the progression of a work-measurement study from preparation to synthesis, starting with consultation with all interested parties. Preliminary observations are made to determine the type of work being performed, and the manner of its execution, as these factors will influence the choice of a suitable method of measurement. Selection of the most appropriate method is dealt with in Section 7.3.

When 'direct observation' methods are used (such as estimating, time study, rated-activity sampling, tape-data analysis) it is essential that the operator selected for study is adequately trained and fully aware of the correct working method, and capable of achieving a reasonable pace of working. This is necessary even when the visualization P.M.T.S. methods are employed to obviate the variations in motion pattern which occur with changes in pace. Should these conditions not be satisfied the supervisor must be consulted with a view to selecting a more suitable subject.

The work-measurement study may be required for the purpose of comparing existing and proposed methods, in which case the study may proceed using the agreed motion-patterns. However, if the purpose be to compute a standard time for forward planning it will be necessary to make a study of a job which does not yet exist; a study which may be carried out only by visualization. Ideally the work-measurement analyst should receive a comprehensive training in work study and have the ability to detect bad methods and ineffective motion-patterns.

The analysis stage is chiefly concerned with the reduction of the complete job

ANALYSIS STAGE

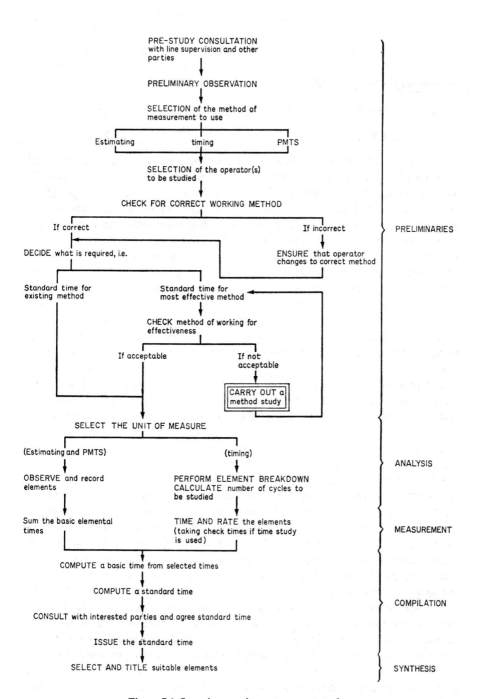

Figure 7.1 Steps in a work-measurement study

into its constituent elements. This breakdown is preceded by a decision on the unit of measure to be adopted. The unit of measure is the basic unit upon which the standard time will be based; examples are 'X s.m.s per part', 'Y s.m.s per box of components', 'Z s.m.s per square yard of floor cleaned', and so on. The selection of a suitable unit of measure is of fundamental importance, and until this has been chosen the measurement stage cannot begin.

The method of analysing a job into elements is described in Section 7.4. Definitions of elements are common to all work measurement, but the characteristics and magnitudes of elements varies with the methods of measurement adopted. The final step in analysis is the determination of the number of cycles to be studied. This may be achieved through the application of statistical theory, as described in Section 7.5.

The greatest differences between the techniques of work measurement are found in the actual measurement processes, but once measurement has been effected the addition of allowances is common to almost all systems.

It is usual for the standard time to be agreed with the interested parties, but this should not be taken as an opportunity for haggling or rate-fixing. The competent analyst should have enough confidence in his work to resist all attempts to debase his standard time. In case of dispute, and where, in fact, there are grounds for suspecting that the standard is not appropriate, a further study (or check study, such as a *production study*) may be requested by management to confirm or amend the original findings.

Finally, the elements of the accepted standard time are examined, and filed under appropriate headings for use as synthetic times, which may be extracted subsequently for incorporation into future standard times. The procedure described in this section will be considered in more detail in this and succeeding chapters.

7.2 DEFINING THE JOB

Classical work study advocates the preceding of work measurement by method study. It is argued that the practitioner should not waste his time measuring jobs which are being performed using the wrong motion-pattern. This should not be taken as an invariable rule to be followed without regard for the circumstances. Some systems of work measurement used in controlling indirect work (described in Chapter 15) are designed for rapid institution of control, after which the more lengthy method studies may be applied as the second phase. Again, the standard time may be required for an existing method to afford a means of comparing this method with an alternative proposed revision.

In all cases, whether existing or new methods are studied, the way in which the job is performed must be clearly defined. This is essential for the settling of arguments, should the standard time be in dispute, or circumstances demand that the job be remeasured. Each standard time is unique to a specified method, and subsequent revision of that agreed method will invalidate the standard.

The revisions to which reference has been made include changes in equipment,

ANALYSIS STAGE

TIME STUDY SPECIFICATION

OPERATION No.	PART No.		STUDY No.		
OP. DESCRIPTION	DESIGN		DEPT./SECT.		
	SIZE		PLANT DETAILS		
JOB DESCRIPTION	MATERIAL		WORKING CONDITIONS		
	STUDY BY		DATE		
JOB SKETCH	JIGS, TOOLS, ETC.		STUDY TIME ANALYSIS	HRS.	MINS.
			EFFECTIVE TIME		
			INEFFECTIVE TIME		
			R.A.		
			CHECK TIME		
			TOTAL RECORDED TIME		
			ELAPSED TIME		
			ERROR		
	FEEDS		A.V. RATING		
	SPEED		OUTPUT		
WORK PLACE LAYOUT			REMARKS		
STANDARD TIME PER					

Figure 7.2 An example of a time-study specification

tools, and materials, as well as alterations to motion patterns and workplace layout. Consequently the work specification should include details of how the job is performed, duties and responsibilities of the operative, descriptions of the equipment, tools, jigs, and fixtures employed in carrying out the task and, of course, a diagram of the workplace arrangement. A suggested form for setting out such details is shown in Figure 7.2.

7.3 SELECTING THE MOST APPROPRIATE TECHNIQUE

Many organizations regard the techniques of work measurement as entirely separate entities, even to the extent of progressing through the levels of sophistication, say from estimating to time study, and finally through to P.M.T.S. The methods of estimating, timing, and P.M.T.S. do differ widely in their form and application, but, nevertheless, may be regarded as different tools which are useful in the particular circumstances to which they are applicable. Micro-P.M.T.S. offers the precision required for measuring the short-cycle manual work encountered on the flow-production line, while the maintenance foreman demands a technique with the qualities of rapid application and simplicity such as comparative estimating.

One factor which influences the practitioner in his choice of technique will be the purpose for which the standard time is required. These purposes will include evaluation of method studies, installation of financial incentive schemes, manning, control of labour through measurement of utilization, production control, costing of labour, estimating, and so on.

If ways and means can be agreed to enable less accurate measurements to be used in appropriate carefully defined circumstances, measurement-time and expense will be saved. There is a very real need, however, for carefully in-built safeguards—the situation must not be allowed out of control—there may well be a need to gain trade-union agreement to new measurement practices.

Measurement need only to be sufficiently accurate to cater for those basic purposes for which the measurements are to be used. To choose the appropriate measurement technique the purpose must be defined and fully appreciated, namely method study, incentives, manning, production control, costing of labour, and estimating. For example, the application for purposes of method study may be to examine present methods, to compare methods, or to estimate savings. Incentives may be concerned with individual, team, group, or ancillary work. All these are related to the variety of work in the broadest sense, i.e. frequency of occurrence, methods variation, length of job-cycle, length of production run. Other factors considered are:

incentive scheme to be employed
whether work is inside or outside the cycle or machine-running time
number of operators in working group
length of payment and calculation period

7.31 DETAIL SITUATION EXAMPLES

Method study. In connection with short-cycle repetitive work, the likely requirement is for P.M.T.S. or time study. Broader situations may be dealt with by activity sampling.

Incentives. For individual incentives a high precision is usually required, say ±5 per cent. The most obvious choices for straight proportion or geared schemes is P.M.T.S., time study, or standard data from these sources.

For group incentives (i.e. for people who are not interdependent but have a common link or purpose), averaging will reduce the accuracy needed for such

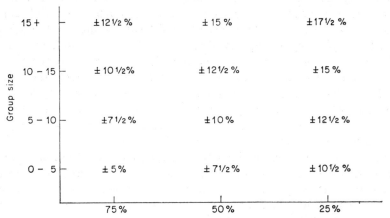

Note: Progressive cumulative averaging or extending bonus calculation period will diminish accuracy need

Figure 7.3 Selecting the required accuracy

schemes, which are usually geared schemes. Further criteria for consideration are the size of the group, the variety of work (i.e. frequency of change of work-type, length of run, total occurrence of one variety per unit of time), and possibly the total estimated standard hours on each variety per time-period; say for example 25 per cent of group on one product in each working week. These points are illustrated in Figure 7.3.

Standards for ancillary staff who supply service requirements must take into account various production levels and requirements. Thus an accuracy of, say, ±20 per cent is acceptable.

7.32 TECHNIQUES AVAILABLE

Some major techniques available, in order of diminishing accuracy (see Figure 7.4), are:

First-level P.M.T.S. (±2½ per cent). Specifically for very short-cycle high production work with cycle times of the order of up to 0·5 min.

Second-level P.M.T.S. (±5 per cent). Should be used for cycle times over 0·5 min. This form should be used as a basis for most standard data forms, for example to establish bench-marks for higher-level systems. Where standard data

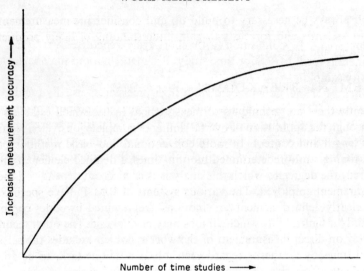

Figure 7.4 The increase of accuracy with the number of studies

are not required bench-marks can be established by techniques of much reduced accuracy, such as method recorded by tape-recorder, as in tape-data analysis.

Time study (± 5 per cent). This technique is basically for the intermediate-length (2·0 to 15·0 min) cycle task, particularly for incentive schemes based on individual operators.

Variants may be:

block timing and rating (± 10 per cent)
overall timing and interval rating ($\pm 12\frac{1}{2}$ per cent)

Estimating. The reliability of this group of techniques depends on the skill of the estimator. It may be used in cases where rapid application is essential, but where the variability in the work precludes the use of precise measurement.

Activity sampling ($\pm$ variable). Generally used for wide spread of tasks and locations with long and often non-repetitive cycles. Used also for method-study overall criteria. Activity sampling cannot be accurate in the absence of pre-methods studies to isolate unrecognized wasteful activities.

7.33 COMMENTS AND CONCLUSIONS

The position may be summarized as follows:

1. The choice of measurement techniques of *appropriate* accuracy radically affects study time and general measurement economics.
2. It is impracticable to try to lay down specific overall rules for the choice of appropriate technique in all possible situations.
3. There are dangers in the too-liberal interpretation and application of techniques of reduced accuracy.
4. It should be borne in mind that method study should help to indicate appropriate measurement technique.

5. It will always be necessary to build up and consider pre-measurement production records and forecasts to gain understanding of likely product-type variations and runs.

7.4 ELEMENT BREAKDOWN

Currently there are techniques of measurement in use which establish basic times through the application of overall timing of complete job cycles, regardless of their length and content. In more conventional and basic methods of estimating, timing, and predetermined motion–times, a detailed elemental analysis is required; the degree to which this analysis is carried out depends on the type of measurement employed. The various systems of P.M.T. have specific numbers of clearly defined elements or 'motions' (represented by codes comprising letters and/or figures) into which all jobs may be analysed. The other techniques which rely on direct measurement of the worker (which excludes the setting of standard times by synthesis) regard each job as a unique task, to be analysed into elements whose size and work-content are carefully selected.

Consequently all trained and experienced P.M.T. analysts studying the same job should produce similar, and often identical, analyses, whereas the time-study practitioners may vary in their interpretations of the job in terms of element breakdown.

Element breakdown is necessary for several reasons. The systems of predetermined motion–times naturally require the job elements to correspond with the standard elements of the particular P.M.T. system. The time-study practitioner and the analytical estimator need to reduce the job cycle to elements which can conveniently be rated and timed. Performance rating is necessary to allow for possible variations in the pace of work, especially during long-cycle work. The analytical estimator finds it easier to estimate the steps of a task separately than to assess the complete job. Moreover the extraction of parts of the job for use as synthetic times may be desired, and analysis into appropriate elements facilitates this selection.

According to B.S.I. Glossary of Terms Used in Work Study an element is:

> A distinct part of an operation selected for convenience of observation, measurement, and analysis (Term number 42003).

A work cycle is defined as

> The sequence of elements which are required to perform a task or to yield a unit of production. The sequence may sometimes include occasional elements (Term number 42001).

The term 'unit of production' in the above definition is a dimension which must be clearly defined in each individual case. Consequently it is necessary to first decide on the unit of production to be adopted before the work cycle, and hence the standard time, may be specified for a job.

Different elements must be recorded and timed in different ways appropriate to their characteristics. To facilitate analysis, measurement, and computation, elements may be classified according to their peculiar qualities. The classifications are:

1. *Repetitive element*. An element which occurs in every work cycle of an operation (Term number 42007).
2. *Occasional element*. An element which does not occur in every work cycle of an operation, but which may occur at regular or irregular intervals (Term number 42006).
3. *Constant element*. An element for which the basic time remains constant whenever it is performed (Term number 42004).
4. *Variable element*. An element for which the basic time varies in relation to some characteristics of the product, equipment, or process, for example, weight, dimensions, quality, etc. (Term number 42005).
5. *Manual element*. An element performed by a worker (Term number 42010).
6. *Machine element*. An element performed automatically by a power-driven machine (or process) (Term number 42011).
7. *Governing element*. An element occupying a longer time than that of any other element which is being performed concurrently (Term number 42008).
8. *Foreign element*. An element observed during a study which, after analysis, is found to be an unnecessary part of an operation (Term number 42009).

The above definitions for elements may be used in combination. Thus one element, for example, may be a repetitive, constant, manual element while another may be occasional, constant, machine element. Some typical elements are depicted in Figure 7.5 and analysed below.

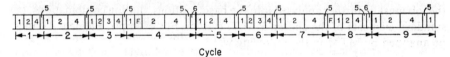

El. No.	Abridged description	Basic time
1.	Collect order and assess the best route	1·62 b.m.
2.	Pick books for order	2·86 b.m. (average)
3.	Enter deficiencies on invoice	2·10 b.m.
4.	Wrap order	3·24 b.m.
5.	Tie order	0·30 b.m.
6.	Place 4 orders on pallet	0·64 b.m.
F.	(Foreign element)	—

Figure 7.5 Diagrammatic representation of several cycles in a job, showing the various types of elements

The example is based on the work of a warehouseman employed by a large publisher. Books on each order are picked, and assembled at a central point for wrapping and despatching.

Element 1. Collect order from supervisor's rack, and study this to assess the most effective route for collecting books.

ANALYSIS STAGE

This element must be performed each time an order is to be filled and takes a constant time to perform. Thus it is a repetitive, constant, manual element.

Element 2. Walk to the appropriate rack and select the book ordered. Place in collecting bag before moving to next rack.

This element is a repetitive element, but the distance walked varies for each book picked. Thus this is a repetitive, variable, manual element.

Element 3. Books which are out of print or not available are entered on the invoice as 'temporarily unobtainable'. This occurs on roughly 35 per cent of the orders. Consequently the element is occasional, constant, and manual.

Element 4. Books are removed from the collecting bag, suitably arranged on a sheet of stout brown paper, and wrapped. Sizes of orders vary according to how many books are ordered. The element is repetitive, variable, manual.

Element 5. Parcels are placed on the tying machine, and automatically tied. This is a repetitive constant element, and mainly machine-performed.

Element 6. Parcels are placed on a box pallet for collection by the dispatch operatives, so the element is occasional (because four parcels at a time are placed on the pallet), constant, and manual.

7.41 COMPARISON OF ELEMENTS

In the foregoing analysis, Element 1 was selected as a suitable element for time study because it was of reasonable duration for rating (i.e., the pace of working would not vary significantly during this element), and there was a natural break-point between perusal of the order and commencing the walk to the rack.

This breakdown would not be suitable for measurement by P.M.T.S., because the job must be further analysed into the basic motions, motion combinations, or standard elements of the chosen P.M.T. system. For example, Element 4 could be analysed by MTM in the following way:

L.H.		t.m.u.	R.H.	
		29·0	B	Bend down
	R–B		R–B	Reach to parcel
		3·5	G1B	Grasp parcel under end
Grasp under end	G1B	3·5		
As R.H.	SC5	2·2	SC5	Take the weight
	M4B5	7·3	M4B5	Lift parcel to carry
		31·9	AB	Arise (straighten body)
		18·6	TBC1	Turn
		51·0	W3P0	Walk to pallet
		29·0	B	Bend down
	M4B5	7·3	M4B5	Place parcel on pallet
	R2E	3·8	R2E	Remove fingers from under
		31·9	AB	Arise
		18·6	TBC1	Turn
		51·0	W3P0	Walk to stack
	Total =	288·6		

The use of MTM-2 reduces this detailed analysis to X elements, consisting of:

	L.H.	t.m.u.	R.H.	
		61	B	Bend-arise
		14	GC5	Grasp under end of parcel
Grasp under end	GC5	14		
As R.H.	GW3	3	GW3	Take the weight
	PA15	6	PA15	Lift to carry
	PW5	1	PW5	ditto
		72	4S	Walk to pallet
		61	B	Bend-arise
	PA15	6	PA15	Place on pallet
		72	4S	Walk to stack
	Total =	310		

MTM-3 reduces the number of elements even further:

	L.H.	t.m.u.	R.H.	
		61	B	Bend-arise
As R.H.	HA15	18	HA15	Lift parcel
		72	4SF	Walk to pallet
		61	B	Bend-arise
	TA15	7	TA15	Place on pallet
		72	4SF	Walk to stack
	Total =	291		

7.42 UNIT OF MEASURE, AND FREQUENCY OF OCCURRENCE

Before any attempt can be made to measure the work it will be necessary to select a unit of measure (or unit of production) upon which the standard time for the job may be established. This is necessary so that in the final compilation of the standard time the basic time for all occasional elements may be apportioned over the unit of production. Apportionment may be achieved by observing the frequency with which the occasional element occurs.

When occasional elements appear at regular intervals (such as Element 6 of the previous example) the frequency may be determined simply, by observing relatively few work-cycles. Elements whose occurrence is irregular need to be investigated over a representative period. The necessary data are obtained through direct observation (by the analyst or worker), from data collected by random-sample observation, or from documentary evidence, such as the invoices in the previous example. Statistics must be collected also for variable elements, so that a representative average may be found which will be appropriate for use in the final analysis. It will be noticed that in both cases the frequencies of occurrence are essentially averages.

To calculate the frequency of occurrence it is necessary to answer the question 'How often does this element occur per unit of measure?' and then to express the frequency as:

ANALYSIS STAGE

$$\frac{\text{number of times the element occurs in a given time}}{\text{number of units produced in this time}}$$

the ratio being reduced to its lowest terms.

The previous example will be used as a vehicle to illustrate the selection of a suitable unit of measure. In this example three obvious choices present themselves as convenient units to adopt. The standard time may be determined: (i) per book picked, (ii) per order filled, or (iii) per pallet of orders completed. The three choices will be analysed separately.

Element 1. This element occurs every time and does not vary, so it is a repetitive, constant element. The first case to be considered is a unit of measure of 'per book'.

The order is collected each time a collection of books is made. The frequency of occurrence is determined from the ratio given above, and to the question 'How often does this element occur per unit of measure ('per book')?' the answer is, 'Once for every five books picked.' This may be expressed as '1 in 5' or 1/5. The basic time allowed 'per book' for this element, therefore, is 1/5 of the actual element basic time, i.e. $1/5 \times 1\cdot62$ b.m., or $0\cdot32$ b.m. Alternatively, suppose the unit of measure chosen had been 'per order filled', the reply to the question would have been 'Once for each order', which may be expressed as 1/1. Hence the basic time remains as $1\cdot62$ b.m.

If the 'pallet of orders' had been chosen as the unit then four orders make up one pallet, and hence the frequency of occurrence is 4/1; the basic time allowed, therefore, is $4 \times 1\cdot62$ b.m., or $6\cdot48$ b.m.

Element 2. This element is repetitive, but varies with each order because the books are situated in different locations and at different distances from the packing table. The basic time for this element must, of necessity, be an average, based on many representative cycles.

Again, if the unit of measure chosen is 'per book' the frequency of occurrence will be 1/1 because a book is picked 'once per book'. A unit based on the *order* will yield a frequency of 5/1 because 'Five books make up one order'. There are 20 books per pallet (5×4), producing a frequency of occurrence of 20/1.

The remaining elements are analysed similarly. Applications of these principles are given later in this chapter under 'Compilation', Section 7.6.

7.5 DETERMINING THE NUMBER OF CYCLES TO STUDY

The number of cycles which must be timed in order to attain a desired level of accuracy depends on the duration of the basic time of the cycle, and on the dispersion of the basic times obtained during the study. The measure of dispersion used is the standard deviation. Although such calculations are usually restricted to the timing techniques (i.e. time study), they are applicable to situations where variable work is measured by P.M.T.S.

There are five notable methods for estimating the number of cycles, but in all cases it is inevitable that the study be interrupted while the necessary calculations are made. Use is made of a nomogram, graph, control chart, or slide-rule calculations.

Of course, with short-cycle jobs it would be impossible for the observer to extend basic times, look up nomograms, or calculate cumulative averages and plot these, while simultaneously performing the study. One method of avoiding interruption is to time ten cycles as a pilot study, and then perform the calculations before starting the study proper.

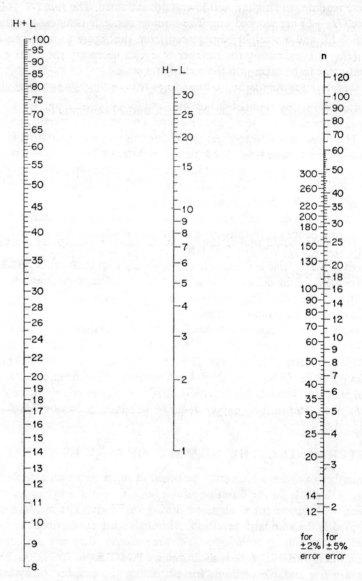

Figure 7.6 A nomogram for the number of cycles to study for a given statistical accuracy

7.51 USE OF THE NOMOGRAM

The most simple method, but one which sacrifices some accuracy to achieve simplicity, is the nomogram. This chart, otherwise known as an alignment chart, is depicted in Figure 7.6. It has been constructed from the formula on page 116.

To use the nomogram it is necessary once more to use basic times extended from the observed times. Ten readings of basic times of cycles are made, and from these readings the highest and lowest are extracted. The sum $(H + L)$ and difference $(H - L)$ are worked out. The nomogram scale is entered at the sum value $(H + L)$, and a straight edge placed from this figure to cross the centre scale at $(H - L)$, indicating the number of cycles necessary for 2 per cent or 5 per cent error to be taken, on the right-hand scale.

For example, suppose the first ten readings from a time study are used:

$$21, 28, 22, 22, 26, 28, 26, 23, 25, \text{ and } 27 \text{ centiminutes}$$

Then
$$H = 28, \text{ and } L = 21$$
$$H + L = 49, \text{ and } H - L = 7$$

The alignment chart indicates that 85 cycles must be studied for ± 2 per cent error, or 14 cycles for ± 5 per cent.

7.52 A GRAPHICAL METHOD

The graphical method makes use of the cumulative average as a means of smoothing out the fluctuations which would result if the individual basic times were plotted on a graph.

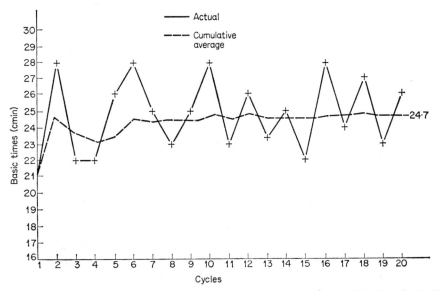

Figure 7.7 A cumulative-mean method for estimating when sufficient cycles have been studied

Figure 7.7 shows the data from similar elements of successive cycles plotted chronologically. It will be seen that there are rather violent fluctuations from cycle to cycle. For purposes of illustration these have been exaggerated, and would probably not occur to this extent in practice. The superimposition of the cumulative average (Figure 7.7) shows how the variation quickly levels off to a steady value. When this appears to happen the indication is that sufficient cycles have been timed.

If the jobs are long-cycle the observer may plot the cumulative averages as they are timed and calculated, and continue until the graph levels off substantially. With short-cycle jobs there will not be time to do this, so as before, ten cycles may be timed, calculated, and plotted as a pilot study. A decision may then be made as to whether further cycles are needed. Should this be necessary ten further observations are taken, plotted, and a further decision made.

7.53 USE OF PROBABILITY AND DISPERSION

A more exact method than those outlined above is based on statistical probability and the particular property of the standard error. In place of the nomogram or graph a formula for n, the number of cycles to be studied, is used as follows:

$$n = \frac{4\sigma^2}{E^2 \bar{X}^2}$$

where σ = the standard deviation of the basic cycle times

E = the error tolerated with 95 per cent confidence

$\bar{X}$ = the average basic time of the ten readings

For example, for an error of ± 5 per cent (95 per cent confidence)

$$= \frac{4\sigma^2}{0\cdot05^2 \cdot \bar{X}^2}$$

$$= \frac{4\sigma^2}{0\cdot0025 \bar{X}^2}$$

In the example used above $\bar{X} = 24\cdot8$ b.m.

$$\sigma = 2\cdot7 \text{ b.m.}$$

Therefore for ± 5 per cent error

$$n = \frac{29\cdot16}{0\cdot0025 \times 24\cdot8^2}$$

$$= 19 \text{ observations (to the nearest whole number)}$$

To halve the error, four times the number of observations must be taken. Compare these results with those obtained from using the nomogram (page 110).

7.531 Derivation of the formula. It was seen in Chapter 6 that individual readings such as basic times were scattered about their mean value; the extent of this

ANALYSIS STAGE

scatter (or dispersion) being measured by the *standard deviation* of the values. Similarly the average of groups of readings (known as sample means) are themselves scattered about their grand average; having their own standard deviation, usually called the *standard error* (of the mean).

If during a time study the first ten cycle basic times are averaged, this mean value might not be equal to the true mean basic time because of the statistical errors which are inevitably present whenever small samples such as ten observations are taken. (Clearly if fifty cycles are timed a much more accurate assessment of the true basic time will result.) However, it is possible to estimate the error to which the average value of the ten readings is subject, by using the standard error as a measure of this uncertainty.

The standard error of mean values is found by dividing the standard deviation (σ) for all cycles possible (i.e. the 'population') by the square root of the sample size (n), or

$$\text{s.e.} = \frac{\sigma}{\sqrt{n}}$$

But clearly it would not be possible to measure all cycles performed over all the years by all workers, so it is inevitable that an estimate must be made. The estimate is based on the small sample of just ten cycles studied. From Chapter 6 the standard deviation is given by:

$$\sigma = \sqrt{\frac{\Sigma fd^2}{n-1}}$$

where d is the deviation of readings from their mean values.

By way of illustration the basic times from Figure 7.7 will again be used. The mean value of these is

$$\frac{248}{10} = 24\cdot8 \text{ b.m.}$$

The standard deviation from Figure 7.7 is 2·7 b.m., and this sample standard deviation must be used as the standard deviation for *all* cycles; there being no other source for this data.

From this, the standard error for means is

$$\frac{2\cdot7}{\sqrt{10}} = 0\cdot85 \text{ b.m.}$$

This is shown graphically in Figure 7.8.

As explained in Chapter 6, roughly 95 per cent of the area under the normal distribution is between the limits of ±2 standard errors each side of the mean, that is, between 24·8 + 1·7 and 24·8 − 1·7 b.m. This means that the true basic time should be between 26·5 and 23·1 b.m., and this statement can be made with 95 per cent confidence.

In general terms the inherent error is 2 standard errors, and this is denoted by $E.\bar{X}$, because it is E per cent of the mean value of $\bar{X}$.

E

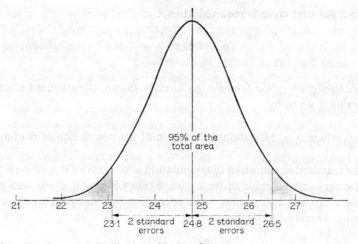

Figure 7.8 A normal distribution with data from the text example superimposed

or
$$E.\bar{X} = \frac{2\sigma}{\sqrt{n}}$$

Then the number of readings to take for a given error (E) is:

$$\sqrt{n} = \frac{2\sigma}{E.\bar{X}}$$

$$n = \frac{4\sigma^2}{E^2.\bar{X}^2}$$

7.54 USE OF THE RANGE

The foregoing method requires the computation of the standard deviation, which is rather laborious in the absence of any mechanical means of calculating. The relationship which the range of readings bears to the standard deviation makes possible the use of a simplified method of determining the number of observations.

Once again the method requires data from a pilot study, such as the first ten observations. From these readings the highest (H) and lowest (L) basic times are extracted and placed in the formula

$$n = 676\frac{(H-L)^2}{(H+L)}$$

for up to 5 per cent error (with 95 per cent confidence).

If n exceeds 10, then additional cycles must be studied to make up the deficiency.

From the data already used above,

$$H = 28, \text{ and } L = 21$$
$$H - L = 7, \text{ and } H + L = 49$$

Thus $\quad n = 14$ approximately.

ANALYSIS STAGE

If up to 2 per cent error is required, then

$$n = 4{,}225\frac{(H-L)^2}{(H+L)}.$$

7.541 Derivation of the formula. As already shown, the standard error of the mean value is given by

$$s = \frac{\sigma}{\sqrt{n}}, \text{ where } \sigma \text{ is the standard deviation of the population of readings} \quad (A)$$

As the standard deviation of the population is not known it must be estimated from the very small sample of ten cycles. From Chapter 4 it was seen that the range of a normal distribution was approximately equivalent to six standard deviations. Therefore:

$$\sigma = \frac{\text{range}}{6} \quad \text{(approximately)}$$

Ten cycles can hardly represent the population of all readings, therefore certain corrections must be made to improve the estimate. For a sample of 10 it has been found that the standard deviation of a population can be estimated by dividing the range (R) of the sample of ten by 3·078 (see standard textbooks on statistics for a proof).

From this
$$\sigma = \frac{R}{3 \cdot 078} \quad (B)$$

and
$$\text{standard error} = \frac{R}{3 \cdot 078 \sqrt{n}} \quad (C)$$

from combining (A) and (B)

Now the range is the highest reading minus the lowest reading of the sample of 10,

or
$$H - L \quad (D)$$

Substituting in (C),
$$s = \frac{H-L}{3 \cdot 078 \sqrt{n}} \quad (E)$$

For 95 per cent confidence two standard-error limits are necessary to give the required error E.

Thus
$$2s = E = \frac{2(H-L)}{3 \cdot 078 \sqrt{n}} \quad \text{(multiplying (E) by 2)} \quad (F)$$

The mean cycle-time may be estimated from the range extremes:

$$\frac{H+L}{2}$$

and the error required may be within 5 per cent of the mean:

i.e. $\quad 5\%$ of $\dfrac{H+L}{2}\quad$ or $\quad\dfrac{0.05(H+L)}{2}\quad$ (G)

Equating (F) and (G) $\quad \dfrac{0.05(H+L)}{2} = \dfrac{2(H-L)}{3.078\sqrt{n}}$

Rearranging for n, $\quad \sqrt{n} = \dfrac{4(H-L)}{0.05 \times 3.078(H+L)}$

therefore $\quad n = \dfrac{16(H-L)^2}{0.0025 \times 9.474(H+L)^2}$

$\quad\quad\quad\quad\quad = 676\dfrac{(H-L)^2}{(H+L)^2}$

7.6 COMPILATION OF THE BASIC TIME

Methods of calculating the basic time from its constituent elements vary according to the technique of measurement used. The basic times yielded by P.M.T.S. need only to be summed (after the application of their respective frequencies of occurrence).

Element times derived through the use of a timing technique require the modification of the rating factor, as described in Section 5.23. Graphical methods are available for the purpose, as described in Chapter 9.

Inevitably the analyst using a timing method will, on occasions, be faced with element basic times which are not representative of the element he is considering. Rating and timing errors will be responsible for variations in 'constant' element times, as will subtle changes in motion patterns, and almost indescernible variations in materials. The analyst must decide which elements' times are to be discarded as unrepresentative and due to error, and which are to be included in the *selected* basic time.* Consider, for example, the following extended time for element 1 of the preceding example:

   Element 1 (centiminutes) 36, 38, 37, 37, 36, 41, 38, 36, 38, 37, 47, 36.

The average is 38 cm, and clearly most of the readings are within 2 cm of this value with the exception of those reading 41 and 47. The analyst may argue that 41 is 'not so bad', but 47 is 'rather a long way out'. But where does one draw the line? What is 'not so bad' and what is 'too bad'? In practice this is usually decided by the analyst purely by inspection, and using his intuition and experience. However, there are statistical methods available for rejecting those readings which have occurred in error. Clearly, whatever method is used, care must be taken to reject only those readings which are definitely erroneous.

The statistical method is based on the theory outlined in Sections 6.13 and

---

* Selected time is defined by B.S. Term No. 35003 as the time chosen as being representative of a group of times for an element or combination of elements. These times may be either observed or basic, and should be denoted as selected observed or selected basic times.

ANALYSIS STAGE

6.14. These deal with the two important concepts of standard error and the normal distribution. Reference to these sections will show that, if the values of the standard error and the mean of the sample element readings be known, then recourse may be had to the theory of probability to differentiate between two occurrences: those readings which are near enough to the mean to be acceptable and representative, and having only chance variation, and those readings which are so far from the mean that chance variation is ruled out and there must be something wrong with the data (i.e. watch error, rating error, foreign element creeping in, and so on). The control chart used by the statistical quality-control inspector serves this purpose.

The statistical theory is based on the distribution of means—not individual readings. It is necessary, therefore, to obtain averages from the individual basic times by first arranging these into equally sized groups.

Control charts are constructed in the following way:

1. The basic times for one element are arranged into groups (for example, into fours).
2. The means ($\bar{X}$) and ranges (W) of each group of four are calculated.
3. The average range ($\bar{W}$) and the grand average ($\bar{\bar{X}}$) of *all* readings is found and used as the centre line of the chart (Figure 7.9).
4. The control lines are computed from the formula: $\bar{\bar{X}} \pm A\bar{W}$
   The value A is obtained from quality-control tables contained in most advanced statistics textbooks. For samples of four the value of A is 0·75 (for 99·8 per cent confidence). The above formula gives control lines for three standard errors (i.e. for 99·8 per cent confidence, which in practice means that any average falling outside the control lines has only one in-a-thousand chance of being there purely by random chance, or put another way, there is almost certainly something wrong and chance is ruled out).
5. The means are plotted on the chart.
6. The study is considered to be satisfactory if all points fall within the control lines.

Suppose, for example, that twenty-four cycles had been timed and the basic times for one repetitive element had been grouped into fours, producing the following means and ranges for the groups of four:

$\bar{X}$ 23·5 25·0 23·5 25·0 20·5 23·8   Grand average $= 23·55 = \bar{\bar{X}}$
W 1   2   1   2   9   3   Average range $= 3·0\ \ = \bar{W}$

$$\bar{\bar{X}} \pm A\bar{W} = 23·55 \pm (0·75 \times 3·0)$$
$$\text{Upper control-line} = 23·55 + 2·25 = 25·80$$
$$\text{Lower control-line} = 23·55 - 2·25 = 21·30$$

It will be seen that the fifth group-mean is below the lower control-line, which indicates that it contains a reading (or readings) which is not representative and the group should be rejected, i.e. not included in the final calculation of basic time.

The foregoing method is useful, but considered by many to be too elaborate to warrant universal application. It is, of course, superior to the purely subjective method.

7.7 BACKGROUND TO ACTIVITY SAMPLING

Work measurement may be said to have two basic functions: the main one being to establish standard times for specified jobs. The second function is the collection of data for such purposes as assessing work-load, and utilization of employees and equipment.

Data for management purposes may be collected over representative periods, either by 100 per cent observation of the situation during that period, or, more economically, by sample studies carried out according to the principles of statistical sampling and the laws of probability. One example of such data collection is to be found in the field of inspection used widely in industry and commerce, and which may be performed on either a 100 per cent or a statistical sampling basis.

The technique in work study most nearly analogous to 100 per cent inspection is the *production study*, described in Chapter 13. The present chapter is concerned solely with the sampling technique known as activity sampling.

Originating in the textile industry in the North of England, activity sampling has become one of the principal vehicles for general data-collection in work study. Development of the technique is credited to L. H. C. Tippett, who in 1935 introduced activity sampling for the investigation of groups of machines, under the title of *snap-reading method*. Activity sampling appears under various other names, including observation-ratio study, random-observation method, and two popular in the United States: ratio delay, and work sampling.

The restrictive B.S.I. definition of work measurement excludes activity sampling from being a technique of work measurement because its purpose is not to 'establish a time for a qualified worker to carry out a specified job at a defined level of performance'. In recent years researchers have augmented the snap attribute-type observations with the addition of rating assessments, which have the effect of producing a better-weighted picture of the manual elements of the work. In so doing they have produced a method of work measurement which more closely resembles time study than its alleged parent technique of activity sampling.

7.71 DEFINITION AND SCOPE OF ACTIVITY SAMPLING

The formal definition of the technique, taken from B.S. 3138: 1969 is:

A technique in which a large number of observations are made over a period of time of one or a group of machines, processes or workers. Each observation records what is happening at that instant and the percentage of observations recorded for a particular activity or delay is a measure of the percentage of time during which that activity or delay occurs (Term number 31007).

The omission of the term *work measurement* from the definition is noticeable, although in many textbooks activity sampling is listed erroneously as one of the techniques for setting standard times.

The production study, or continuous-observation study, is of course very time-consuming and requires the observer to be aware of, and to note, every important change or happening. Inevitably he is restricted to observing one or perhaps two operators or machines. The activity-sampling study offers greater flexibility and allows the observer to move about and so cover a very much larger number of employees or machines. On the other hand the production study gives a reliable account of the situation during the period of the study, whereas sampling will always be subject to an inherent statistical error due to the very nature of its format. However, this does not detract from the usefulness of the method. The error may be predetermined and even chosen by the observer, and may be reduced to negligible proportions by taking a sufficiently large number of observations. Furthermore, human error will always ensure that 100 per cent observation and 100 per cent accuracy are not synonymous. Fact-finding surveys such as these merely indicate the situation as it exists at the time of the study. If this is supposed to represent the continuing situation, great accuracy is meaningless in the face of the not inconsiderate day-to-day variations to which the situation is prone.

Activity sampling may be used with advantage in the following circumstances:

1. To observe several employees over a working period using the minimum number of observers;
2. To assess the proportion of time spent on a particular activity or activities;
3. To assess the proportion of idle, ineffective, or waiting time;
4. To assess the amount of machine-breakdown time;
5. To observe the situation over a long period, which would be uneconomical to do by continuous observation.

The effect of spreading the study over a longer period often produces a more representative and realistic result. Activity sampling may be carried out by less highly trained staff, such as the supervisor of the section, who, being in attendance much of the time, is well placed to make the observations over a long period.

7.72 PROCEDURE

The general procedure for carrying out an activity-sampling study is common to all applications, although the detail will be affected by local conditions and the type and purpose of the study.

1. Definition of the problem

The design of the study will be influenced by the type of problem to be investigated. Problems range from simple utilization assessment (i.e. working versus idle time), to a detailed analysis of the proportion of time spent on various tasks. In addition to this 'static' type observation, the practitioner may be required to travel a route to observe mobile equipment such as fork-lift trucks or tea trolleys.

2. Consultation

The degree of consultation will depend on general management policy or 'custom and practice', but at the very least common courtesy demands that the practitioner informs the supervisor in charge and explains to him the purpose of the study. Consultation is fully explained in Chapter 4.

3. Orientation

This stage is one of collecting initial information on which to base the full study. From the information assembled an element or job breakdown is made, and the accuracy or limits of error agreed. The representative period over which the study is to extend is decided at this stage.

Further information which will enable an assessment of the number of observations to be made is gleaned from a *pilot study*, which is often carried out prior to the main study. Such a study indicates the probable proportion of time spent on various activities, and this rough estimate is used in a formula derived for the purpose of calculating the number of observations.

4. Final preparations

By this stage all the initial information has been collected and the number of observations determined. Using random time-tables or by other suitable means the appropriate random times are generated and listed in chronological order on an observation record-form, on which is shown the element breakdown. Where appropriate, routes are planned, and other preparations are made.

5. Observation

Observations are made, marking the respective columns against the random times according to the activity observed at the particular random or systematic time. A production count is made at the same time.

6. Calculations

Results of the study are analysed simply as the sum of the number of times the activity was observed, expressed as a percentage of the total number of observations in the study. Where more than one activity (or inactivity) are involved the calculation is performed for each activity in turn.

Sampling techniques produce results which are subject to statistical error. Thus the percentages just calculated are merely estimates, and must be qualified by the limits of error to which they are subject. The accuracy of the study is assessed additionally by some practitioners through the medium of the control chart, which indicates those results which are not appropriate and should be eliminated from the studies.

7. Conclusions

The study may be written up as a report on the findings of the observer together with his interpretation of the statistical data. The report is presented to management for consideration and appropriate action.

7.8 SOME MATHEMATICAL CONCEPTS

Before illustrating the technique with an example some of the basic concepts on which activity sampling is based will be considered.

7.81 NUMBER OF OBSERVATIONS

Activity sampling is based on the theories of sampling and probability. In Chapter 6 a formula was derived for calculating the limits of error to which a binomial situation such as this is subject. The reader who wishes to gain a thorough understanding of activity sampling, and in particular the statistical theory behind the technique, is advised to read and appreciate the principles outlined in Chapter 6.

The pilot study gives an estimate of the proportion (p) of time spent on a particular activity being studied. The limits of error are set two standard deviations (or, more correctly, two standard errors) from this estimate in order to ensure that the observer may be 95 per cent confident that the estimated error is correct.

The standard-error formula for a binomial distribution is (from page 77)

$$\sqrt{\frac{p(100-p)}{n}}$$

where n is the number of observations which must be made to ensure a certain

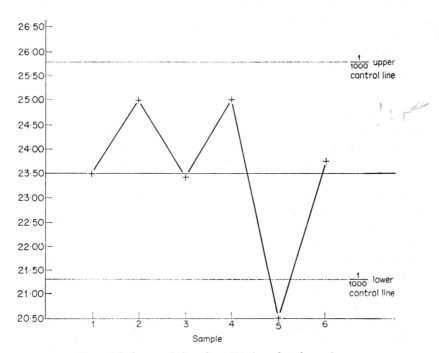

Figure 7.9 A control chart for validation of study results

required accuracy. The limits of error are to be set at two standard errors, therefore:

$$\text{limits of error } (L) = 2\sqrt{\frac{p(100-p)}{n}}$$

This formula is fundamental to activity sampling.

Rearranging this formula produces one for estimating the number of observations required to attain a required error:

$$n = \frac{4p(100-p)}{L^2}$$

In the present example the pilot study indicated a machine utilization of about 86 per cent. During the course of one day the department manager paid twelve visits to the machine section and noted that Machine No. 1 was idle on two occasions, Machine No. 2 on one occasion, and Machine No. 3 was idle on two occasions, giving an average of 1·7 in 12, or 14 per cent idleness per machine. This rough estimate of 14 per cent is used in the formula to determine the number of observations.

The observer decides the limits of error which can be tolerated consistent with economy of time. A result within ½ per cent of the true value may be desirable in certain circumstances, but in the present example this would require making a phenomenal 19,300 observations. A more realistic error is, say, 5 per cent, when the number of observations would be:

$$n = \frac{4 \times 14 \times 86}{5^2}$$

$$= 193 \text{ observations (approximately)}$$

It will be seen that to reduce the error to one-tenth of its value (i.e. from 5 to 0·5 per cent) it is necessary to increase the number of observations one-hundred-fold (193 to 19,300). This is the 'inverse square-law' effect in the relationship, because n is proportional to $1/L^2$. Figure 7.10 shows how the number of observations decreases with L. In this instance the observer decided that 193 observations was excessive and compromised on ± 7 per cent error, which required approximately 100 observations.

The formula for n describes a familiar shape when plotted against p, for a fixed value of L

$$n = \frac{4p(100-p)}{L^2}$$

$$n = \frac{400p}{L^2} - \frac{4p^2}{L^2}$$

$$n = \frac{-4}{L^2}p^2 + \frac{400}{L^2}p$$

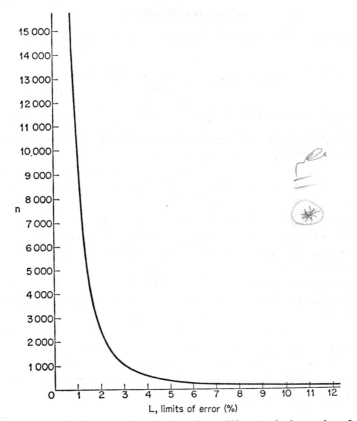

Figure 7.10 A graph of n plotted against L to show the rapid increase in the number of observations required for increased statistical accuracy

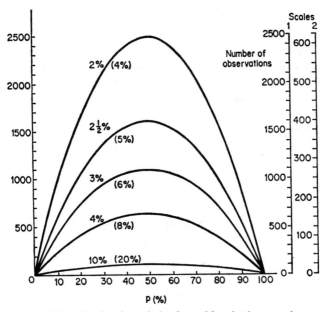

Figure 7.11 A family of parabolas formed by plotting n against p

To the mathematician this is the formula for a parabola, the maximum value of which varies with L.

The maximum or 'peak' of the activity-sampling parabola always occurs where $p = 50$, as shown in Figure 7.11, which illustrates a family of parabolas. This may be proved quite simply using calculus.*

The implication of the fact that $p = 50$ is a maximum for a given value of error is that if the number of observations is determined for $p = 50$ this ensures that whatever the actual value of p the result must be *more accurate* than required. The value $p = 50\%$ gives the greatest error. This is best illustrated by way of example. Suppose an activity-sampling study is being performed without a pilot study being carried out, but a result better than ±5 per cent is required. If p in the formula be set at 50, then:

$$n = \frac{4 \times 50 \times 50}{25}$$

$$= 400 \text{ observations}$$

On completion of the study using 400 observations, if p in fact came out to be, say, 70 per cent, the error in this result would be:

$$L = \pm 2\sqrt{\frac{70 \times 30}{400}}$$

$$= \pm 4 \cdot 58 \text{ per cent, which is better than the required 5 per cent.}$$

Again, if the result were, say, $p = 10$ per cent:

$$L = 3 \text{ per cent, which, again, is better than 5 per cent.}$$

It must be appreciated that the formula used above was derived from the standard error for a binomial distribution, for which there are two forms (*see* Chapter 6, p. 78).

The absolute form is given by $\pm\sqrt{p(1-p)n}$, or from this the error may be given as $\pm 2\sqrt{p(1-p)n}$ for 95 per cent confidence. This formula gives the standard error in terms of *observations*, which is not a convenient form for the purposes of activity sampling, which requires it in the form of a proportion of observations to total observations (p).

The standard error of the proportion of time the machines are idle (or working) gives an estimate of the error inherent in this estimated proportion. This

---

* Differentiating the above equation for a maximum:

$$\frac{dn}{dp} = -2\frac{4}{L^2}p + \frac{400}{L^2}$$

$$0 = -2\frac{4}{L}p + \frac{400}{L^2} \quad \left(\frac{dn}{dp} = 0 \text{ for a maximum}\right)$$

$$2\frac{4}{L^2}p = \frac{400}{L^2}$$

$$2p = 100$$

$$p = 50$$

parameter, known as *standard error of proportion*, is given by $\pm\sqrt{\dfrac{p(1-p)}{n}}$ or, when p is a percentage, by $\pm\sqrt{\dfrac{p(100-p)}{n}}$, with the corresponding formula for error at 95 per cent confidence level.

When the standard error of proportion formula is used the error is applied (added and subtracted) directly to the observed percentage (p). In the example used above (i.e. idleness = 14 per cent, and 100 observations) the full result would be written as

$$p \pm 2\sqrt{\dfrac{p(100-p)}{n}}$$

or $\quad 14 \pm 2\sqrt{\dfrac{14 \times 86}{100}} = 14 \pm 6{\cdot}94$ per cent

i.e. 20·94 to 7·06 per cent, within which limits the true actual value for p will lie, with 95 per cent confidence. Note that the error is applied directly to p and not as a *percentage* of p: it is not 6·94 per cent of 14.

7.811 *A general formula.* By convention, the usually accepted level of confidence for activity sampling is 95 per cent. From the theory of Chapter 6 it will be seen that for a normal distribution (to which the binomial distribution in this case approximates) 95 per cent of the area under the curve lies between the limits set by two standard deviations each side of the mean. In the case of probability, if the total area beneath the normal distribution is equated to unity, areas under specific portions of the curve are proportional to probabilities. Thus the area between the limits of two standard deviations each side of the mean represents the probability (95 per cent, or 0·95) that any value taken at random, from this distribution will in fact occur between these limits. For this reason the inherent error is taken as *two* standard deviations. It will be recalled that the standard deviation (or standard error) formula is

$$\sigma = \sqrt{\dfrac{p(100-p)}{n}}$$

giving rise to the general formula:

$$\text{Error} = X\sqrt{\dfrac{p(100-p)}{n}}$$

where X is the number of standard deviations (or standard errors), to give the desired confidence level. Some values for X are given below (values have been rounded):

X	Confidence level
1	68·0
1½	87·0
2	95·0
2½	98·8
3	99·7

It will be noticed that as the confidence level increases the limits of error increase, assuming that *p* and *n* remain constant. For example, in the calculations made in the previous section 14 per cent idleness with 100 observations produced an error of ±6·94 per cent on the 95 per cent confidence level. Raising the confidence to 99·7 per cent has the following effect:

$$L = \pm 3\sqrt{\frac{14 \times 86}{100}}$$

$$= 10{\cdot}41 \text{ per cent}$$

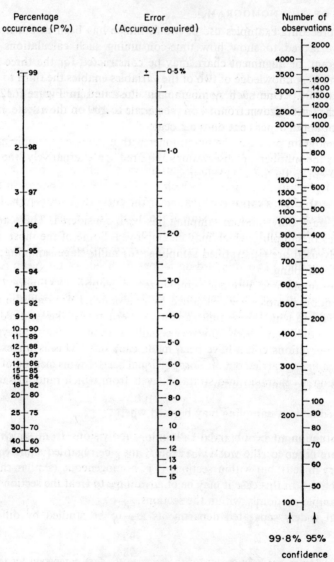

Figure 7.12 A nomogram for the number of observations

From these values we may be 95 per cent sure that the true value of this activity lies between 86 + 6·94 per cent and 86 − 6·94 per cent, but we are 99·7 per cent sure that it lies between 86 + 10·41 and 86 − 10·41 per cent. This is logical because the wider the limits are opened the more sure we are that we are right. In the extreme it could be said that we are *absolutely* certain that the true value lies between 0 and 100 per cent! Clearly this last statement is useless and meaningless, so a compromise which has practical usefulness is chosen. Such a compromise is the 95 per cent confidence level.

7.82 USE OF THE NOMOGRAM

In the preceding examples use of the formulae has been demonstrated and these have served to show how time-consuming such calculations may be. A nomogram, or alignment chart, may be constructed for the three variables of p, L, and n. A knowledge of two of the variables enables the third to be found from the chart.* One such nomogram is illustrated in Figure 7.12. In this example the line is drawn from 14 on the p scale to 100 on the n scale, indicating an error of about 7 per cent on the L scale.

The nomogram provides the practitioner with a quick and simple method of estimating the number of observations required, or alternatively, the limits of error inherent in the result of his study.

7.83 METHODS OF SAMPLING

At this stage only random sampling has been considered. There are certain other types of sampling which may be employed. Some of the most important and widely used are (i) stratified sampling; (ii) multi-stage sampling; and (iii) systematic sampling.

In the example used here and summarized in Section 7.9 (Figure 7.13) completely random samples were obtained by randomizing not only the times but also the days to which these times applied. Alternatively, the required 100 random times could have been allocated equally to each of the five days, that is, twenty observations could have been made each day. This is an example of _stratified sampling_ carried out on a proportional basis. In this method of sampling the situation is separated into strata or levels from which random samples are drawn.

Stratified activity-sampling may be used where:

1. The situation to be observed falls into subdivisions from which separate data are required: the work in a factory may be regarded as heterogeneous (or very mixed), but within sections it is homogeneous (similar throughout the section). In this case it may be of advantage to treat the sections as strata and sample randomly within the sections.
2. Several widely separated departments are to be studied by different observers.

* J. M. Allderige has been credited with designing the first nomogram for this purpose (*Factory Management and Maintenance* Vol. 112, March 1954).

128 WORK MEASUREMENT

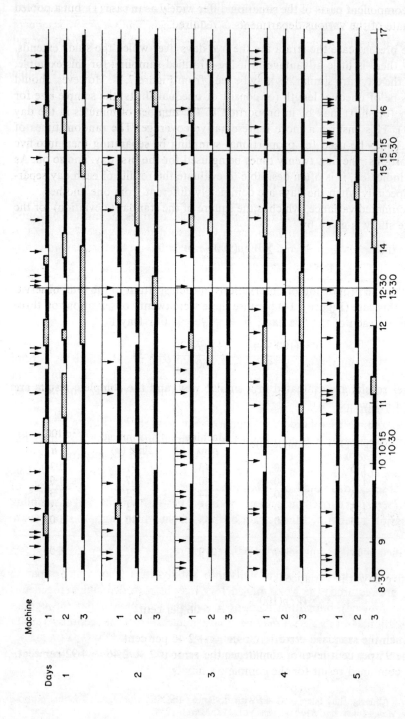

Figure 7.13 A diagrammatic representation of five days' activities, with random observations shown by the arrows

ANALYSIS STAGE

3. The component parts of the situation differ widely as in case (1), but a pooled estimate of the various departments is required.

In the present case the strata are the five days over which the study extends. Possibly there is little advantage in using stratified sampling for this example, because the days are similar, as may be seen from Figure 7.14. However, should the days be of unequal length (or consist of unequal shifts) the sample size for each day (or shift) may be in proportion to the number of minutes in the day (or shift). This ensures a more representative coverage. The random times of Table 7.1 may be used for proportional sampling by separating them into five equal portions: the first twenty times being used for the first day, and so on. As already indicated, it is often desirable to evaluate the results of each day separately, especially when the days are completely different from one another.

The estimated variance (which is the square of the standard deviation) for the complete study is given by

$$\frac{\sum W^2 p(100-p)}{n}$$

where W is the proportion of observations in each stratum to the total observations. In this case (Figure 7.14) because there were twenty observations on three machines on each of the five days, $W = \frac{60}{300} = \frac{1}{5}$. For Day 1

$$\frac{W^2 p(100-p)}{n} = \frac{(\frac{1}{5})^2 \, 71 \cdot 29}{60} = 1 \cdot 34$$

The other results are evaluated in a similar way, and the complete results are tabulated below:

Stratum	Weight (W)	Sample (n)	Machine runs	% running time (p)	$\frac{W^2 p(100-p)}{n}$
Day 1	1/5	60	43	71	1·34
2	1/5	60	40	67	1·03
3	1/5	60	44	73	1·31
4	1/5	60	44	73	1·31
5	1/5	60	48	80	1·07
Totals	1	300	219		6·06

For all five days the variance is

$$\frac{\sum W^2 p(100-p)}{n} = 6 \cdot 06 \text{ per cent}$$

from which the standard error is $\sqrt{6 \cdot 06} = \pm 2 \cdot 46$ per cent

On the 95 per cent level of confidence the error is $2 \times 2 \cdot 46 = 4 \cdot 92$ per cent, and the combined result for the complete study is

$$\frac{219}{300} \pm 4 \cdot 92 = 73\% \pm 4 \cdot 92\%$$

WORK MEASUREMENT

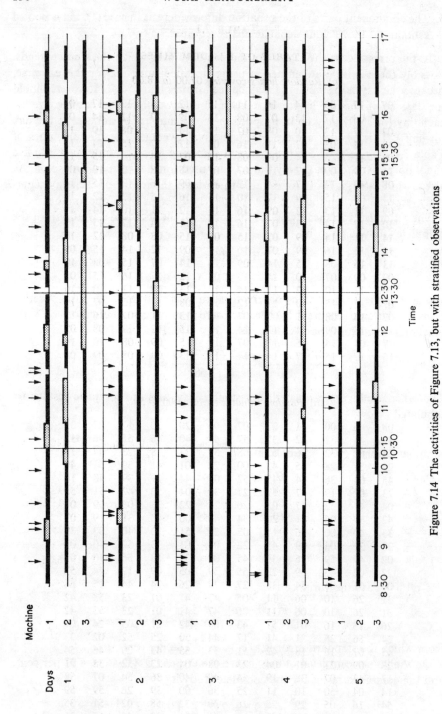

Figure 7.14 The activities of Figure 7.13, but with stratified observations

TABLE 7.1

A TABLE OF RANDOM TIMES

(a) *Random hours covering* 07.00 *to* 18.00

07	17	09	14	16	11	15	12	11	13	17	07
17	08	17	13	07	08	13	09	17	11	14	08
07	17	13	10	08	07	14	14	13	08	10	13
10	14	08	07	11	16	08	17	17	15	14	15
09	11	15	14	09	08	16	07	14	16	15	12
11	09	09	12	11	07	07	15	08	10	10	09
08	13	12	17	14	12	16	09	16	13	15	11
13	09	15	09	17	10	15	16	15	13	12	10
09	11	13	12	09	16	15	12	11	10	14	16
10	17	14	13	07	16	10	11	14	17	13	17
14	11	14	09	07	15	07	15	08	08	12	10
12	15	16	09	07	16	13	09	12	14	15	09
13	16	08	15	14	09	16	14	07	15	08	10
09	09	07	10	14	13	08	12	17	16	15	08
10	12	10	12	11	14	17	13	10	11	10	16
07	12	15	10	17	12	10	08	13	15	07	16
07	14	08	11	07	07	12	13	11	07	16	17
16	16	09	14	12	14	14	15	12	17	08	09
17	07	14	15	13	07	12	15	09	08	10	16
15	17	15	17	14	14	15	13	09	08	09	08

(b) *Random minutes covering* 00 *to* 59

08	46	00	11	27	07	05	20	30	17	27	31
42	48	33	22	37	57	41	40	46	55	46	12
24	42	32	36	05	58	37	33	18	23	44	13
25	37	28	55	42	00	22	01	41	39	51	40
45	04	36	14	45	52	03	44	49	39	48	16
23	02	19	47	59	27	30	04	50	42	51	55
04	47	32	17	03	26	24	22	33	03	59	07
42	39	06	41	20	34	49	00	48	31	19	07
33	40	00	36	14	03	39	51	33	40	03	39
28	21	04	09	45	22	03	01	33	52	40	07
02	49	59	26	39	45	02	32	18	56	10	03
22	02	43	39	23	55	39	05	09	17	44	52
00	01	03	34	38	27	29	26	24	44	15	54
31	26	10	06	11	05	47	41	01	23	53	42
31	26	10	06	11	05	47	41	01	23	53	42
26	13	16	01	51	43	07	42	21	05	24	07
34	56	36	31	41	17	44	50	25	52	02	03
45	07	16	18	42	51	33	55	43	20	46	54
23	09	21	01	04	23	08	04	23	52	38	01
17	43	07	38	19	54	40	40	36	54	07	58
14	04	50	10	11	25	36	00	39	26	59	59
54	16	05	29	25	28	30	32	58	02	51	35
35	08	52	47	52	10	05	36	02	43	17	31

In this example stratified sampling does not have any particular advantage. When the length of the day varies the sample size (n) may be in proportion to the number of minutes in the day. The weighting (W) will reflect this proportion: otherwise the calculation is identical with that above. If, for example, Department A works on a 12-hour day, B works 10 hours, and Department C is restricted to an 8-hour working day, the weighting (W) for Department A would be $\frac{12}{30} = 0.4$, while for B and C, W would be 0.33 and 0.27 respectively.

Results may be analysed by machine instead of by days, using proportional sampling as summarized below. The data are taken from Figure 7.14. The reader should check these, using this Figure and Table 7.2.

TABLE 7.2

THE FIRST STAGE IN RANDOM-TIME GENERATION

One hundred random times, with random days added from random numbers

Day	Random time	Day	Random time	Day	Random time	Day	Random time
1	09.02	3	09.01	5	15.51	4	08.39
1	14.22	1	13.31	4	09.23	2	14.29
3	16.00	5	14.18	5	10.54	4	15.47
1	11.31	4	16.38	3	10.10	5	15.07
3	11.45	5	16.10	2	15.12	2	09.44
3	09.17	4	16.47	5	12.39	4	11.33
1	13.34	5	09.23	5	08.47	1	15.08
3	09.35	5	14.38	4	13.07	4	16.23
2	08.49	5	08.51	5	15.44	3	09.36
3	09.01	3	15.41	1	09.08	1	08.30
3	11.26	1	09.42	1	11.40	1	09.18
5	09.13	2	16.04	3	15.17	5	11.09
1	13.56	5	12.19	1	16.30	4	11.01
2	09.07	5	11.11	5	11.05	4	09.21
3	11.09	4	08.52	3	09.26	5	16.43
2	15.08	1	08.45	5	15.41	5	08.36
3	09.59	5	16.55	4	16.42	1	16.39
2	12.45	2	11.05	4	11.55	5	15.35
1	10.36	1	13.43	4	10.04	3	08.42
5	11.21	3	11.17	5	08.40	5	16.04
3	10.07	2	08.51	3	16.00	4	11.19
5	11.52	4	14.23	2	13.32	5	15.40
1	09.26	1	08.55	1	09.54	3	13.54
4	11.19	1	16.05	1	11.10	3	14.19
2	11.34	1	11.43	2	16.12	5	16.07

Machine	Runs	Idle (mgt)	Idle (worker)
1	68 (78.1)	16 (9.3)	16 (12.6)
2	80 (78.1)	9 (11.8)	11 (10.6)
3	70 (77.6)	17 (13.3)	13 (10.6)

Comparison of these results with the actuals (in brackets) show wide discrepancies from the true values.

A third method which appears to produce very reliable results is *systematic sampling*. This method completely abandons the idea of randomness in the observation times and relies on the inherent irregularity of the work for the random element. In most cases breakdowns and stoppages are unpredictable and create a sufficiently random situation which admits the use of systematic sampling.

Sample observations are made at regularly spaced intervals which gives an even coverage of the day's activities. In the example already quoted 100 observations are to be made over five days, which requires 20 evenly spaced observations to be made each day. Thus the spacing of the observations should be 430 min divided by 20, or 21·5 min between consecutive observations. The next factor to determine is the starting time. If this is to be 8·30 on each day, the second observation will be at 8·51½, and the last at 16·18¼, but anything which happens after this time will never be seen. There are similar disadvantages of course, with starting at 8·51½ and finishing at 16.40. One solution to this is to increase the time-interval to 22·5 min (i.e. 430 min ÷ 19). A better solution is to select the starting time each day from random times. The study using systematic sampling, with random starting times, is shown diagramatically in Figure 7.15, and the results are summarized below:

Machine	Runs	Idle (mgt)	Idle (worker)
1	78 (78·1)	11 (9·3)	11 (12·6)
2	79 (78·1)	11 (11·8)	10 (10·6)
3	79 (77·6)	15 (13·3)	6 (10·6)

TABLE 7.3

A COMPARISON OF OBSERVED RESULTS AND TRUE RESULTS

Machine	Random	Systematic	Stratified	Actual
Machine running				
1	80	78	68	78
2	78	79	80	78
3	73	79	70	78
Idle time (management fault)				
1	11	11	16	9
2	9	11	9	12
3	16	15	17	13
Idle time (worker fault)				
1	9	11	16	12
2	13	10	11	11
3	11	6	13	11

When all results are compared (Table 7.3) it is clear that systematic sampling produces a result which is more representative of the true situation than random

134 WORK MEASUREMENT

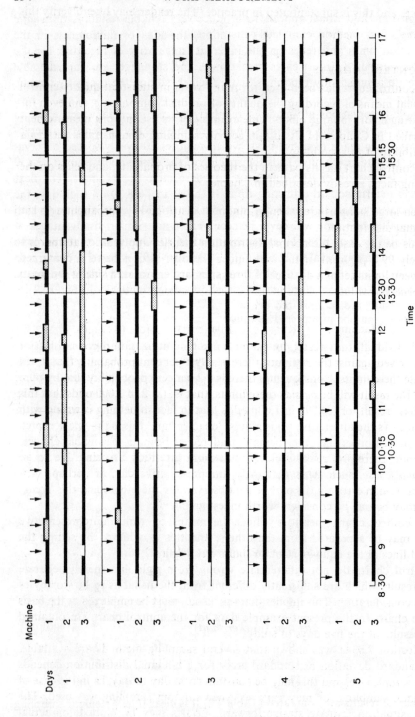

Figure 7.15 The activities of Figure 7.13, but with systematic observations

The limits must be set at 78·7 + 15·8 = 94·5 per cent (upper limit)

and at 78·7 − 15·8 = 62·9 per cent (lower limit)

On plotting the five results on the chart (Figure 7.16) it will be seen that all lie within the control-lines and are acceptable. Inevitably on some occasions charts will contain results which fall outside the control-lines. Those results which are out of control are rejected and a new value for p and fresh control-lines are computed which omit the out-of-control readings.

The situation depicted in Figure 7.14 presents a somewhat different problem. A varying sample size (n) means that no longer can the control-lines

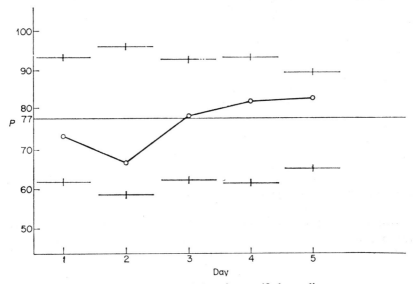

Figure 7.17 A control chart for stratified sampling

be placed a constant distance from p. The total number of observations is, as before, 300, and p is 77·3, but the calculation for L must be made for each day:

Day	n	Machine running observ.	percentage	$\pm 3\sqrt{\dfrac{p(100-p)}{n}}$	Upper line	Lower line
1	69	50	72·4	16·2	93·5	61·1
2	39	26	66·7	18·9	96·2	58·4
3	60	47	78·3	16·0	93·3	61·3
4	51	42	82·4	16·2	93·5	61·1
5	81	67	82·7	12·5	89·8	64·8

The chart, together with the results, are plotted in Figure 7.17.

A simplified, but in some respects inferior, method which overcomes the objection of varying control-lines, is to obtain an *average sample-size* which may be used once only in the formula to produce 'average' control-lines. In this example the average sample-size is $\frac{300}{5} = 60$, the calculation being based on $p = 77 \cdot 3$ per cent:

$$L = \pm 3 \sqrt{\frac{77 \times 23}{60}} = \pm 16 \cdot 3 \text{ per cent}$$

Constant control-lines are set at $77 \cdot 3 + 16 \cdot 3 = 93 \cdot 6$ per cent

and at $77 \cdot 3 - 16 \cdot 3 = 61 \cdot 0$ per cent

A superior method for dealing with variable sample-sizes is the *stabilized standard error** chart. In effect this chart performs a significance test on each

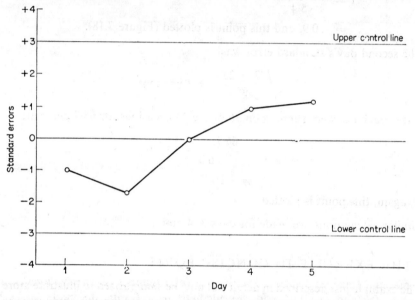

Figure 7.18 An alternative control chart, scaled in standard errors

result by measuring the deviation of each day's result from the overall study result (p) in terms of standard errors (σ). A deviation of more than 3σ indicates an assignable cause with a significant difference on the 99·8 per cent confidence level.

The chart is scaled in standard errors (Figure 7.18) and control-lines drawn at plus 3σ and minus 3σ. Points are first converted to standard errors using the formula:

$$t = \frac{\text{(particular day's result)} - \text{(study result } p)}{(\sigma \text{ for the day's result})}$$

* *See* Duncan, Acheson J., 'Detection of non-random variation when the sample size varies', *Industrial Quality Control*, January 1948; also Sastry, A. S., 'Control charts for work sampling studies with variable sample sizes', *Work Study and Management Services*, April 1968.

The standard error for the day's result is computed in the usual way, i.e.

$$\sigma = \sqrt{\frac{p(100 - p)}{n}}$$

For the first day's result

$$\sigma = \sqrt{\frac{77 \times 23}{69}}$$

$$= \pm 5.4 \text{ per cent}$$

The machines were running on 50 out of the 69 occasions they were observed, or 72·4 per cent of the day, therefore

$$t = \frac{72.4 - 77.3}{5.4}$$

$= -0.9$, and this point is plotted (Figure 7.18).

The second day's standard error was

$$\pm \sqrt{\frac{77 \times 23}{39}} = 6.3 \text{ per cent}$$

and the machines were running on 26 of the 39 occasions, or 66·7 per cent; thus

$$t = \frac{66.7 - 77.3}{6.3}$$

$$= -1.7$$

and, again, this point is plotted.

Similar calculations are made for days 3, 4 and 5.

7.9 THE EXERCISE IN CONCISE FORM

The example just described in detail will now be summarized to illustrate more concisely how a study is performed in practice. The steps in this study may be compared with those outlined in the procedure set out in Section 7.72.

The present problem is concerned with the utilization of three similar machines in a small section of a machine shop. The study was made at the request of the department manager, who wished to obtain reliable data on the performance of the three machines. The problem appeared to be one of excessive breakdown time, but this needed confirming.

It had been decided that idle time should be analysed as 'management-responsible' and 'worker-responsible'; the former to include breakdowns, lack of work, shortage of material, interruptions, and machine changeovers, and the latter to cover the worker taking relaxation allowance (at the workplace or away), chatting to other operatives (if his machine were stopped during this conversation time), and other stoppages caused by the worker.

The pilot study performed after the usual preliminaries revealed an apparent utilization for the machines of 86 per cent on average. This figure resulted from 12 snap observations made during the course of one day by the department manager (see Section 7.81).

It had been decided that a study spread over one week would be representative of the situation, and that random times on random days during the week would be employed. From the nomogram it was clear that the number of observations which would be required to achieve an error of better than ± 1 per cent was too large to be practicable (4,816 observations). On the other hand 10 per cent was not sufficiently precise. A compromise figure of ± 4 per cent, on the 95 per cent confidence level, was reached. To realize these limits it was necessary to make 100 observations during that week on each of the machines (300 observations in all).

The required number of random times for these observations was extracted from the appropriate tables (Table 7.1) to cover the periods 8.30 to 10.15, 10.30 to 12.30, 13.30 to 15.15, and 15.30 to 17.00. These are set out in Table 7.2. Any random times shown on the table which were outside these periods were discarded, and the number made up to 100 with other random times. In addition to the random times a *random number table* was used to provide the day code against each random time (i.e. Monday = 1, Tuesday = 2, and so on). Only numbers 1 to 5 were retained. Finally, when 100 random times had been collated they were reassembled in chronological order on an appropriate observation record-sheet (Figure 7.19).

On the first day of the study the observer positioned himself so that he could see all three machines. At precisely 8.30 he observed the machines, noticing that none was in action because the operatives were not in attendance. Consequently he recorded these observations in the columns 'machine idle (worker fault)' for all three machines.

The next random time was 8.45, which allowed him nearly 15 min to pursue other tasks. Again at 8.45 precisely he noted that all three machines were operating, which was duly recorded. In this manner the study continued for the rest of the week.

It will be seen from the chart that random times may occur consecutively (e.g. Day 3, 9·36 and 9·37), or are sometimes widely spaced out (e.g. Day 2, 9.44 and 11.05). This is a feature of the use of random times which often perturbs the practitioner and can give rise to greater errors than if regularly spaced times are used. Nevertheless this has no effect on the statistical limits and confidence calculated for the study, which are still valid.

Often it is argued that workers will anticipate the arrival of the observer and will start working (or shut the machine down) deliberately to bias the result. It has been found through experience that workers cannot be bothered with such practices, but if this is considered to be a real problem the first day's results may be omitted from the study.

On completion of the study, analysis of results may be made. In the present example nine totals (3 machines × 3 columns) were obtained on summing the 'ticks' (Figure 7.19). In the first column for 'machine No. 1 runs' there were 80

ACTIVITY SAMPLING

ANALYSIS CHART

Operation: *Machine section IV* **Observer:** *D.H.J.* **Date:** *6 October*

Time	Machine 1			Machine 2			Machine 3		
	1	2	3	4	5	6	7	8	9
Day 1/ 8·30			✓			✓			✓
8·45	✓			✓			✓		
8·55	✓			✓					✓
9·02	✓			✓			✓		
9·08		✓		✓			✓		
9·18		✓		✓			✓		
9·26	✓			✓			✓		
9·42	✓			✓			✓		
9·54	✓			✓			✓		
10·36		✓		✓			✓		
11·16	✓				✓		✓		
11·31	✓			✓			✓		
11·40	✓			✓			✓		
11·43	✓			✓			✓		
13·31			✓			✓		✓	
13·34	✓					✓		✓	
13·43	✓			✓				✓	
13·56			✓		✓			✓	
14·26	✓			✓			✓		
15·05			✓	✓			✓		
16·00	✓			✓					✓
16·30	✓			✓			✓		
16·39	✓			✓			✓		
Day 2/ 8·49	✓			✓			✓		
8·51	✓			✓			✓		
9·07	✓			✓			✓		
9·44	✓			✓			✓		
11·05	✓			✓			✓		
11·34			✓	✓			✓		
12:23	✓			✓				✓	
13·32			✓			✓		✓	
14·29	✓				✓		✓		
15·03	✓				✓		✓		
15·12	✓				✓		✓		
16·04		✓			✓			✓	
16·17	✓				✓			✓	
Day 3/ 8·30	✓					✓	✓		
8·42	✓					✓	✓		
9·01	✓			✓			✓		
9·17	✓			✓			✓		
9·20	✓			✓			✓		
9·35	✓			✓			✓		
9·36	✓			✓			✓		
9·58			✓	✓			✓		
10·07	✓			✓			✓		
10·10	✓			✓			✓		
11·09	✓			✓			✓		
11·17	✓			✓			✓		
11·26	✓			✓			✓		
11·45		✓			✓				✓
Total	38	5	7	36	8	6	38	8	4
	M/C 1 runs	M/C 1 idle (management)	M/C 1 idle (worker)	M/C 2 runs	M/C 2 idle (management)	M/C 2 idle (worker)	M/C 3 runs	M/C 3 idle (management)	M/C 3 idle (worker)

Time	Machine 1			Machine 2			Machine 3		
	1	2	3	4	5	6	7	8	9
13·32			✓			✓			✓
13·54	✓			✓			✓		
14·19		✓				✓	✓		
15·27			✓	✓			✓		
15·43	✓			✓			✓		
16·02	✓			✓				✓	
Day 4/ 8·39	✓			✓					✓
8·52	✓			✓			✓		
9·21	✓			✓			✓		
9·23	✓			✓			✓		
10·03			✓	✓			✓		
11·09	✓			✓			✓		
11·19	✓			✓			✓		
11·33		✓		✓			✓		
11·55		✓		✓				✓	
12·13	✓			✓				✓	
14·23		✓		✓				✓	
15·43	✓			✓			✓		
15·47	✓			✓			✓		
16·03	✓			✓			✓		
16·15	✓			✓					✓
16·47	✓			✓			✓		
16·57	✓			✓			✓		
Day 5/ 8·16	✓			✓			✓		
8·40	✓			✓			✓		
8·47	✓			✓			✓		
8·51	✓			✓			✓		
9·13	✓			✓			✓		
9·23	✓			✓			✓		
10·54	✓					✓	✓		
11·03	✓			✓			✓		
11·09	✓			✓				✓	
11·11	✓			✓				✓	
11·21	✓			✓				✓	
11·52	✓			✓			✓		
12·19	✓			✓			✓		
12·39	✓					✓	✓		
14·10		✓		✓				✓	
14·33	✓				✓		✓		
15·07	✓			✓			✓		
15·35	✓			✓			✓		
15·40	✓			✓			✓		
15·41	✓			✓			✓		
15·44	✓			✓			✓		
15·51	✓			✓			✓		
16·04	✓					✓			✓
16·06	✓					✓			✓
16·10	✓			✓					✓
16·43	✓			✓			✓		
16·55	✓			✓					✓
Total	80	10	10	79	9	12	73	16	11
%									
Error	8	6	6	5	5·6	3·7	8·9	3·9	6·2

HWS 11

Figure 7.19 A completed observation sheet for the exercise in Section 7.9

ticks which, on 100 observations, represents 80 per cent running time, leaving 20 per cent idle time made up of 10 per cent management-responsible and 10 per cent worker-responsible. Use of the nomogram or formula will show that for $p = 80$, and $n = 100$, the inherent error (on the 95 per cent confidence level) is ± 8 per cent, so that the actual running time for this machine must lie within the limits 72 to 88 per cent. Of course, in practice it is not possible to check this result because the actual value will not be known. However in the synthesized example it is fortunate that actual results are available from detailed analysis of Figure 7.13. A full list of results together with actual true results for comparison are given in Table 7.3. It will be seen that in all cases the actual results lie well within the stated limits of error. Control charts for this study are shown in Figures 7.16, 7.17, and 7.18, and fully described in Section 7.84.

7.91 USE OF COMPUTERIZED STUDY BOARDS

The foregoing exercise was an example of a manually performed study, and this was provided in order that the reader could better understand the principles involved.

With the prevalence of computers it is only natural that such equipment should be enlisted to simplify the work of the practitioner. Suitable hardware is available in the form of computerized study boards. These are used principally for time studies, and are fully described in Chapter 17. Software has been written for the various models, which will analyse activity sampling studies, and provide print-outs of the recordings and analyses.

8 Predetermined Motion–Time Systems

8.1 INTRODUCTION

Predetermined motion–time systems (P.M.T.S.) form one of the three main divisions of work measurement, in company with the others, the techniques of timing and estimating. All are designed to establish the time required to complete a job at a defined level of performance, although they are based on differing principles and all have different characteristics.

Predetermined motion–time systems may be regarded as forming a subdivision of *synthesis*, because the standard elements which compose the system are used to build up a basic time for the job being studied. The difference is a matter of degree. The term *synthetic data* may be reserved for significant and distinguishable parts of operations which may be common to many other operations. Predetermined motion–time data are for basic and fundamental human motions and simple mental elements of general application (or, with derived systems, of specific application) and may be used in combination to build up synthetic data. The distinction becomes less obvious with the higher-level systems of P.M.T., and this is especially so with those which are designed for use in a specific area such as clerical or maintenance work (for example, Clerical Work Data, and Universal Maintenance Standards). The differences between the various levels of data are shown diagrammatically on page 152.

The British Standards Institution in its Glossary of Terms Used in Work Study and Organisation and Methods (B.S.I. 3138: 1979) defines P.M.T.S. as 'work measurement techniques in which times are established at implicit rates of working for classified human movements and mental activities. These are used to build up the time for an operation or task. P.M.T.S. data have been developed for common combinations of basic human movements or mental activities'.

The implication of the definition is that P.M.T. systems are based on the basic human motions of which all work is composed. However, many of the systems derived for use in specific areas of work include elements which are far from being '*basic* human motions', for example erecting a ladder, or inserting documents into a folder. These are examples of P.M.T. elements from derived, higher-level systems. It is useful to regard the basic systems of predetermined motion–times as *micro-P.M.T.S.*, and the derived systems as *macro-P.M.T.S.*

The definition also appears to preclude the mental elements of work, concentrating on 'motions'. Many systems, particularly the Work-Factor Mento Factor System, include mental elements in their data tables, and as long ago as World

War I, Frank Gilbreth was thinking of manual work as including mental elements.

The first principle behind the definition is that manual work-cycles are composed of basic elements or motions that can be identified in a consistent manner by the analyst. The first classification system was produced by Frank Bunker Gilbreth, who identified seventeen basic movements which he called *therbligs*. Later he

SYMBOL	NAME	COLOUR
	Search	Black
	Find	Grey
	Select	Light grey
	Grasp	Red
	Hold	Gold ochre
	Transport loaded	Green
	Position	Blue
	Assemble	Violet
	Use	Purple
	Disassemble	Light violet
	Inspect	Burnt ochre
	Pre-position	Pale blue
	Release load	Carmine red
	Transport empty	Olive green
	Rest for overcoming fatigue	Orange
	Unavoidable delay	Yellow
	Avoidable delay	Lemon yellow
	Plan	Brown

Figure 8.1 Gilbreth's original 'therbligs'

added a further therblig to his list to cover the mental action 'plan'. Therblig symbols and associated colour-codes are shown in Figure 8.1.

The second principle is that time standards can be applied to these basic motions. The principle was stated first by A. B. Segur in 1927: 'within practical limits the time required for all experts to perform true fundamental motions is a constant'. The total time required for the job is the sum of all the individual basic motion-times. Lately the validity of this last statement has been questioned.

Almost without exception the total task-times obtained by such summation are *basic* times; they do not include any allowances. One of the very few exceptions to this is found in a higher-level system known as the Clerical Work Improvement Programme, whose data times include a relaxation allowance equivalent to an addition of 16⅔ per cent.

By the mid 1920s the principle of measuring work with the aid of a stop-watch had become firmly established in the United States. Unfortunately time study in its primitive form suffered from many deficiencies, and even the introduction of rating in the 1930s did little to improve its image. The lack of consistency between standard times for different but similar jobs was obvious to the astute representatives of the emerging trade unions. Work-Factor was developed in the early 1930s because of union opposition to the stop-watch. The situation was not helped by opposition of the American Senate, which prohibited the use of any part of the U.S. Government Appropriation to be used as funds for the purpose of timing an employee of the government by means of a stop-watch. These problems put pressure on the early practitioners to search for means of determining standard time other than by direct timing. Two pressing requirements were for consistency, and for greater reliability.

There are only two major international systems in use in this country at present. These are Work-Factor and Methods–Time Measurement. Many other systems currently are in use, but all, with few exceptions such as Mulligan Standards, are derived from Methods–Time Measurement.

It is generally held that A. B. Segur of Oak Park, Illinois, developed the first commercially used system. It was completed about 1925 and called Motion–Time Analysis (M.T.A.). Segur used much of Gilbreth's research material in developing the system which is now operated as a proprietary system by the A. B. Segur Company. M.T.A. motion values are given in minutes to the fifth decimal place. The elemental motions themselves are based directly on therbligs.

One of the most widely known international systems is Work-Factor. The property of the Wofac Company (a Division of Science Management Corporation), Work-Factor was developed by a team of researchers under the direction of Joseph H. Quick in America during the 1930s. Details of the system have been published in textbook form.*

Work-Factor Time Data were developed through original motion–time studies using six-second and three-second sweep stop-watches, photo-electric timers, and motion-picture analyses. Data were collected from an extremely large sample of workers and operations. Seven major Work-Factor systems have been developed, which may be used separately or together.

Detailed Work-Factor is intended to produce very accurate time standards, and is used as a basis for the other, higher-level systems. The system is based on a Work-Factor Time Unit of 0·0001 min. Ready Work-Factor (a second-level system) was instigated to simplify the training of engineers, foremen and supervisors in Work-Factor, and to simplify Work-Factor applications for standard data. Ready Work-Factor Time Data were developed by simplifying Detailed

---

* J. H. Quick, J. A. Duncan, J. A. Malcolm, Jr., *Work-Factor Time Standards* (New York: McGraw-Hill, 1962).

Work-Factor Time Values, during the period 1946 to 1961. The system is based on a time unit (t.u.) of 0·001 min. A third system, Abbreviated Work-Factor, based on a time unit of 0·005 min, was developed to meet the particular needs of small-quantity jobbing-shop operations, maintenance, and construction work. There are two forms of Abbreviated Work-Factor called A-100 and B-100.

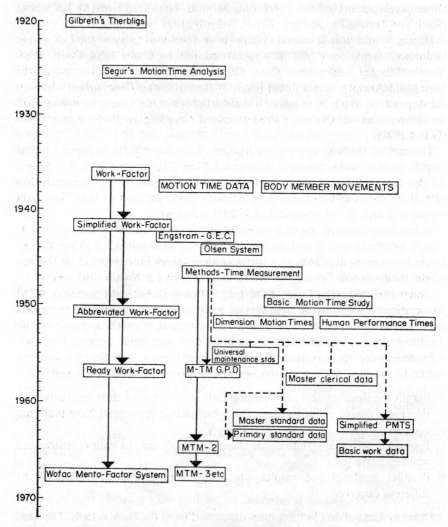

Figure 8.2 A chronology of predetermined motion–time systems

The fifth system, the Wofac Mento-Factor System, is probably unique in that it deals exclusively with mental-process times such as are encountered in office work and inspection tasks. It is a first-level system and fully compatible with the other Work-Factor systems. Wocom is a computer system for making analyses from certain input data. It is described in Chapter 17. The remaining systems are Work-Factor Standard Data and Work-Factor Programs.

The General Electric Company of the United States operates several systems of predetermined motion-times, the first of which was developed in 1938 by Harold Engstrom and H. C. Geppinger. The system was known as Motion-Time Data for Assembly Work, and was described by Ralph M. Barnes in 1940.* Other developments followed, including Motion-Time Standards (M.T.S., about 1950), and Dimension Motion-Times, roughly three years later.

During World War II several systems were developed to assist the U.S. Army Ordnance. Work on P.M.T.S. was carried out by Capt. John Olsen, Prof. Marvin Mundel, and others. Capt. Olsen's method, designed for use at the Springfield Armory, was a direct result of the American Government's ban on the stop-watch, to which reference has already been made. One system developed for this purpose was Ordnance Predetermined Approximate Performance Times (O.P.A.P.T.).

The end of the war saw the emergence of a system from Western Electric Company Inc. under the title 'Elemental Time Standards for Basic Manual Work'. Basic Motion Timestudy (B.M.T.) was designed by the Canadian firm of J. P. Woods and Gordon Ltd of Toronto, and first used in 1951. This data system is a revision of Methods-Time Measurement.

In the latter part of the 1940s details of the other major system of P.M.T. known as Methods-Time Measurement (MTM) were published in the United States,† and made available to industry as a non-proprietary method. MTM was originally devised for use at the Westinghouse Brake & Signal Corporation.

Under the guidance of the MTM International Directorate, national MTM Associations are formed in appropriate geographical areas in order to protect the MTM system against incorrect use, to co-ordinate research, and to supervise the correct training of instructors and practitioners through approved courses.

In developing their systems, Maynard, Stegemerten, and Schwab were required to produce a system of predetermined motion-times which would:

1. Permit the development of good methods in advance of their installation.
2. Provide a means of measurement in which employees could have faith and confidence.
3. Provide a universal system of measurement applicable to both repetitive and non-repetitive jobs.
4. Produce analyses and results which could be consistently repeated by different observers.

Other systems of MTM have been developed from the basic system. The need for higher-level systems produced, in the 1950s, General Purpose Data (G.P.D.) developed in the United States, and about the same time special-purpose systems of Universal Maintenance Standards (from the H. B. Maynard Company) and

* R. M. Barnes, *Motion and Time Study* (New York: Wiley, 1940).
† H. B. Maynard, J. G. Stegemerten, J. L. Schwab, *Methods–Time Measurement* (New York: McGraw-Hill, 1948).

Master Clerical Data (M.C.D.), devised by the Serge Birn Company of America made their appearances.

In the 1960s other systems derived from MTM originated in the U.K. Previously all work on P.M.T.S. was carried out in the North American continent, and Master Standard Data was one of these. The United Kingdom methods included Primary Standard Data (P.S.D.), and a British Rail–I.C.I. venture produced the first-level data system of Simplified P.M.T.S., and from this the second-level system of Basic Work Data (for maintenance work), and Clerical Work Data. Responsibility for the care and development of these latter techniques has now been entrusted to the Management Economics Division of the Production Engineering Research Association (P.E.R.A.).

Most of the systems named above were 'unofficial' developments, that is, they were not part of the research programme of the MTM International Directorate. Apart from G.P.D., the official systems, developed mainly in Sweden and the United Kingdom, include MTM-2, which was completed about 1965, and MTM-3 (*c.* 1970). Concurrently with these developments work was completed on Tape Data Analysis. These systems are described later in this chapter.

During the 1950s interest began to focus on the largely unmeasured but potentially rewarding field of indirect work. Development of higher-level systems of P.M.T. was stimulated by the inadequacy of existing micro-P.M.T.S. to cope with highly variable work in an economic or practical way. These systems are mainly for use in specific areas such as maintenance or clerical work, and include those mentioned above, derived from MTM, and others which were independent of that parent system, such as the Clerical Work Improvement Programme (C.W.I.P.), a system of W. D. Scott & Company from Mulligan's Standards. These data were developed by Paul B. Mulligan in the United States from film analyses.

Other innovations which were tried and introduced with a measure of success in the 1960s were Modapts and Office Modapts, derived from MTM by the Australian Association for Predetermined Motion–Time Systems and Research.

8.2 DEVELOPING THE SYSTEMS

Gilbreth's original therbligs, eighteen in number, were sufficient to describe the jobs performed by various manual workers for the purposes of motion study. Gilbreth, and later Segur, realized that in order to adapt such a system of classification to suit the measurement of work the therbligs must be refined and subdivided, not only to describe motions more precisely but also to allow for the different variables such as distances moved and degrees of difficulty. For example, the therblig *transport loaded* covers the action of an operator who is moving an object from one place to another, but the analyst using a contemporary system of first-level micro-P.M.T.S. would need to know how far it was moved, how much it weighed, if there was any resistance to movement, if care needed to be exercised to avoid damage to operator or object, if the object started or finished at rest, and many other facts about the variables involved.

The particular information to be collected would depend upon the particular system used, but this serves to illustrate the amount of thought which goes into the analysis of just one simple motion.

The amount of research, data, and validation study necessary to produce a reliable system of predetermined times is prodigious. The Work-Factor System, for example, took twelve years to perfect* during which time, from 1934, about a half-million man-hours of engineers' time were used in the research, validation, and application testing of the data, which included more than 10 million individually recorded time-values.

When a time study is made in the course of a work-measurement study the observer usually studies one, two, or perhaps three operatives to obtain his data. The Work-Factor researchers studied no less than 1,100 factory and office employees during the 1930s and 1940s. It is the use of such large samples which gives a system of P.M.T.S. like Work-Factor its statistical validity and its credibility.

Data for P.M.T.S. tables may be collected from laboratory studies, or directly from the areas in which they are to be subsequently applied. The latter approach has many advantages, and this type of clinical data was used in many of the major systems.

The timing methods which Gilbreth employed in many of his studies as long ago as World War I included using such advanced equipment as motion-picture cameras and other forms of still photography. The systems of P.M.T. are required to measure in extremely small units of time, the smallest being the Work-Factor Time Unit, which is only 0·0001 minute in length, and MTM's time-measurement unit (t.m.u.), which is somewhat larger, at 0·00001 hour (0·0006 min). Decimal minute stop-watches would be of little use where such precision is demanded. The Work-Factor researchers resorted to paper-tape electronic timers to measure motions of very short durations, while 16 mm ciné-cameras were used to record complex motions for subsequent analyses. Time study was used extensively, for which stop-watches with three-second or six-second sweep hands and calibrations in thousandths of a minute were employed. Joseph H. Quick† reveals that while data were collected from the working situation, some data were verified in the laboratory using stroboscopic photography.

Barnes‡ reports the use of photography and high-speed timing methods by Engstrom and Geppinger in the development of their Motion–Time Data System, and the originators of Methods–Time Measurement also had recourse to such facilities.

Several of the major researchers used film analyses. For example, in the development of MTM, once the elements and their variables had been defined it remained to apply time standards to them. This was accomplished by a procedure

---

* This refers to the Detailed Work-Factor System. Research continued, and still continues into further developments of the system.

† J. H. Quick, J. A. Duncan, J. A. Malcolm, Jr., *Work-Factor Time Standards* (New York: McGraw-Hill, 1962).

‡ R. M. Barnes, *Motion and Time Study* (New York: Wiley, 1962).

known as frame count. The speed of the cameras at the time of filming was 16 frames per second. By counting the number of frames occupied by each motion it was possible to determine the length of time taken by each motion. The time for one frame is 1/16th sec., which is 1/16 ÷ 3,600 hours, or 0·00001736 hours. This proved a very cumbersome value to work with, so the originators adopted a unit of 0·00001 hour which they called the *time measurement unit* (t.m.u.).

Not all film analyses were carried out with camera speeds of 16 frames per second. Higher speeds (micro-motion) were often employed to produce more accurate values for the more complex and shorter elements.

Some form of levelling or rating was used to refer the observed times to a particular performance standard. The reference level in no cases corresponded to the recognized standard ratings of, say, 133 (A.S.M.E.), or 60 (Bedaux) in these early systems, although some methods derived from MTM have been converted, for example Simplified P.M.T.S., and Clerical Work Data. Mostly the element times were selected to average out at a 'fair day's work' level, although individual elemental times may be set at paces which deviate from this rating.

Work-Factor data were levelled by at least two, and in some cases, up to five experienced engineers who made simultaneous but independent assessments. Rating was based on the evaluation of the skill and effort of operators, using a system developed under the direction of Joseph H. Quick. The reference performance level was taken as that of the 'average experienced operator, working with good skill and good effort'. With the MTM system the operators were rated at the time of filming, using the Westinghouse Leveling System. This procedure was developed at the Westinghouse Corporation by Lowry, Maynard, and Stegemerten, and a description of this method of levelling is given in Chapter 5. All MTM data are expressed at a performance level of MTM 100.

Attempts have been made to refer the performance levels of various systems to the recognized rating scales. In the case of Work-Factor the originators have never been disposed to quote officially a reference level in such terms, although it has been put just below 100 rating on the B.S.I. recommended scale (75 Bedaux).* Several investigators have attempted to establish the average rating level at which the MTM motion–times have been set. A general agreement has been reached as a result of recent research,† that the most appropriate value is equivalent to U.S. day-rate, or 83 rating on the B.S.I. (0–100) scale. This allows practitioners to extend basic times derived from MTM analyses to the B.S.I. 100 (standard) rating by multiplying by a factor 83/100. This, of course, applies to most of the MTM-derived systems, although some, notably Simplified P.M.T.S., Basic Work Data, and Clerical Work Data have rejected the t.m.u. in favour of milliminutes and the B.S.I. standard rating, to which all element times have been converted.

* J. A. Malcolm, Jr., 'Work-Factor Merits and Comparisons' (a paper given at the Work-Factor Associates Conference, October 1968).
† F. J. Neale, 'Conversion Factor for MTM', *Work Study and Management*, March 1966, p. 121; P. M. Burman *et al.*, 'MTM and the B.S.I. Rating Scale', *Work Study and Management Services*, February 1969; J. H. D. Kerse, 'The t.m.u. and the B.S.I. Rating Scale', *Work Study and Management Services*, January 1969.

8.3 CHARACTERISTICS

The major characteristics of all these systems are similar; all systems of P.M.T. are used for measuring physical work, and to some extent mental work. They cannot be used for operations where the time is governed by the running time of a machine, nor for a process which is beyond the control of the operator. The times are sufficient for normal manual work, but are not suitable for tasks which involve an unusually high degree of mental effort. For example, P.M.T.S. could be used to analyse the removal of a tightly fitting pin from a hole with a pair of pliers. The data would not necessarily cover the extraction of a tooth by a dentist unless the time required for the extraction includes due recognition of the degree of thought, skill, and care involved in addition to the physical actions employed.

P.M.T.S. provide an extremely fine analysis of the job. Time–study data treat the job in terms of elements of work which involve times of the order of 0·1 to 0·5 min to perform. For measuring elements below 0·1 min duration the stopwatch becomes both unreliable and physically impossible to operate. P.M.T.S. can analyse far smaller elements of work. Furthermore the build-up of the element time is accounted for motion by motion. A P.M.T.S. analysis shows not only how long the job should take, but also why it is performed in this way. This can be of great use in improving the methods used and in reducing the job time. Detailed analyses may be of value when queries and complaints are raised about the time allowed.

Unfortunately this same characteristic makes the detailed P.M.T. systems extremely time-consuming to use. The time required to analyse a job using the MTM-1 system can be as much as 150 times as long as the job takes to perform. Two approaches are in current use. One solution is to employ a computerized system such as Wocom, described in Chapter 17. The other is to develop simplified systems from the parent, first-level system. Abbreviated Work-Factor and Ready Work-Factor are examples of the latter approach.

MTM has corresponding levels of data. The detailed form is known as MTM-1 and qualifies as a first-level system. The second-level development, MTM-2, consists of a simplification of MTM-1 and combines the basic motions of MTM-1 to achieve the objective of simplifying the data and increasing the speed of application.

The objectives of all higher-level systems are similar, but each follows its own individual paths of development to attain these objectives. MTM-2 was derived from the basic parent MTM-1 by (i) combining, (ii) averaging, (iii) substituting, and (iv) eliminating the various MTM motions. For example, the combination of 'release', 'reach', and 'grasp' of the basic system produced the category 'get' of the second-level MTM-2, while 'sit-stand' are eliminated and analysed as 'bend-arise' plus other relevant motions.

In the development of the Work-Factor second- and third-level systems of Ready Work-Factor and Abbreviated Work-Factor, simplification was achieved more by grouping data into larger sets. Thus the eight Standard Elements of

Detailed Work-Factor are preserved in Ready Work-Factor, and not combined as in the case of MTM-2. This is also true of Abbreviated Work-Factor A-100, although some combinations are evident in Abbreviated Work-Factor B-100, principally in the Standard Element 'Pick-up', which is composed of 'Reach', 'Grasp', and 'Move'. These Standard Elements are not superseded by the Standard Element 'Pick-up'.

In Ready Work-Factor the individual Motion Distance categories (in inches) are reduced from twenty-seven in the Detailed Work-Factor to just five groups of Motion Distances. By further simplifying the Arm, Leg, Trunk, Finger–Hand, and Foot-Transport Motion–Time-Tables and integrating these into one Table, the 410 individual Motion–Times of Detailed Work-Factor are reduced to just twenty-five in Ready Work-Factor. Comparable simplification of the 'Grasp', 'Pre-position', and 'Assembly' Motion–Time-Tables are achieved.

The basic MTM-1 motions of 'reach', 'grasp', and 'release' are combined to form the MTM-2 element 'get', but disappear as individual motions, unlike the case of the Standard Element 'Pick-up' in Abbreviated Work-Factor. Similarly the basic MTM-1 motions 'move' and 'position' are replaced by the MTM-2 motion 'put'. The original elemental times for these MTM-1 movements, which number 239, have been reduced to thirty in the 'get' and 'put' Tables.

MTM-3 is a third-level simplification using even fewer classifications than MTM-2. 'get' and 'put' combine to form 'handle', and thus all five MTM-1 motions are now reduced to a single category. With each stage of simplification there is a loss of detail, and to some extent accuracy, though the latter is minimized by statistical methods of averaging and simplifying the detailed data. The simplification is illustrated below:

1st-level MTM-1	2nd-level MTM-2	3rd-level MTM-3	higher MTM-V U.S.D. M.C.D. U.M.S.
Basic Motions			
Release ⎫ Reach ⎬ Grasp ⎭	get ⎫		Combinations to give simple and
Move ⎫ Position ⎭	put ⎭	handle	complex elements

Systems in the last category (higher-level) are not only quicker to apply than the first three, they are intended also for specific kinds of work. MTM-V is used for machine-shop work, U.S.D. (Universal Standard Data) is designed for general maintenance, as is U.M.S. (Universal Maintenance Standards); and M.C.D. (Master Clerical Data) is intended for use in clerical work. These systems are not suitable for use outside the areas for which they were intended, whereas the first-, second-, and third-level data are for general application.

Many companies developed simplified systems from MTM-1; the I.C.I. Ltd technique of Simplified P.M.T.S. is one well-known example. These systems are numerous and have been compiled with various degrees of logic and accuracy.

P.M.T. systems have a great advantage over other systems of work measurement in that they minimize the effects of personal judgment in the setting of time standards, although they do not eliminate this factor completely. The motion categories, the rules of application, and the motion–times themselves are based on extremely large quantities of research material, as described in an earlier section of this Chapter. The rating factor is built into the system and does not have to be applied each time the study is made, as it does with time study and rated-activity sampling. Differences in rating concepts between analysts are eliminated.

A very high degree of consistency is possible in setting the standard, but it must be realized that individual analysts may put slightly different interpretations on the rules of application in certain motions, particularly when using a 'visual' technique such as MTM-2. It was to solve problems of consistency and personal judgment in work measurement that Work-Factor was originally developed so many years ago.

There is a footnote to the B.S.I. definition quoted above, which refers to the methods application of P.M.T.S. This has been criticized on occasions on the grounds that it does not stress sufficiently the importance of method. The time taken to perform a task depends on the sequence of basic motions employed by the operator. It is not possible to use a system of P.M.T. without first establishing that sequence. Further, the information is presented in extremely fine detail, with little or no ambiguity. Because of this, P.M.T.S. provide a useful means of establishing effective working methods upon which the standards may be based. The qualifying of each motion in terms of time enables the practitioner to establish the standard time for the job and the method simultaneously, thereby obviating the need for lengthy time studies and experiments on the new method.

Predetermined motion–time systems are useful for developing standards for existing jobs, but they are even more valuable as a means of determining methods and predicting times for new work. The old techniques of motion study can be used to find an effective method, but they have no means of determining the time for such a method. A time study cannot be carried out on a job until production has started unless expensive experiments are carried out with special 'mock-up' models of the task. Standard data and synthetic times can give some guidance as to time, but it is in P.M.T.S. that detailed, precise, work descriptions exist.

These are the chief characteristics of P.M.T.S. They provide a very detailed analysis of the job, useful for both method study and work measurement. It is possible to develop correct methods and time standards in advance of production with a precision and flexibility not afforded by any other systems. Times can be developed for work-cycles which are too short for normal stop-watch studies. Personal judgment problems are minimized by clearly defined rules for applying motion categories, and by the use of levelled times, which eliminate the need for rating.

Against these advantages it may be said that P.M.T.S. can only analyse manually controlled operations. Process times and machine-running times must be studied by other means. Work which requires an abnormally high degree of control cannot be covered by P.M.T.S. either. The application of the systems where the work is predominantly mental and involving decision-making is limited. The production of detailed analyses is time-consuming, often to the extent of rendering the systems uneconomic to use. This problem has been alleviated by the introduction of second, third, and higher levels of data, and by the development of computerized systems.

8.31 SPEEDS OF APPLICATION

It is difficult to compare the application speeds (or ratios) of the different systems of P.M.T.S., and even more difficult to compare these with other methods of work measurement, because of the problem of knowing what to include in the calculation. For example, should the generation of the data be included or merely the application? In the case of, say, comparative estimating, an extremely favourable but misleadingly high speed of application will be obtained.

Some estimates have been given by the custodians of various systems, for example:

Work-Factor: Application ratios for 'an intricate 2-handed assembly with many parts, after methodizing is complete, and including making a right- and left-hand study':

Detailed: 1 hour of engineer's study time per min of operation, i.e. 60:1
Ready: 30 min of engineer's study time per min of operation, i.e. 30:1
Abbreviated: 3 to 10 min of engineer's study time per min of operation, i.e. 3:1 to 10:1
Wocom: 18 min of engineer's study time per min of operation (this does not include input of computer data) 18:1

These ratios may be reduced by as much as 50 per cent by using the special pre-printed analysis forms.

There are corresponding application ratios for the systems of Methods–Time Measurement:

MTM, 200:1; MTM-2, 70:1; Taped MTM-2, 11:1
Direct comparisons with Work-Factor are not possible because of the differences in methods of assessing the ratios.

8.32 PREVALENCE OF SYSTEMS

Often the need for so many systems is questioned by the layman. Originally, when Work-Factor was developed in the 1930s, the only system of any significance then available was Segur's M.T.A. Work-Factor, developed quite independently, arose from a need for an alternative and reliable means of measuring work to replace, or at least to augment the use of time study. Other systems followed, but these were mainly intended to be used locally within the originating organizations. Later Methods–Time Measurement emerged as a result of research by H. B. Maynard and others to produce methods formulae primarily for use in machine-shops.

When MTM was made available to industry as a non-proprietary system many large industrial concerns as well as management consultancies were stimulated to produce their own systems based on the MTM standards, and during the 1950s and 1960s new systems proliferated.

The basic first-level and higher systems of Work-Factor, MTM, Basic Motion Timestudy, Primary Standard Data, Master Standard Data, and others, persist because of the understandable confidence of the originating organizations in their own systems. By owning the copyright to a system they are independent of any other company, owe no royalties, and are able to develop and apply their standards as they will.

The most widely used system in the United Kingdom is Methods–Time Measurement, but on a global scale it is almost impossible to assess the relative importances of MTM and Work-Factor, because the latter takes pride of place in some European as well as some Asian countries. Furthermore, many companies who employ practitioners trained by these organizations have themselves trained others in the art of applying the techniques, and many of these are not officially registered as practitioners or instructors with the originating bodies.

Of the many systems of macro-P.M.T.S. the most widely used (in the U.K.) appear to be Clerical Work Improvement Programme, Basic Work Data, Clerical Work Data, and Universal Maintenance Standards.

8.33 VALIDITY

The originators of the various systems claim many advantages for predetermined motion–times over other methods of work measurement, some of which have been mentioned earlier. Advocates also praise the 'accuracy' of their systems as compared with time study and estimating. Inevitably their characteristics have been criticized over the years, and it is necessary for users to appreciate the shortcomings as well as the advantages of each.

One of the most important merits claimed for P.M.T.S. is the *universality* of the data, which allegedly will give comparable results wherever they are applied. The claim is subject to the reservation that data are valid within the area from which they were derived originally. In the case of lower-level micro-P.M.T.S., data may be subject to more liberal interpretation and application, but at the other extreme, craft data and data-blocks may be valid only within the confines of a particular shop or department of the organization. It may be necessary to generate individual sets of data for each section studied, and indeed the originators warn users against attempting to transfer data-blocks from one area or company to another, even though apparently the work is identical in both areas. Local differences in conditions may invalidate such standards.

Clearly the more basic and fundamental the elements, the more universally applicable are the data. But even the first-level P.M.T. systems may be affected by national characteristics and anthropometric differences, and should therefore, be validated before use.

Another important consideration is that of *accuracy*, or more precisely *errors* which may be evident and inherent in any system. Most errors arise from five main sources.

1. Errors may be statistical and the inevitable result of sampling.
2. They may be human errors caused by incorrect analyses of jobs, or different interpretations of methods or rules of application.
3. They may result from combining or simplifying data to produce higher-level systems.
4. Least significantly, errors may result from 'additivity' effects.
5. Finally, errors may be caused by the failure of the system to account for all the variables which significantly affect the elemental times.

It is not meaningful to speak of the 'accuracy' of any system because this implies comparison with some yardstick or unique and absolute standard, and, of course, none exists. On the other hand it is possible to compare a standard set by P.M.T.S. against one for the same job set by time study (or any other method). Clearly this will not produce an absolute value of accuracy because no one can say that the time-study time is absolutely correct. Thus absolute accuracy is a myth, and comparisons must be made in terms of *relative* accuracy of one technique compared with another.

An elemental time is inevitably the average of many observations made by several observers. Any sample average is subject to a statistical error, and this, as described in Chapter 6, is known as the *standard error of the mean*, defined by the ratio:

$$\text{standard error} = (\text{standard deviation}) \div \sqrt{(\text{sample size})}$$

Suppose, for example, that the mean time to perform a 'reach' of 250 mm was found from a sample of 900 observations to be 6·5 milliminutes with a standard deviation of 0·8 milliminutes, and that the standard could be taken as representing that of the population of reaches, then the standard error of the mean is: $0·8 \div \sqrt{900}$, or 0·027 milliminutes. Thus, with 95 per cent confidence (± 2 standard errors) the mean of 6·5 milliminutes is subject to a statistical error of $\pm 0·054$ milliminutes. In a job sequence of elements it is likely that the individual statistical errors will be compensating, and consequently cancel, but it is possible that in a particular job the errors will be biased in one direction.

The problem is more serious in the case of macro-P.M.T. systems designed for highly variable work, where often very large standard deviations are quoted. For example, one large public utility using MTM-2 Motion Data quotes mean job-times of the order of 3,000 t.m.u. with standard deviations of over 2,000 t.m.u. Even after performing the task as many as 50 times, the mean time is still subject to an error (with 95 per cent confidence) of ± 600 t.m.u., which represents a relative error of ± 20 per cent. In some cases the standard deviation even exceeds the mean time.

From the above it is clear that the user must employ such times with care, first discovering the parameters, and hence the errors to which his standards are subject.

The *percentage error* depends on the length of the cycle being studied, so relative errors may be quoted only for given cycle-lengths. The accuracy of MTM-2 is stated to be within 5 per cent of MTM-1 values (with 95 per cent confidence) for cycle durations of one minute. It is instructive to compare such first-, second-, and third-level systems, and a full analysis is given in Figures 8.8 and 8.13 and explained in the accompanying text.

Several reasons prompted the development of systems of P.M.T., one of which was the prohibition of the use of the stop-watch as a means of measuring work. Such a situation once obtained in the United States. Another was the need for consistency and fairness between standards set by work measurement. Champions of P.M.T.S. use consistency as a major weapon when advocating use of the systems. The assumption is made that trained practitioners would make identical analyses of a given job and hence arrive at identical basic times for doing that job. There are two reasons why this ideal may not be realized in practice, apart from those discussed elsewhere. First there is the *human error*; the analyst may misinterpret the situation, or he may miss certain subtle elements such as 'apply pressure', or even fail to include parts (or repetitions) of the work. Secondly the job may be prone to misinterpretation or ambiguity, in which case the analyst must make a decision as to the most appropriate element to include. This is especially true of the observational systems such as Ready Work-Factor or MTM-2.

From the foregoing it may be concluded that although the data sheets ensure relatively good consistency from correctly compiled analyses, inconsistencies can result from incorrect analyses.

Errors arise when second- and higher-level techniques are developed from basic detailed systems. The use of a measuring system which is less precise than the system from which it is derived will inevitably introduce errors due to rounding. In a basic system, arm movements may be measured to the nearest inch (2·5 cm), whereas in a higher-level system the same motions will be measured to the nearest 6-in. (15 cm), the same basic time being assigned to all arm movements between, say, 15 and 30 cm, and another basic time for movements between 30 and 45 cm. One example of this effect, discussed later in Section 8.6, is MTM-2. In this case not only are moving distances grouped into larger ranges, but also several basic elements are combined to form new, single elements, as in the MTM-2 element 'get', which is comprised of the MTM-1 elements 'reach', 'grasp', and 'release'.

Recently the validity of the straightforward adding of P.M.T. elements has come under investigation. The criticism is based on the allegation that it is not possible to incorporate in any particular element sufficient consideration of the preceding and succeeding elements which must inevitably affect it. Thus something must be lost (or perhaps, gained) when three totally independent times are added to represent the overall motion consisting of three dependent and linked events. The many standard elements of a system may be arranged in an extremely large number of combinations, and to cover all of these with individually derived standard elements would increase the number of standards to an impracticable size, thereby defeating the objective of P.M.T.S. to provide a workable, reliable, and compact system of work measurement.

The systems do, of course, allow for preceding and succeeding elements as well as they are able. An example of this is found in the classes of the element 'move' in MTM (*see* page 168), and there are many others.

Most systems have been developed through many years of research and subjected to validating studies, some more rigorously than others. In spite of the thoroughness of the research it is inevitable that some factors or variables are omitted. These may include mental elements such as simple split-second decisions which must be made before a notion is made, or different environmental conditions under which similar jobs may be performed in different places.

Small errors may result from levelling differences. The major systems of P.M.T. based their elemental times on the pace of 'a fair day's work'. Levelling to a definite rate (such as 100 B.S., or 60 on the Bedaux scale) was not used. This means that, although it has been found that MTM time standards are generally equivalent to 83 rating (B.S.I.), individual elemental standards will be different from this. On very short-cycle jobs it is possible that the rating differences of individual elements will not cancel, but will introduce a bias. Generally the error will be negligible. However, it may be necessary for the company using the system to translate the basic times of the particular system to suit the standard rating level which it has adopted.

For the reader who wishes to consult published literature the article by H. Schmidtke aud F. Stier, 'An Experimental Evaluation of the Validity of P.M.T.S.', *Journal of Industrial Engineering*, May 1961, Vol. 12, is recommended.

8.4 ELEMENTS AND ANALYSES

The original therbligs depicted in Figure 8.1 were not sufficiently sensitive to provide precise analyses and subsequent measurement. A. B. Segur refined the therbligs and added standard times. Others have since enlarged the list with their own additions.

The major divisions of work are the standard elements, of which the therbligs are examples. These standard elements are subject to several sub-divisions and variables. Typical P.M.T.S. elements are shown in Figure 8.3, and some of the variables are explained below.

Most jobs are predominantly composed of hand operations. Obtaining objects (tools, materials, etc.) and placing them in specific positions constitute a great proportion of all work. Typical standard elements associated with picking up and moving objects are identified by the titles 'reach', 'grasp', 'move', 'position' (or 'pre-position'), and 'release'. It has been found in MTM-2 research, for example, that these elements comprise about 90 per cent of all usual work. Variations of these titles are 'get', 'put', 'place', and 'obtain'. In order to analyse all work other elements are needed to cover walking, eye movements, assembling, disassembling, and some mental processes.

In some systems, including MTM and S.P.M.T.S. (Simplified P.M.T.S.), times for 'reaches' and 'moves' are contained in separate tables, and may be made only with the hands and fingers, but in Work-Factor, Reaches and Moves (between which no distinction is made in the Transport Motion–Time-Table)

Figure 8.3 Typical P.M.T.S. elements

may be executed using any of the Body Members. Basic Motion–Timestudy also has a common 'Reach or Move' Table.

Variables of P.M.T.S. standard elements increase the times necessary for the performance of basic elements, and include the following:

1. *Distances* moved by the fingers, hands, arms, trunks, and feet in performing the work have a direct bearing on the time taken to make the move. This parameter is measured in inches or centimetres. Twisting motions may be measured in degrees.
2. *Weight*. This variable has at least three forms:

    (i) the weight of the object when lifted or moved affects the time required to execute the movement;
    (ii) resistance to motion may exist such as is experienced when pulling open the drawer of a filing cabinet, or when sliding a heavy box along the floor;
    (iii) torque is apparent when an operative turns a handle or a nut against strong resistance to motion.

    Often jobs require the operator to grip an object and exert a force on it, there being little or no discernible movement. This occurs when a final tightening is given to a nut. This form of 'weight' is treated in different ways by the various systems of P.M.T.: in B.M.T. and in MTM it is called 'apply pressure' (*see* page 171), and separate rules and tables are provided, while Work-Factor designates this as a 'Time Equivalent' and computes the action as a 1-in. motion of the Body Member involved (or a 45° Turn for Forearm Swivel).
3. *Types of Motion*. Classifications of this variable include 'U' motions, and the state of the hands at the start and/or finish of the motion, i.e. whether the hand is in motion or at rest at the beginning or end.
4. *Care and Precaution*. These are time-increasing factors. Jobs may include movements performed in dangerous circumstances in which injury may be caused to the worker, or damage to materials or equipment. Work with soldering irons, power saws, or the carrying of full, open, pots of paint all require care, and must be given due time-allowance. In Work-Factor this is one of the four *Work-Factors* composing the 'Major Variable—Manual Control'.
5. *Difficulty and precision*. Some motions are 'basic' such as tossing aside an object into a large bin, but when specific placing is needed a slight hesitation to allow the operative to gain control before the final insertion, or in the case of very precise placing a careful steering motion, must be paid for in terms of extra time-allowance. Parts to be assembled or picked up may be very small or of an awkward shape. They may be entangled, nested, or sticky. The 'target' size and the closeness of fit between mating parts cause difficulties when assembling. In Work-Factor the part being assembled is known as the Plug, and the space into which it is assembled is the Target. The Plug/Target ratio

decides the degree of difficulty. Moreover Plug/Target ratios which are close to unity (i.e. a very tight fit) require the Plug to be moved into a perpendicular position before it is possible to achieve the assembly. This is known as 'Upright'. Generally the degree of difficulty in all systems is assessed by the sizes of objects handled or dimensions of eventual resting-places.

8.41 ANALYSES

The methods of compiling analyses vary from system to system. The main differences between the MTM-based systems and the Work-Factor system is in the use of a common time-scale in the former, as opposed to separate columns in Work-Factor analyses. A simple comparison is given below.

8.5 METHODS-TIME MEASUREMENT—MTM-1

The following sections are intended to provide an appreciation of a typical system of P.M.T. It is stressed that this is not a manual of MTM, and potential users must seek instruction from an approved source of training.

Methods–Time Measurement is one of many predetermined motion–time systems. It was developed in the 1940s by H. B. Maynard, G. J. Stegemerton and J. L. Schwab. Details of the system were published in 1948, in a book entitled *Methods–Time Measurement* (McGraw-Hill). Most predetermined motion–time systems have been developed by firms either for use within their own organization, or for use in consultancy. Details of the systems are not usually published. MTM data, however, have been made completely available to the general public—including the original research. As a result of this the system has achieved world-wide popularity.

Areas of training, qualification and research into MTM are controlled and co-ordinated all over the world by a body known as the International MTM Directorate. This body is elected triennially by a General Assembly of the co-operating National MTM Associations. The International Directorate consists of a President, Co-ordinating Regional Trustees, Secretary, Treasurer, and a number of other General Trustees. There are national associations in the following major industrial nations: U.S.A./Canada; France; Germany; Japan; the Netherlands; Norway; Sweden; Switzerland; United Kingdom; Finland; and Belgium. These bodies are formed and held together by the common interest of promoting and co-ordinating the proper use of MTM within their designated areas.

Standards of training, including qualification of instructors, marking of examinations, duration and content of training courses, and so on, are supervised throughout the world by a single co-ordinating body, the International MTM Directorate. A consistent and high standard of application and practice is then promoted and maintained.

Since its original publication in 1948, the system has been refined and extended by additional research done by the National MTM Associations under the auspices of the International Directorate. Many other P.M.T. systems have been developed from the basic MTM data, notably MTM-2 and MTM-3, which form the subject of the next section. In this section we will examine the basic MTM data, which is officially known as MTM-1. The information given will provide a fairly full appreciation of the technique, but is no substitute for the full training course necessary to qualify a practitioner in MTM.

The MTM system is based upon the two principles that all human work can be performed by using a limited number of basic motions in various combinations; and that time standards can be established for these motions. MTM is defined as 'a procedure which analyses any manual operation or method into the basic motions required to perform it, and assigns to each motion a predetermined time standard which is determined by the nature of the motion and the conditions under which it is made'.

The basic motions and standard times of MTM are derived from research carried out by the originators at the Westinghouse Electric Corporation. Dr Maynard and his team studied a wide variety of manual tasks, beginning with drill-press work.

Some of the data were collected on film and subjected to a frame-by-frame analysis to isolate the basic motions. Other methods, including time study, were used to produce these motions, whose code letters are:

Reach	R
Grasp	G
Release	RL
Move	M
Position	P
Disengage	D
Turn	T
Apply Pressure	AP
Crank	C
Eye Action	ET & EF
Foot Motion	FM
Leg Motion	LM
Side Step	SS
Bend, Stoop or Kneel on One Knee	B, S, KOK
Arise	AB, AS, AKOK
Kneel on floor Both Knees	KBK
Arise	AKBK
Sit	SIT
Stand	STD
Turn Body	TB
Walk	W

These actions are either single, distinct movements of the limbs performed for a given purpose, called 'basic motions', or combinations of basic motions called 'basic elements'. Reach, for example, is a single movement of the hand or fingers to a destination (usually made to an object), and is described as a basic motion. Grasp, on the other hand, is a combination of very small reaches with the fingers and various other motions in order to get hold of the object. It is therefore described as a basic element.

With few exceptions, the time required to perform a given motion or element varies with the conditions under which it is performed. It was thus necessary to sub-classify these motions, to provide a suitable way of assessing the factors which alter the time. These factors or 'variables' will be discussed later in the chapter.

As explained in Section 8.2, the time-unit adopted by the originators was the time-measurement unit (t.m.u.).

$$1 \text{ t.m.u.} = 0\cdot00001 \text{ hours}$$
$$= 0\cdot0006 \text{ minutes}$$
$$= 0\cdot036 \text{ seconds}$$

All times on the MTM Data Card are expressed in t.m.u. Conversion to minutes at 100 B.S. performance can be made by extending by the 83 B.S. rating, or by adjusting the factor proportionally:

$$\text{mins @ 100 B.S.} = \text{t.m.u.} \times 0\cdot0005$$

The remaining dimensions—distance and weight—can be expressed either in Imperial or in metric units. Time values exist for both.

8.51 THE ELEMENT 'REACH' (R)

The most important characteristic of 'reach' is the predominant purpose, of moving the hand or fingers to a destination. Movements for any other purpose would be analysed by different motions; for example movement of the hand to lay aside a screwdriver on a bench is classed as 'move'.

Three factors or variables affect the time taken to perform a reach. The first of these is the distance travelled by the hand or fingers. The distance is measured in inches and is written immediately after the 'reach' symbol R.

e.g. R10B = Reach 10 inches (25·4 mm) to an object in a location which may vary slightly from cycle to cycle (*see* case and description in Reach Table)

When very long reaches are made, the body usually assists the hand by moving at the same time. Only part of the total distance reached is contributed by the hand and arm. The rest is produced by the body action—which is called 'body assistance'. The actual time taken for the whole operation will be the time needed for the hand and arm motion alone, since the body movement is simultaneous.

There are five kinds of reach, called 'cases'. These form the second variable of the motion. They are distinguished by the symbols A to E. The case of reach employed depends largely on the nature of the object to which the reach is made.

The first case, A, is a reach to an object in a fixed location or to an object held in the other hand, or on which the other hand rests. Case A reaches are performed automatically. Very little visual or mental control of the muscles is necessary. For this reason they are the fastest of all reaches.

Case B is a reach to a single object in a location which may vary slightly from cycle to cycle. R-B is the commonest case of reach. It requires some visual control—it is necessary to glance at the object, but the eyes are not required for the whole motion. Reaching for a tool lying on a bench would be case B.

Case C is a reach to an object, jumbled with others in a group so that 'search' and 'select' occur. The important characteristic is that 'search and select' must occur. R-C is made to a single object in the group. Therefore, while the reach is being performed it is necessary for the eyes and mind to single out the object which will finally be picked out from the group. *The objects must be jumbled together*. R-C does not apply to neatly stacked objects. Reach to a nut from a pile of nuts in a box would be classed as case C. Case C reach requires a high

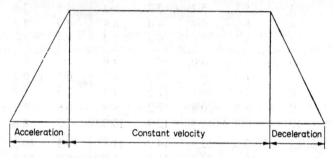

Figure 8.4 A type 1, or standard 'reach'

degree of visual and mental control. For this reason it takes longer to perform than Cases A or B.

Case D is a reach to a very small object, or where an accurate grasp is required. Where the object is fragile or dangerous, or where access to it is restricted, Case D reaches also apply, because of the care involved: for example, reach to a small piece of broken glass.

Case E is a reach to an indefinite location to get the hand in position for body balance, or for the next location, or out of the way. This is the only case where the hand is not reaching to an object. It is simply a movement of the hand to get it to a general location—with no intention of doing anything when it arrives.

The third variable of reach is called type of motion. There are three types of motion. The first is the normal reach action. The hand starts at rest in one place and travels to its destination, where it again comes to rest. This is called a Type I or Standard Reach (*see* Figure 8.4).

A Type I reach consists of the following components:

1. A period while the hand is accelerating from rest at the start of the motion.
2. A period while the hand is travelling at a steady speed towards the destination.
3. A period while the hand is slowing down in order to stop at its destination.

Type II and Type III reaches frequently occur in highly repetitive work, where the hand performs a sequence of motions, one after the other, without stopping.

A Type II motion is one where the hand starts off from rest, but does not come to a standstill at the end. Rather, it carries on smoothly into the next motion without stopping. An example of this would be pushing a light object

TABLE 8.1

TABLE OF VALUES FOR 'REACH'

Distance moved inch	Time t.m.u.				Hand in motion	
	A	B	C or D	E	A	B
¾ or less	2·0	2·0	2·0	2·0	1·6	1·6
1	2·5	2·5	3·6	2·4	2·3	2·3
2	4·0	4·0	5·9	3·8	3·5	2·7
3	5·3	5·3	7·3	5·3	4·5	3·6
4	6·1	6·4	8·4	6·8	4·9	4·3
5	6·5	7·8	9·4	7·4	5·3	5·0
6	7·0	8·6	10·1	8·0	5·7	5·7
7	7·4	9·3	10·8	8·7	6·1	6·5
8	7·9	10·1	11·5	9·3	6·5	7·2
9	8·3	10·8	12·2	9·9	6·9	7·9
10	8·7	11·5	12·9	10·5	7·3	8·6
12	9·6	12·9	14·2	11·8	8·1	10·1
14	10·5	14·4	15·6	13·0	8·9	11·5
16	11·4	15·8	17·0	14·2	9·7	12·9
18	12·3	17·2	18·4	15·5	10·5	14·4
20	13·1	18·6	19·8	16·7	11·3	15·8
22	14·0	20·1	21·2	18·0	12·1	17·3
24	14·9	21·5	22·5	19·2	12·9	18·8
26	15·8	22·9	23·9	20·4	13·7	20·2
28	16·7	24·4	25·3	21·7	14·5	21·7
30	17·5	25·8	26·7	22·9	15·3	23·2

Case and description

A Reach to object in fixed location, or to object in other hand or on which other hand rests.

B Reach to single object in location which may vary slightly from cycle to cycle.

C Reach to object jumbled with other objects in a group so that search and select occur.

D Reach to a very small object or where accurate grasp is required.

E Reach to indefinite location to get hand in position for body balance or next motion or out of way.

out of the way—to clear a space perhaps. Type II motions can also start with the hand in motion and finish with the hand at rest.

A Type III motion is one in which the hand is in motion both at the beginning and at the end. This type of motion consists solely of a constant-velocity period, and lacks both acceleration and deceleration. It is the fastest of the three types. The symbol 'm' is used to denote 'hand in motion', for example:

$$\begin{aligned}\text{mR10B} &= \text{R10B with the hand in motion at the start (Type II)}\\ \text{R10Bm} &= \text{R10B with the hand in motion at the finish (Type II)}\\ \text{mR10Bm} &= \text{R10B with the hand in motion at the start and finish (Type III)}\end{aligned}$$

Type I reach times are given in Table 8.1 in the first four case-columns. In cases A and B, the values for Type II motions are given in the 'hand in motion' columns. Type II reaches for cases C, D and E all have to be calculated.

8.52 THE ELEMENT GRASP (G)

Grasp—symbol G—is the element used to get hold of an object, and follows naturally from reach, which is used to get the hand to the object. It is defined as 'the basic element employed to secure sufficient control of an object with the hand or fingers to permit the performance of the next basic motion'. Grasp is only performed by the hand or fingers. It does not occur when contact is made by the feet, as in pressing a foot button, nor does it occur when control is established over the object using tweezers or tongs. The former case is analysed by body motions and the latter is treated as a move motion. The time required to perform a grasp varies with the type of grasp used and the conditions prevailing at the time (Table 8.2). The MTM system recognizes eleven cases of grasp, in five main groups. These groups are:

        G1—pick-up grasp
        G2—regrasp
        G3—transfer grasp
        G4—select grasp
        G5—contact grasp

G1—pick-up grasps—are used to pick up single objects that are lying by themselves, or which are placed in an orderly fashion in relation to their surroundings.

The third type of grasp is called 'transfer grasp'. The code is G3. Transfer grasp occurs when an object is passed from one hand to the other. Control must be completely transferred to the second hand. Merely grasping the object with the second hand, in order to hold it in both, would not qualify for a transfer grasp.

'Select grasp', G4, is used to grasp an object jumbled with others so that 'search and select' occur. There are many examples of G4s, especially in assembly work, where parts are presented to the operators in a box or tote-pan. For this reason it is often called the tote-pan grasp. Selection of a nut from a box of nuts, or of a tool from a toolbag, are usually G4s. G4 grasp is subdivided according to the

TABLE 8.2

TABLE OF VALUES FOR 'GRASP'

Case	Time t.m.u.	Description
1A	2·0	*Pick up grasp*—small, medium or large object by itself, easily grasped.
1B	3·5	Very small object, or object lying close against a flat surface.
1C1	7·3	Interference with grasp on bottom and one side of nearly cylindrical object. Diameter larger than $\frac{1}{2}$ inch.
1C2	8·7	Interference with grasp on bottom and one side of nearly cylindrical object. Diameter $\frac{1}{4}$ inch to $\frac{1}{2}$ inch.
1C3	10·8	Interference with grasp on bottom and one side of nearly cylindrical object. Diameter less than $\frac{1}{4}$ inch.
2	5·6	*Regrasp*.
3	5·6	*Transfer grasp*.
4A	7·3	Object jumbled with other objects so search and select occur. Larger than $1 \times 1 \times 1$ inch.
4B	9·1	Object jumbled with other objects so search and select occur. $\frac{1}{4} \times \frac{1}{4} \times \frac{1}{8}$ inch to $1 \times 1 \times 1$ inch.
4C	12·9	Object jumbled with other objects so search and select occur. Smaller than $\frac{1}{4} \times \frac{1}{4} \times \frac{1}{8}$ inch.
5	0	Contact, sliding or hook grasp.

sizes of the object being grasped. The size ranges given are intended as a guide to which subdivision should be allowed:

G4A—objects larger than $1'' \times 1'' \times 1''$
G4B—$\frac{1}{4}'' \times \frac{1}{4}'' \times \frac{1}{8}''$ to $1'' \times 1'' \times 1''$
G4C—objects smaller than $\frac{1}{4}'' \times \frac{1}{4}'' \times \frac{1}{8}''$

The last category of grasp is G5. This is a contact grasp, used when sufficient control is gained over the object by touching its surface with the fingers or hand. G5 takes no time to perform—in fact it is the break-point between 'reaches' and 'moves'. It is recorded in an MTM analysis to show where control has been established over the object.

8.53 THE ELEMENT RELEASE (RL)

'Release' is the basic element used to relinquish control of, or contact with, objects by the hand or fingers. The symbol for release is RL. Release is the action of letting go of something. It is the opposite of 'grasp'. Like grasp it does not apply to letting go of an object which is held by a tool (for example a pair of tweezers). This situation is analysed as *move*—normally M*f*B to open the jaws of the tool (*f* being the distance moved of ¾ in. or less).

There are two kinds of release. The first—RL1 is performed by a simple opening of the fingers. The second kind—RL2 does not involve opening the fingers. It is simply a removal of the hand or fingers from the object. RL2 is the opposite of G5. Like G5 it does not take any time to perform, because it is simply a break-point, and is included in an analysis to show the point at which control of the object is lost.

8.54 THE ELEMENT MOVE (M)

'Move' is the basic motion employed when the predominant purpose is to transport an object to a destination with the hand or fingers. The symbol for 'move' is M. Move has many of the same characteristics of 'reach'. It employs the same body members—the hands and fingers. The rules governing distance are the same. The same kinds of body assistance and the same three types of motion occur. The rules for dealing with them are similar to 'reach'.

The chief distinction between 'reach' and 'move' is in the predominant purpose. The predominant purpose of 'move' is to move an object to a destination, whereas the purpose of a 'reach' is to move the hand. Reaches can in fact be performed with an object palmed in the hand if the purpose is to move the hand and not the object—i.e. for the purpose of picking up a second object.

There are three cases of 'move' (Table 8.3). The cases are related to the type of destination involved and to the accuracy required. There is no relationship between the cases of 'move' and those of 'reach'. Case A move is used to move an object to the other hand, or against a stop. These moves require little or no conscious control. Movements to the other hand are guided by kinetic sense in a similar manner to 'reach'. Movements of objects against a stop of some kind eliminates the need for accurate location of the object at the end, and can thus be done very quickly. Case B is the most common case of 'move'. It is used to move an object to an approximate or indefinite location. M-B covers such motions as laying aside tools, finished products, and so on. Case C moves are very highly controlled. Case C is used to move an object to an exact location, which means to a tolerance of about $\frac{1}{2}$ in (13 mm). If two hands are used, the effect of the weight, in slowing down the motion, is effectively halved. When the object is slid, part of the weight is supported by the surface on which the object rests, and the effective resistance is reduced. This effective resistance is called effective net weight (E.N.W.). There are rules in the MTM Manual which describe the calculation of E.N.W. Effective net weights of 2·5 lb (1 kg) or less do not significantly effect move times. A move with weight is split into two parts, the static component and the dynamic component.

8.55 THE ELEMENTS POSITION (P) AND DISENGAGE (D)

8.551 *The Element 'Position'* (*P*). Position is the basic element employed to align, orient and engage one object with another with motions so minor that

Table 8.3

TABLE OF VALUES FOR 'MOVE'

Distance moved inch	Time t.m.u.				Weight allowance		
	A	B	C	Hand in motion B	Weight (lb) up to	Factor	Constant t.m.u.
¾ or less	2·0	2·0	2·0	1·7			
1	2·5	2·9	3·4	2·3	2·5	1.00	0
2	3·6	4·6	5·2	2·9			
3	4·9	5·7	6·7	3·6	7·5	1·06	2·2
4	6·1	6·9	8·0	4·3			
5	7·3	8·0	9·2	5·0			
					12.5	1.11	3·9
6	8·1	8·9	10·3	5·7			
7	8·9	9·7	11.1	6·5	17·5	1·17	5·6
8	9·7	10·6	11·8	7·2			
9	10·5	11·5	12·7	7·9	22·5	1·22	7·4
10	11·3	12·2	13·5	8·6			
					27·5	1·28	9·1
12	12·9	13·4	15·2	10·0			
14	14·4	14·6	16·9	11·4			
16	16·0	15·8	18·7	12·8	32·5	1·33	10·8
18	17·6	17·0	20·4	14·2			
20	19·2	18·2	22·1	15·6			
					37·5	1·39	12·5
22	20·8	19·4	23·8	17·0			
24	22·4	20·6	25·5	18·4	42·5	1·44	14·3
26	24·0	21·8	27·3	19·8			
28	25·5	23·1	29·0	21·2	47·5	1·50	16·0
30	27·1	24·3	30·7	22·7			

Case and description

A Move object to other hand or against stop.
B Move object to approximate or indefinite location.
C Move object to exact location.

they do not justify analysis as independent basic elements. The 'position' symbol is P (Table 8.4).

Position has three components—'align', 'orient' and 'engage'. 'Align' is to line up the two parts on a common axis, called the axis of engagement. Most positions involve fitting one object into another, and the axis of engagement is

TABLE 8.4
TABLE OF VALUES FOR 'POSITION'

Class of fit		Symmetry	Easy to handle	Difficult to handle
1. Loose:	no pressure required	S	5·6	11·2
		SS	9·1	14·7
		NS	10·4	16·0
2. Close:	light pressure required	S	16·2	21·8
		SS	19·7	25·3
		NS	21·0	26·6
3. Exact:	heavy pressure required	S	43·0	48·6
		SS	46·5	52·1
		NS	47·8	53·4

the path along which the object must travel during insertion. 'Orient' consists of all the adjustments that have to be made to match the shape of the object to the shape of the hole so that they will fit together. The third component—'engage'—consists of inserting the object into the hole. Position values allow for engagements of up to 1 in. depth of insertion.

8.552 The Element 'Disengage'. 'Disengage'—Symbol D—is defined as the basic manual action employed to break contact between one object and another, and is characterized by recoil caused by the sudden ending of resistance (Table 8.5).

TABLE 8.5
TABLE OF VALUES FOR 'DISENGAGE'

Class of fit		Easy to handle	Difficult to handle
1. Loose:	very slight effort, blends with subsequent move	4·0	5·7
2. Close:	normal effort, slight recoil	7·5	11·8
3. Tight:	considerable effort, hand recoils markedly	22·9	34·7

It is performed by the hand and fingers, and is used to analyse situations where there is friction between two objects that are being separated. When separation occurs the hand jumps or recoils. Involuntarily this recoil is the chief characteristic of 'disengage', and must occur for disengage to be present. Mere separation of objects without recoil would be analysed by 'move' or 'turn'.

There are two variables of disengage:

    1. Class of fit
    2. Ease of handling.

8.56 THE ELEMENTS 'TURN' (T), 'APPLY PRESSURE' (AP), AND 'CRANK' (C)

8.561 The Element 'Turn' (T). 'Turn' is the basic motion employed to rotate the empty or loaded hand about the long axis of the forearm. The symbol is T. This motion is performed by twisting the arm on its own axis, and it is this rotating action that distinguishes it from other motions such as 'reach' and 'move'. When an object is to be rotated, the hand will perform turning motions only if the arm is within 30° of the axis of rotation. There are numerous examples of turn motions. They can occur during the removal and fitting of light bulbs into bayonet sockets, or rotating of control knobs, dials, and so on (*see* Table 8.6).

8.562 The Element 'Apply Pressure' (*AP*). 'Apply pressure' is an application of muscular force during which object-resistance is overcome in a controlled manner accompanied by essentially no motion (½ in. = 13 mm or less). 'Apply pressure' was first noticed during the final tightening of a screw with a screwdriver, during which pressure was applied without causing any appreciable movement of the

TABLE 8.6

TABLE OF VALUES FOR 'TURN'

	Weight lb	\	Time t.m.u. for degrees turned				
		30°	45°	60°	75°	90°	105°
Small	0 to 2	2·8	3·5	4·1	4·8	5·4	6·1
Medium	2·1 to 10	4·4	5·5	6·5	7·5	8·5	9·6
Large	10·1 to 35	8·4	10·5	12·3	14·4	16·2	18·3
		120°	135°	150°	165°	180°	
Small	0 to 2	6·8	7·4	8·1	8·7	9·4	
Medium	2·1 to 10	10·6	11·6	12·7	13·7	14·8	
Large	10·1 to 35	20·4	22·2	24·3	26·1	28·2	

screw. Two important facts emerge from this. Firstly, force is applied to the object. This force is quite noticeable and is normally preceded by a slight hestitation, while the muscles are tensed up in order to deliver the force to the object. The second factor is that little or no movement results. In fact, movement is not the prime intention. In the case of the screwdriver, the purpose is to ensure that the screw is tight. This use of AP is very common where things are to be fastened together. It frequently occurs when screw caps are fitted to bottles and containers. It can also happen when a push-fit cap is fitted over the nib of a pen. In this case, force is applied to overcome the resistance of the retaining ring or clip. In practice, though movement is not intended, it usually does occur when AP is used. Minor movements of up to ¼ in. (6.4 mm) are allowed for in the times for 'apply pressure'. 'Apply pressure' has been the subject of considerable research since its original discovery. In January 1973 the International MTM Directorate officially adopted a system of revised apply-pressure data which replace all previous data on the subject. The appropriate decison model for selecting the correct values for apply-pressure elements is shown in Figure 8.5, and the Table of Values in Table 8.7.

172 WORK MEASUREMENT

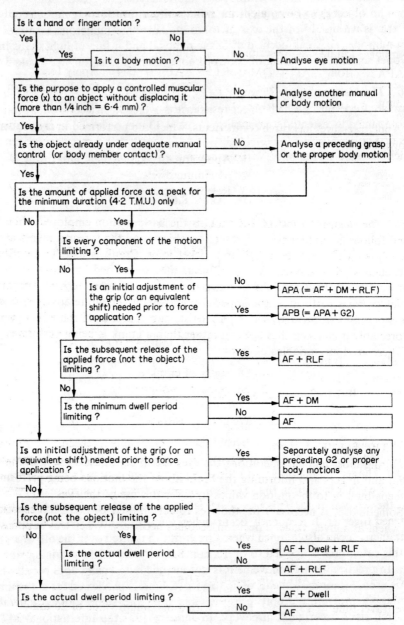

Figure 8.5 'Apply pressure' decision diagram

TABLE 8.7

TABLE OF VALUES FOR 'APPLY PRESSURE'

	Full cycle			Components	
symbol	t.m.u.	description	symbol	t.m.u.	description
APA	10.6	AF + DM + RLF	AF	3.4	Apply force
			DM	4.2	Dwell, minimum
APB	16.2	APA + G2	RLF	3.0	Release force

'Apply pressure' has three components, which are performed in the sequence shown below:

1. Apply force
2. Minimum dwell
3. Release force.

8.563 The element 'crank' (C). 'Crank' is the basic motion employed when the hand follows a circular path to rotate an object, and the forearm pivots about the elbow, which acts as a ball-and-socket joint. Crank is used to describe a particular kind of physical action. Objects may be rotated by moves or turns, but neither of these motions displays the particular characteristics of crank. When crank is performed, the upper arm and elbow remain relatively still, and the work is performed by the hand, wrist, and forearm. If the elbow moves appreciably it is a sign that a move rather than a crank is being performed.

There are four variables of crank:

1. Diameter of crank
2. Number of revolutions
3. Continuity of motion
4. Weight or resistance.

8.57 THE ELEMENT 'EYE ACTION'

During most manual operations the eyes are in continuous use, guiding the hand and fingers and informing the brain about the motions being performed. Time is built in to the motion values to cover this use of the eyes.

Sometimes it is necessary for the eyes to act on their own. This is especially so when, at a given stage, a decision becomes necessary. When this happens all

TABLE 8.8

TABLE OF VALUES FOR 'EYE ACTIONS'

Eye-travel time = $15.2 \times \dfrac{T}{D}$ t.m.u., with a maximum value of 20 t.m.u.

where T = the distance between points from and to which the eye travels,

D = the perpendicular distance from the eye to the line of travel T.

Eye-focus time = 7.3 t.m.u.

actions by other parts of the body have to stop, until the eyes have seen, and the decision has been made. At times like this the special element 'eye action' is used (Table 8.8). There are two kinds of eye action, eye focus and eye travel.

Eye focus

Eye focus is the basic action performed by the eyes and mind to recognize a readily distinguishable characteristic of an object within the area of normal vision.

Eye travel

Eye travel is the basic eye-action employed to shift the axis of vision to a new location. Like eye focus, it is only used when all other motions must cease whilst the eyes do their work.

8.58 THE ELEMENT 'BODY MOTION'

Body motions (Table 8.9) are made with the feet, legs, and body itself. They are deliberate actions, and not made incidentally to some other motion. This distinguishes them from body assistance. A foot motion is a movement of the

TABLE 8.9

TABLE OF VALUES FOR 'BODY MOTIONS'

Description	Symbol	Distance	Time t.m.u.
Foot motion:			
Hinged at ankle	FM	Up to 4 inches	8·5
With heavy pressure	FMP		19·1
Leg or foreleg motion	LM—	Up to 6 inches	7·1
		Each additional inch	1·2
Sidestep:			
Case 1—Complete when leading leg contacts floor	SS-C1	Less than 12 inches	Use REACH or MOVE time
		12 inches	17·0
		Each additional inch	0·6
Case 2—Lagging leg must contact floor before next motion can be made	SS-C2	12 inches	34·1
		Each additional inch	1·1
Bend, stoop, or kneel on one knee	B, S, KOK		29·0
Arise	AB, AS, AKOK		31·9
Kneel on floor—both knees	KBK		69·4
Arise	AKBK		76·7
Sit	SIT		34·7
Stand from sitting position	STD		43·4
Turn body 45 to 90 degrees:			
Case 1—Complete when leading leg contacts floor	TBC1		18·6
Case 2—Lagging leg must contact floor before next motion can be made	TBC2		37·2
Walk	W-FT	Per foot	5·3
Walk	W-P	Per pace	15·0
	W-PO	Per pace	17·0

foot hinged at the ankle. It covers movement of the foot in one direction only. The symbol is FM. A second category—foot motion with pressure (FMP)—allows for application of force by the feet. It is the equivalent of FM plus an AP(A). Leg motion—LM—is a movement of the leg. It covers movements of the whole leg hinged at the hip, and those of the lower leg only, hinged at the knee. Distance is its only variable.

$$LM12 = \text{leg motion of 12 in. (300 mm)}$$

Walking is a series of leg motions with the predominant purpose of moving the body forwards or backwards to a new location. Walking can be either obstructed or unobstructed. Unobstructed walking covers walking on a smooth level surface with no load. It does not cover negotiating obstacles, sharp corners, slopes, stairs, and so on. 'Obstructed walking' covers all normal industrial conditions and should therefore be used in analysis in preference to 'unobstructed'. Obstructed walking covers walking with a load, walking up and down stairs and slopes; walking on difficult or slippery surfaces, walking backwards, and so on.

The code for walking is W. 'Per foot' analyses are shown as W-FT with the number of feet travelled. Unobstructed per-pace analyses are shown as W-P with the number of paces taken. Obstructed per-pace values are coded W-PO.

8.59 SIMULTANEOUS MOTIONS

When two or more MTM motions are performed simultaneously, the total time required for the combination of motions will be the time for the longest or limiting motion. There are two ways in which motions can be performed simultaneously.

In the first instance, the motions are performed concurrently by different body-members. They are written on the same line, and the one which takes the lesser time is ringed to show that it is 'limited'. Time is allowed for the motion that occupies the most time. If the motions are identical, then no circling is necessary. For example:

(R8A)	11·5	R8C
(G1A)	7·3	G4A
(M8B)	12·2	M10B
RL1	2·0	RL1
	33·0	

In the second case, the motions are performed by the same body-member. Regrasps and turns are very often made at the same time as moves. The limiting motion is written first, the remainder follow, preceded by a bracket, then limited motions are struck out.

```
        13·4  ⎧ M12B
              ⎨ G2
              ⎩ I905
        ─────
         13·4
```

Motions frequently occur in both kinds of combination at once.

```
   (R10A)    13·4  ⎧ M12B
                   ⎨ I903
                   ⎩ G2
             ─────
              13·4
```

Sometimes a whole series of motions in one hand limits out a series of motions in the other. The ease, or difficulty, with which two or more motions can be performed simultaneously is influenced by the amount of mental and visual control required. Normally it is not possible to perform two highly controlled motions at the same time. One hand tends to overlap the other. In these cases an overlapping motion is recorded.

```
Towards location   (M−)    13·5    M10C  ⎫
                           16·2    P2SE  ⎬ Position object
Overlap            MfC      2·0          ⎭
Position object    P2SE    16·2
                   RL1      2·0    RL1
                         ──────
                          49·9
```

Another influencing factor is practice. As the operator gains his experience, it becomes easier to perform motions simultaneously. Some motion combinations can be performed simultaneously when they are carried out within the area of normal vision, but not when they are outside it. Others depend upon ease or difficulty of handling. Table 8.10 may be used to recognize those motions which may reasonably be expected to be performed simultaneously.

8.6 MTM-2

Because of the exceptionally fine degree of analysis, the basic MTM system (MTM-1), described in the previous chapter, is time-consuming and therefore expensive to use. For some work areas, simpler and therefore cheaper systems are more appropriate, and a number of them have been developed from MTM-1. Some of these have been developed to cover special fields and others are of a general nature. Several of these, and their authors, are listed in Section 8.9.

In 1964 the International MTM Directorate issued instructions for a new

TABLE 8.10

TABLE FOR 'SIMULTANEOUS MOTIONS'

Reach				Move						Grasp					Position						Disengage				Motion.
A,E	B	C,D		A,Bm		B		C		G1A G2 G5	G1B G1C		G4		P1S		P1SS P2S		P1NS P2SS P2S		D1E, D1D	D2		Case	
		W*	O*	W*	O*	W*	O*	W*	O*		W*	O*	W*	O*	E**	D**	E**	D**	E**	D**		E**	D**		
								●	●									●	●	●				A, E	Reach
			●				●	●	★			●	●	★	●	●	●	★	★	★			●	B	
			●	●	★	●	★	★	★		●	★	★	★	★	★	★	★	★	★	●	★	★	C,D	
																		●	●	●				A, Bm	Move
												●	●	★	●	●	●	★	★	★			●	B	
								●	★		●	★	★	★	★	★	★	★	★	★	●	★	★	C	
																		★	★	★		★	★	G1A, G2, G5	Grasp
											★	★	●	★	★	★	★	★	★	★	★	★	★	G1B, G1C	
													★	★	★	★	★	★	★	★	★	★	★	G4	
															●	★	★	★	★	★	★	★	★	P1S	Position
																	★	★	★	★	★	★	★	P1SS, P2S	
																			★	★	★	★	★	P1NS, P2SS, P2NS	
																								D1E, D1D	Disengage
																								D2	

☐ Easy to perform simultaneously

● Can be performed simultaneously with practice

★ Difficult to perform simultaneously even after long practice. Allow both times

* W = Within the area of normal vision
O = Outside the area of normal vision

** E = Easy to handle
D = Difficult to handle

Motions not included in above table

Turn – normally EASY with all motions except when TURN is controlled or with DISENGAGE

Apply pressure – may be EASY, PRACTICE, or DIFFICULT. Each case must be analysed

Disengage – any class may be DIFFICULT if care must be exercised to avoid injury or damage to object

Disengage – Class 3 – normally DIFFICULT

Release – always EASY

Position – class 3 – always DIFFICULT

second-generation system to be developed. This was to be a system at the general level—that is, it was to apply to all types of work. It was to be the internationally accepted second level of MTM data. The data were to comply with a strict specification—they were to be:

1. Consistent between analysts and areas of application.
2. Fast to handle.
3. Universally named.
4. Easy to understand.
5. Descriptive of the method.

6. Combinable with other MTM data.
7. Based on MTM.
8. Specified in relation to speed of application and accuracy of results.

The development work was carried out by the International Standing Committee for Applied Research. Much of the research information was supplied by the Swedish MTM Association. Data was also collected in the U.K. and in the United States. This was compared with the Swedish data to ensure that the system was universally applicable. The project was finished in May 1965 and approved by the managing board of the International MTM Directorate in Munich, in the June of that year. The system was called MTM-2.

MTM-2 is a system of synthesized MTM data and is the international second general level of MTM data. It is based exclusively on MTM and consists of:

1. Single basic MTM motions, and
2. Combinations of basic MTM motions.

The data were developed from MTM-1 analyses provided by the Swedish MTM Association and Swedish industry. These analyses were taken from widely differing industrial situations. The total number of motions in the analyses is over 14,000.

These analyses were carefully audited to ensure that the principles of MTM-1 had been adhered to. They were then analysed by computer to provide the data that were used to compile MTM-2. At the same time, tests were carried out to find and examine any basic differences between the motion sequences used in the different areas from which the initial data had been taken. From these tests it was concluded that the differences between various areas of application are slight, and that MTM-2 is independent of such factors. Based as it is on actual motion-sequences performed, it is oriented towards the operator rather than to his equipment. That is to say, MTM-2 analyses the actual movements of the operator, rather than making deductions about the motions necessary from the relationships between parts and equipment, classes of fit, and so on.

The demands on the data—especially speed of handling—indicate that simplicity of the system must be a fundamental quality. This simplification was achieved by MTM-technical means and by statistical means. Technical simplification was used to minimize the number of alternatives and variables in the elements. It simplifies the decision-making processes by the use of rules, and decision models that enable the analyst to identify the difficult alternatives by exception.

Statistical simplification was achieved by examining the motion sequences, distances, weights, etc., and using modern probability theories to reduce the large amount of data available in MTM-1 in a way that minimizes loss of accuracy. This form of simplification was carried out in several ways.

Some MTM-1 motions were combined to give a single MTM-2 category. Thus 'reach', 'grasp' and 'release' were combined to form 'get'. Distances were averaged to give a limited number of distance classes, and some elements were averaged also, for example AP(A) and AP(B) were averaged to give a single value for 'apply pressure'. Other elements were substituted by others—'turn' is substituted by the 'reach' part of 'get' or the 'move' part of 'put'—and yet more were eliminated altogether. 'Sit' and 'stand' have been eliminated. They are analysed as 'bend' and 'arise' in MTM-2.

The study of motion sequences showed that the majority of tasks are performed using the sequence 'reach-grasp-move-release'. There is also a high incidence of repeating 'move' motions, and of 'move' followed by 'position'. The predominance of this basic sequence, R, G, M, RL, suggests an apparent

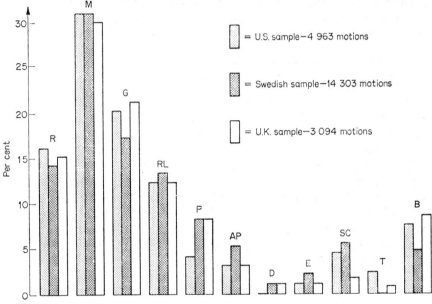

Figure 8.6 The distribution of MTM motions

solution of a four-motion sequence in MTM-2. It is not easy to analyse two-handed over-lapping motions using this sequence, however, and the final solution was to use two main sequences:

Release, reach and grasp = get
Move and move plus position = put

The importance of these five MTM-1 categories is shown in the bar chart, which shows the distribution of the MTM-1 motions in test samples from the different countries. More important than this, the chart shows that the distribution of the motions is very similar in Sweden, the United Kingdom and the United States. There are slight differences, and this is to be expected, just as there are slight differences between industries and work areas. However, the averages in MTM-2 have been carefully constructed to achieve minimum deviation over a large sample of data, and the evidence indicates that valid results are achieved in all but the most highly repetitive areas. This is shown in the bar chart, Figure 8.6.

The close similarity between Swedish and U.K. data for distance is shown by the cumulative frequency graph for 'reach' and 'move' distances (Figure 8.7). The distribution of distances is practically identical.

The system was carefully tested by the Swedish MTM Association to establish the accuracy of results and the speed of application of MTM-2 compared with MTM-1. These tests were made in several Swedish companies and on tasks that

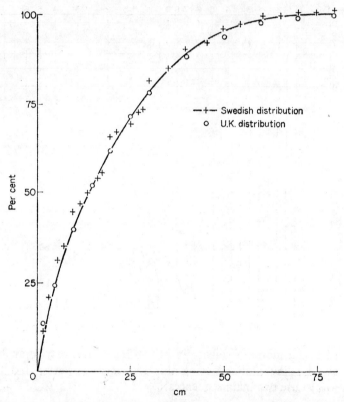

Figure 8.7 Cumulative frequency-distribution of distances of 'reach' and 'move'

ranged from light assembly to foundry work. The tasks were simultaneously, but independently studied by analysts in MTM-1 and MTM-2. The results were then compared.

The results show that MTM-2 has an average bias of zero per cent. That is, the results are neither tight nor loose as compared with MTM-1. The deviations between the two are such that MTM-2 analyses are within ±5 per cent, with 95 per cent confidence, from MTM-1 at cycles of one minute or more. These figures take into account both the differences between the systems themselves and variations that can be expected to exist through the imperfections of the analyst. The averages can be improved on or otherwise depending on the skill of the application. The relative accuracy of MTM-2 compared with MTM-1 is shown in the graph, Figure 8.8. The speed of analysis with MTM-2 was almost twice that of MTM-1. Subsequent experience indicates that even higher speeds may be achieved.

The results of the tests indicate that MTM-2 is suitable for analysis of work-cycles greater than about one minute. Deviations from MTM-1 can be expected to be considerably more than 5 per cent for shorter cycles. This indicates

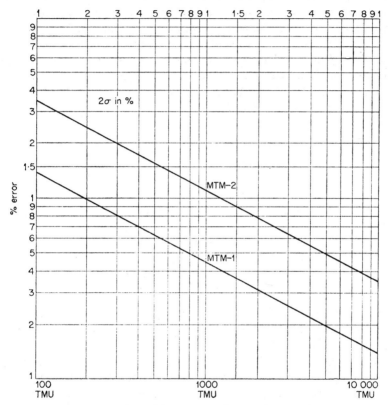

Figure 8.8 Accuracy of MTM-2

that MTM-1 should be used for shortwork. Simplification of the data must result in a certain loss of accuracy for each of the various motion-categories. For example, the MTM code GA15 consists of a reach followed by a G5. The reach can be either case A or case B, and the distance can be anything from 2 to 6 inches. The time-value for any action falling within the broad definition would be 6 t.m.u. In MTM-1 there are ten values to cover the same range of possibilities. MTM-2 analyses should contain a wide variety of motion cases and distances, so that the errors can be balanced. It is not suitable for highly repetitive situations which involve only a limited number of motion categories and distances. MTM-2 provides a simple method of analysing combined and simultaneous motions. Combined and simultaneous motions, on the other hand, are very often far from simple. When complex simultaneous motions occur, and are critical to the analysis, MTM-2 does not provide a sufficiently detailed answer. MTM-1 should therefore be used.

It should be pointed out, at this stage, that MTM-2 has been designed so that it can be combined with other MTM systems. It is possible, therefore, to make an analysis in MTM-2, only reverting to MTM-1 for those parts of the study that warrant its use. The final result would be a mixture of the two systems, and would avoid the inconvenience and expense of making a detailed MTM analysis of a complete job, simply to achieve required accuracy in a few small places.

The MTM-2 system has eleven motion categories and thirty-nine time-standards, ranging from 1 to 61 t.m.u. The system has been so devised that each motion category is derived directly from MTM-1 motions. They do not overlap each other, and it is not possible to replace any element in MTM-2 by other MTM-2 elements. The MTM-2 elements are:

Get	G
Put	P
Regrasp	R
Apply pressure	A
Eye action	E
Foot motion	F
Step	S
Bend and arise	B
Get weight	GW
Put weight	PW
Crank	C

The element codes are used internationally and should not be translated into other languages. The symbol for 'get' is always G, even though the word for 'get' might begin with a different letter in other tongues.

The time-data for MTM-2 are contained in the data card illustrated in Figure 8.9. The top part of the card refers to 'get' and 'put'; 'get weight' and 'put weight' are shown in the centre, and the remaining elements are shown on the bottom of the card. The card shows the elements, their variables, and the time standards for each. It does not show the rules that govern their use. These rules and their characteristics will be discussed in the rest of this chapter. As with the previous section, and the following one on MTM-3, the information given is at the appreciation level only. MTM analysis is a matter of skill as well as theory, and neither this nor any other book, can provide the reader with all that he needs to undertake a successful MTM project.

8.61 THE MOTION—'GET'

'Get' is an action with the predominant purpose of reaching with the hand or fingers to an object, grasping the object, and subsequently releasing it. 'Get'

MTM 2

CODE	GA	GB	GC	PA	PB	PC
−5	3	7	14	3	10	21
−15	6	10	19	6	15	26
−30	9	14	23	11	19	30
−45	13	18	27	15	24	36
−80	17	23	32	20	30	41
GW 1−1 Kg.				PW 1−5 Kg.		
A	R	E	C	S	F	B
14	5	7	15	18	9	61

WARNING: Do not attempt to use this data unless you have been trained and qualified under a scheme approved by the International MTM Directorate.

Figure 8.9 The MTM-2 data-card

starts when the hand begins to move towards the object. It covers all motions necessary to gain control of the object and all those used to let go of it. It is built up of three MTM-1 motions—'reach', 'grasp' and 'release'. The symbol for 'get' is G.

There are three variables to 'get'. These are:

1. Case.
2. Distance.
3. Weight or resistance to motion.

1. *Case*

The cases of 'get' are distinguished by the type of grasping motion used by the operator. There are three cases. For ease in analysis the algorithm or decision

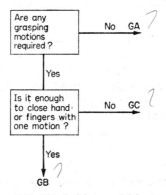

Figure 8.10 The decision model for 'get'

model given in **Figure 8.10** has been prepared. All decisions about the cases of 'get' should be made by using this model.

2. *Distance*

MTM-2 is a metric system. Distances and weights are measured in centimetres and kilograms. A set of British equivalents does exist, and is used in the United States and Canada (Table 8.11). In the United Kingdom, however, the metric system has been adopted.

TABLE 8.11
MTM-2 DISTANCES

inches		centimetres		code
over	not over	over	not over	
0	2	0	5	5
2	6	5	15	15
6	12	15	30	30
12	18	30	45	45
18	upwards	45	upwards	80

3. *Weight or Resistance to Movement*

Get is affected by the static component of weight. As with MTM-1, the effective net weight is used to determine the time allowance. Simplified rules have been produced for determining the E.N.W. If the object is to be moved spatially by one hand, the E.N.W. is equal to the total weight of the object. If it is moved by both hands, the E.N.W. is half the total weight. For sliding movements the E.N.W. is taken as 40 per cent of the weight when one hand is used, and 20 per cent of the weight for both hands. Weights of less than 2 kg (4 lb) per hand are ignored.

The weight component of 'get' is analysed as a separate action—'get weight' (GW). 'Get weight' is defined as the action required for the muscles of the hand and arm to take up the weight of the object. It starts as soon as the fingers have closed on the object in the preceding 'get', and must be completed before any actual movement can take place.

8.62 THE MOTION 'PUT'

'Put' is an action with the predominant purpose of moving an object to a destination with the hand or fingers. It starts with the object grasped and under control at the starting point and includes all the motions necessary to move the object to its destination and seat it in place. It is built up from the MTM-1 sequences 'move' and 'move plus position'.

'Get' and 'put' are distinguished by their predominant purpose. It is possible to perform a 'get' with an object palmed in the hand provided that the predominant purpose is to move the hand and not the object. The symbol for 'put' is P, and the variables are case, distance, and weight.

1. Case

The three cases of 'put' are distinguished by the amount of correction needed at the end of the 'move' stage. Corrections consist of stops, hesitations, and changes in direction necessary to locate the object. They are generally caused by difficulty of handling, close fits, lack of symmetry between the parts, or uncomfortable working positions. They are unintentional movements and should not be confused with short PAs, which are purposive, and usually of easily discernible length.

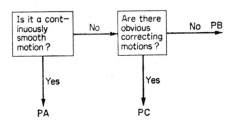

Figure 8.11 The decision model for 'put'

The decision model is used in the same manner as the model for 'get'. The questioning procedure begins with the left-hand box. The characteristics of PA and PC are clearly covered by the two questions. PB emerges as the correct answer if the put does not fit either of the other two cases. Where there is doubt even then, it is permissible to assign the higher case.

2. Distance

Distance is handled in a similar manner to 'get'. When one part is inserted into another, the 'put' includes an engagement of up to 2·5 cm (1 in.). If the insertion is greater than this, an additional 'put' must be allowed.

3. Weight

The effect of weight on a put is handled by 'put weight' (PW). This is an addition to a put motion, depending on the weight of the object moved. As with 'get weight', PW is only allowed when the effective net weight exceeds 2 kg. 'Put weight' starts when the object begins to move, and consists of the additional time over and above the 'move' time in 'put', to compensate for the differences in time required in moving heavy and light objects over the same distance. It ends when the 'move' ends.

One t.m.u. is allowed per 5 kg of effective net weight. Between 2 and 5 kg the code PW5 is used. Between 5 and 10 kg the code is PW10, and so on. 'Get weight' is normally related to 'get', and 'put weight' to 'put'. There are occasions when 'get weight' can be associated with 'put' also. When a spanner is used to turn down a bolt against resistance, a 'get weight' may be necessary to overcome the initial resistance of the bolt after the spanner has been put to it.

Example
A spanner is used to make 8 turns of the bolt against 5 kg resistance. The analysis would be:

Spanner to bolt 8 times	8PC30
Take strain	8GW5
Turn bolt	$\begin{cases} \text{8PA30} \\ \text{8PW5} \end{cases}$

8.63 REGRASP

Regrasp is a hand action with the purpose of changing the grasp on an object. It starts with the object held in the hand and includes up to three fractional movements of the fingers, to readjust the object. It ends with the object in a new position in the hand. If more than three finger-movements are used, more than one regrasp is assigned. The action takes place with the object held in the hand. If the hand lets go and then secures another grasp on the object, the action will be 'get' and not 'regrasp'. The code for 'regrasp' is R. There are no variables.

8.64 APPLY PRESSURE

'Apply pressure'—symbol A—is an action with the purpose of exerting force on an object. 'Apply pressure' in MTM-2 is just a simplified form of the MTM-1 element. The time-value is a weighted average of the times for AP(A) and AP(B), and it covers the same actions. 'Apply pressure' can be performed by any body-member. It starts with the member in contact with the object and includes all three components of 'apply pressure' that were included in MTM-1. That is:

1. Application of controlled increasing muscular force on the object.
2. A minimum dwell or reaction time, to permit the reversal of the force.
3. The release of force.

It ends with the body-member in contact with the object, but with the muscular force relaxed. The dwell or reaction time included in the 'apply pressure' value is only sufficient to cover the minimum time necessary to reverse the force. If the dwell is prolonged in any way, the 'apply pressure' value alone will not be sufficient. This situation arises when force has to be maintained upon the object —for example when somebody presses a doorbell button, the pressure is held for some time to produce a long ring. This sort of extended-dwell time must be analysed as process-time. Apply pressure in MTM-2 has no variables. As with AP in MTM-1, the primary purpose is to exert force and not to move the object though in practice 'apply pressure' can involve minor movements of up to 6 mm ($\frac{1}{4}$ in.). Where the intention is to cause movement, the action should be described by 'get weight', 'put' and 'put weight'.

8.65 CRANK

Crank—symbol C—is a motion with the purpose of moving an object in a circular path of more than half a revolution with the hand or fingers. 'Crank' is obviously used for turning cranks and hand wheels. It can also be used in cases where a crank handle as such is not present, for example dialling a telephone, or winding up a ball of wool.

There are two variables of 'crank'. These are:

1. Number of revolutions
2. Weight or resistance

For example: 8 Cranks with 5 kg weight.

1. Continuous	5	GW5
	120	8C
	8	8PW5
	133	TMU
2. Intermittent	40	8GW5
	120	8C
	8	8PW5
	168	TMU

8.66 EYE ACTION

'Eye action' is an action with the purpose of either recognizing a readily distinguishable characteristic of an object, or shifting the aim of the axis of vision to a new viewing area. It is a simplification of the MTM-1 actions 'eye focus' and 'eye travel'. Only one value is given—7 t.m.u.—and the symbol is E.

As in MTM-1, 'eye actions' are only allowed when all other motions must cease while the eyes perform their task. Use of the eyes to perform other motions is accounted for in these motions. For example, the need for constant visual attention is one of the factors that prolongs the time required for a PC. 'Eye action' ends when other motions can begin again.

Two distinct actions are covered by this element. The first is to focus the eyes on the object and to recognize an easily distinguishable characteristic. This involves making a simple 'yes' or 'no' decision, as when an electric circuit is being tested, with a bulb, for continuity. The tester looks at the bulb to see if it lights up when he makes the contact. First he focuses his eyes on it, and then makes a simple decision: Is the bulb alight? Yes or No? All these visual and mental processes are covered by one eye action. The second action is to move the eyes in order to 'look' from one place to another. This occurs when the tester in the example moves his eyes to look at the bulb after concentrating on making the contact. Notice that, in both cases, all other motions have to cease. There is nothing he can do with his hands until he has seen the light bulb and made his decision.

8.67 FOOT MOTION, AND STEP, BEND, AND ARISE

With 'bend', these two motions, 'foot motion' and 'step', account for all body-motions used in MTM-1. 'Foot motion' has the code F and 'step' is S.

Foot motion is a short foot or leg motion when the purpose is not to move the body. Depression of a pedal would need one F, but releasing it would require another.

'Step' is a leg motion with the purpose of moving the body, or a leg motion longer than 30 cm (12 in.). It starts with the leg at rest and includes any movement of the leg—of any length—that is designed to move the body. This can be a step taken in walking, a step to the side, or one of the small leg-movements used in turning the body round. The predominant purpose of displacing the trunk identifies these as 'step'. It ends with the leg in a new location. Walking is a series of steps and can be quickly evaluated by counting the number of times the foot hits the floor. Step also includes those motions of the leg which are not intended to move the body, but are longer than 30 cm (12 in.). Decisions between 'foot motion' and 'step' should be made using the decision model depicted in Figure 8.12.

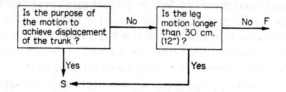

Figure 8.12 The decision model for 'foot motions' and 'step'

8.671 Bend and arise. 'Bend and arise' is a lowering of the trunk followed by a rise. The code for 'bend and arise' is B.

'Bend and arise' includes a number of other actions. Stooping, in which the knees are bent as well as the trunk, so that the hands reach to the floor, and the subsequent 'arise', is included in the time for B. Sit and the subsequent stand, and also kneeling on one knee and rising, are included. Kneeling on both knees is analysed as two B's.

8.68 MOTION COMBINATIONS

When two or more motions are performed at the same time, the time allowed is the time for the motion which has the highest value on the MTM-2 data-card. The rules and procedures are very similar to those of MTM-1. The limited motion is ringed and its time-value ignored, because it is overlapped by the limiting motion. For example,

	LH		TMU	RH	
Get wood-block	(GB15)		14	GB30	Get hammer
Get washer	GC15		19	(GB15)	Get nut

Motions can also be performed simultaneusly by one hand.

$$\left.\begin{array}{lr}\text{Centre punch to mark} & \text{PB15} \\ \text{Regrasp punch} & \cancel{R}\end{array}\right\}15$$

For speed of analysis, minor time-limited motions are not always shown.

8.69 TAPE-DATA ANALYSIS

Tape-Data Analysis was developed by the U.K. MTM Association as a solution to problems encountered in its research programme into the measurement of maintenance work. Maintenance jobs are non-repetitive. The same task rarely happens twice, and even when it does, the methods used vary from occasion to occasion. Several studies may be necessary to establish a representative motion-pattern, but each of these studies must be based on a single observation. Furthermore, the work-cycles tend to be much longer than in production work, and these two factors present great problems to the analyst. Because of the length of the cycle, the analyst is forced to take notes in some form. The usual form is in writing, but this has proved unsuccessful because the writer's attention was distracted from the job and he missed parts of the operation. It has also proved impossible to keep up with the operator without shortening the notes to the point that they are useless for future reference.

A system was needed that would enable the analyst to record the activity as quickly as it was performed, and in such a way that the information could be used to reconstruct the motion pattern that was used. Also the means of recording must leave the analyst free to observe 100 per cent of the time. The system must ensure that speed and accuracy of analysis are within acceptable limits. Finally, it had to be simple, easily understood by the average industrial worker, and easily learnt by MTM-2 analysts in order to minimize training costs.

Tape-Data Analysis has been designed to overcome these problems. The analyst is trained to make a verbal recording, using a small portable tape-recorder. This enables him to watch the operation for the whole of the time. Even using this method it is not possible to record all the motions, in terms of MTM-2 codes, as fast as they occur. GB15, for instance, takes 10 t.m.u. to perform, but about 28 t.m.u. to record. To cope with this, the commonly occurring MTM-2 motion sequences have been isolated and re-coded. The finger motions involved in turning a nut with the fingers ten times are 10 GB5, plus 10 PA5. These have been re-coded, since they are a very common combination. The code for these finger-turns is F5—thus 10 F5. Body-motion codes and other elements that are repeated (for example, 10 PA15 for wiping something dry) are fairly easy to record as they stand. No special codes are needed for them. To complete the picture the analyst is trained to record information about the objects and conditions to help in reconstructing the analysis of the remainder of the motions. Once the record has been completed it can be played back to the analyst, who can then write the MTM-2 pattern for the task in the conventional way, translating the codes and his notes on the conditions.

The motion sequences used in T.D.A. are a record of actual MTM-2 motions used by qualified tradesmen, in the performance of their normal work, in typical industrial conditions. The observer judges the motions that he sees, in terms of conformity to the standard pattern, and records in terms of the variation from the standard—if any.

The first sequence concerns turning objects with the fingers. This applies to tightening and loosening nuts, screws, wing-nuts, and so on. It also applies to some screwdriver operations. The code is F. Finger motions can fall into either the 5- or the 15-distance classes. The former is written F5 and the latter F15. All times are given in t.m.u.

$$F5 = \begin{array}{rl} 7 & GB5 \\ \underline{3} & PA5 \\ 10 & \end{array}$$

If there is appreciable weight (over 2 kg) the fact is recorded—F5 weight (written F5W). An F15W is similar to F5W, but the distance code is larger. The value is 20 t.m.u. It is permissible to use the T.D.A. codes in the subsequent MTM-2 analysis, provided the actions follow the standard sequence. Variations in the grasp are also recorded during analysis. If more than one grasping motion were involved, the record would be '5F5-GCs'.

The second sequence concerns spanner-turns. In this case the spanner is put to the nut or bolt with a PC and turned with a PA. The distances now fall into the 5, 15 and 30 categories. The code is S. If weight is present it is again recorded, verbally as 'weight' and in writing as W.

S15 and S30 follow similar patterns with different distance-codes. The values are $S15 = 32$ and $S30 = 41$. The weight normally experienced when resistance is met with a spanner sequence is 10 kg.

Variations in distances are recorded as they are with F. The analyses for transfers involving different distances, cases of get, and so on, are written out in full in the MTM-2 analysis.

'Wipe' is used to describe cleaning an object with a cloth. The cloth is held in the hand, which then grasps the object—thus wrapping it in the cloth. The actual wiping action is covered by the PA15. The verbal code is 'wipe' and the written code is W.

'Rub' consists of a backward and forward motion across the surface being rubbed. The normal distance is 15, and weight may be involved. The code is 'rub'—both verbal and written. Variations in distance are recorded, as in RUB-5.

$$RUB = \begin{array}{rll} 6 & PA15 & forward \\ \underline{6} & PA15 & back \\ 12 & & \end{array}$$

Each pair of motions is counted as one action. The weight involved in rubbing is usually about 3 kg.

RUB W = 3 GW3 ⎫
         6 PA15 ⎬ forward
         1 PW5 ⎭
         3 GW3 ⎫
         6 PA15 ⎬ back
         1 PW5 ⎭
        ───
         20

'Joggle' is used to describe moving an object to and fro in order to fit or loosen it. It was devised to cover the special case where something, especially something large or heavy, is being fitted into a confined space. The object frequently jams against the sides of the space and has to be wriggled free. This happens frequently when a drawer sticks. Joggle consists of two movements to free the object—one forward and one back. Sometimes both hands are used. Distances vary and are recorded—J5, J15, J30 etc. Three categories of weight are also recognized—negligible, medium and heavy. Negligible weights have no weight codes, and are simply recorded by using the 'joggle' symbol J and distance. Medium weights are called 'joggle weight' e.g. J15W:

J15W = 5 GW5 ⎫
        6 PA15 ⎬ forward
        1 PW5 ⎭
        5 GW5 ⎫
        6 PA15 ⎬ back
        1 PW5 ⎭
       ───
        24

Heavy weights are referred to as 'joggle heavy'—code HW. This sort of action is often found in heavy to-and-fro levering with a crowbar.

'Ratch' is a special case of 'joggle', usually performed with a ratchet spanner or ratchet screwdriver. Again it consists of a backward and forward motion; but force is applied in one direction only. The written symbol is RTCH.

'Scrape', like 'ratch', consists of two motions, with force applied in one direction only. It applies when tools are used to scrape a surface, for example when scraping paint. The tool is placed fairly carefully on the object and then run along it with a smooth, continuous scraping action. The usual distance is 15, but variations in distance should be recorded. Two conditions exist—with and without weight. These are distinguished in the usual way. The verbal codes are 'scrape' and 'scrape weight', and the written form is SCR.

Hammering sequences are used for a wide variety of purposes. There are two kinds of sequences. The first sequence—tap—is used for locating taps, and where careful tapping is necessary. Locating taps are used as a preliminary to heavier blows. When a nail is hammered into a plank of wood, one or more very light taps will be made to 'start' the nail. These are locating taps. The code, verbal and written, is TAP.

The second category, 'blow' (written BLW) covers less accurate hammering. The blows are heavier and vary in distance. The distances are normally either 15 or 30. BLW 30 follows the same pattern, with a larger distance. The value is 22 t.m.u.

The last sequence involves 'bend and arise'. For convenience' sake this element has been split into two: 'bend down' (BD), and 'arise from bend' (AB).

These sequences do not cover all the possibilities, but they do comprise a large number of the commoner actions used in maintenance. A good MTM-2 analyst can construct new sequences to suit himself, but it should be remembered that the more sequences he creates, the more he has to remember, and the more has to be learnt by other analysts. The purpose of these motion sequences and codes is only to permit economical recording of actual MTM-2 motions. The object should be to have the minimum number of sequences necessary to achieve this.

8.7 MTM-3

MTM-3 is the third general MTM system. It was developed by the Swedish MTM Association, and formally approved by the International MTM Directorate in 1970. The intention was to provide a system that would be faster to use than MTM-2, and would answer to a definite specification as regards its applicability, accuracy, and speed. These criteria are very similar to those laid down for MTM-2.

The elements in MTM-3 are related to the actual movements used by the operator, rather than the kind of work, workplace layout, or to the nature of the objects handled. Because of this, MTM-3 can be used for any kind of work. The definitions of the elements and their break-points are so designed that MTM-3 can be used in conjunction with MTM-1 or MTM-2 elements in an analysis. It is thus combinable with other MTM systems.

The element definitions are unambiguous and clear. This makes the system easy to learn. It also reduces the possibility of errors, and thus makes MTM-3 easy to use. Like all MTM-based systems, MTM-3 provides a description of the method. This is of value, both for method study and for training. It is based upon MTM, and its accuracy and speed have been tested against MTM-1 to provide quantitative information about its reliability and the economics of its use in any given situation. Its speed is shown to be twice as fast as MTM-2. Analysis time should be between 25 to 50 times the cycle time of the job under examination. MTM-3 has no significant bias compared with MTM-1—it is not consistently tight or loose.

The random deviations average out at about plus or minus 5 per cent cycle times at 15,800 t.m.u. with 95 per cent confidence. These random deviations, or errors, can be caused by the errors that exist in the system—MTM-3 is very much simplified in relation to MTM-1; and so there will be errors in the times assigned to the motions actually performed during a task, even though the values given in the data-card may be correct average times for such motions in general. The accuracy of the system will depend largely on the length of the study and on the variety of motions performed. MTM-3 gives a reliable result at cycles around ten minutes and above. At cycles less than this, the error increases rapidly. At a cycle of 2·5 minutes it is about ±10 per cent—which is rarely acceptable for industrial engineering purposes. The error also rises if there is a high degree of repetition in the cycle. The movement of an object through a distance of 16 cm is given the same time as a similar movement through a distance of 20 cm. An error must therefore be present, slight though it may be. If there is a sufficient variety of distances and other variables, these errors will balance each other and the result will be reliable. If there is not, however, the errors will be multiplied to such an extent that the result will be very inaccurate. The cycle must, therefore, be long enough to produce a variety of motions, and

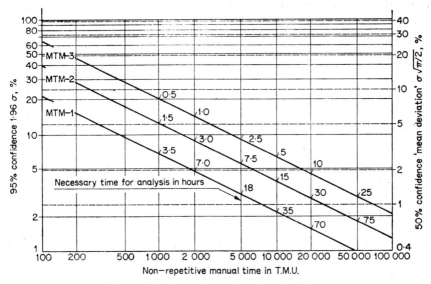

Figure 8.13 Comparison of random deviations

highly repetitive elements must be excluded. Comparisons between the three systems are shown graphically in Figure 8.13.

The figures produced to show the accuracy of the system take into account the second source of errors—mistakes in observation by the analyst.

The MTM-3 system uses four elements, and assigns to them ten time-values ranging from 7 to 61 t.m.u. There are data available in British and metric units. The metric system is used by the U.K. MTM Association, and will be used in this text. The data-card for use with MTM-3 analyses is illustrated in Figure 8.14.

The MTM-3 elements and their codes are:

Handle	H
Transport	T
Step	SF
Bend and arise	B

Other elements such as 'apply pressure', 'regrasp', and weight factors have been allowed for in the time-values for 'handle' and 'transport'; they are not analysed separately.

MTM-3 is not designed to analyse motions with a frequency greater than 10. For example, the element 'Handle' is used for screwing down a bolt on a nut (once the threads have been engaged). If more than 10 handles would be required to complete the running down of the nut, the action must be analysed with MTM-1 or MTM-2.

Metric units

MTM-3				
CODE	HA	HB	TA	TB
−15	18	34	7	21
−80	34	48	16	29
	SF	18	B	61

Figure 8.14 Data card for MTM-3

8.71 HANDLE

'Handle' is a motion sequence with the purpose of getting control over an object with the hand or fingers and placing the object in a new location. 'Handle' may be regarded as a combination of 'get' and 'put'. The sequence begins when the hand or fingers start to reach towards the object. It includes the reach, the grasping of the object, its subsequent movement, and where applicable, the positioning and release of the object. It ends either with the release, or, if the object is not released, with the end of the movement. Where the object is inserted into another, 'handle' allows 2·5 cm (1 in.) of insertion.

'Handle' also applies when an object is grasped and pressure is applied. The application of force is looked upon as a short motion.

8.72 TRANSPORT

'Transport' is a motion with the purpose of placing an object in a new location with the hand or fingers. This is, in fact, a simplified version of 'put'. It starts when the hand starts to move the object, and includes the movement, and where they apply, the positioning and release of the object. As with 'handle', it ends when the object is released, or with the cessation of movement, if the object is not released immediately. If engagement takes place, 2·5 cm of insertion is allowed for. If the object is placed in the other hand, the transport stops when the object arrives in the other hand. Transports in a circular path stop when the hand has completed one revolution.

There is a clear distinction between 'handle' and 'transport'. Handle occurs when the hand has no control over the object and must first pick it up before any movement can take place. Transport occurs when the hand is already in control of the object and movement can start immediately.

8.721 Variables of 'handle' and 'transport'. There are two variables of 'handle' and 'transport'; case and distance.

The two cases, A and B, depend on whether correcting motions are necessary at the end of the element. Correcting motions in MTM-3 are exactly the same as in MTM-2. That is, a correction is an unintentional stop, hesitation, or change in direction at the terminal point of the element. They are caused by the high degree of accuracy needed in positioning the object. Highly controlled grasping actions in 'handle' are ignored. Only the 'put' part of the motion is considered. Case A 'handle' and 'transport' require no corrections. Case B 'handle' and 'transport' contain corrections.

Handle and transport codes are written in the following form:

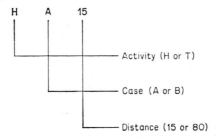

8.73 APPLICATION OF PRESSURE

A single application of pressure is allowed for in the times for handle and transport. This applies when an apply pressure is required in seating an object, and also when the hand grasps an object in order to apply pressure (special case of handle). If several applications of pressure take place, the first of them is included in the element time for the handle or transport, and the rest are analysed as transport—TA15.

The following decision model (Figure 8.15) is designed to assist in analysing this sort of situation:

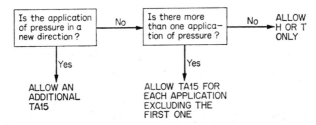

Figure 8.15 Decision model for 'Apply pressure'

8.74 STEP

'Step' is a leg motion with the purpose of moving the foot or leg in one direction. It starts with the foot or leg at rest and ends with it in a new location.

It covers all movements of the feet and legs regardless of the purpose. It can be a movement of the foot alone, hinged at the ankle, as in pressing a pedal. It covers all movements of the leg, both those that are hinged at the knee and those that are hinged at the hip. It also covers all movements of the legs that are designed to move the body. This includes walking, stepping to the side, and turning the body round.

Steps are single movements of one limb, in one direction only. Depressing a foot pedal would be one step, and releasing it would be covered by a second step. Walking is evaluated as in MTM-2, by counting the number of times the foot hits the floor.

8.75 BEND AND ARISE

'Bend and arise' is a lowering of the trunk followed by a rise. It is exactly the same element as 'bend and arise' in MTM-2, and covers all the actions that are covered by that element. 'Kneel on both knees' is covered by 2B in the usual way.

8.76 MOTION DISTANCES FOR HANDLE AND TRANSPORT IN CONJUNCTION WITH STEP AND BEND

Body motions are usually made in conjunction with 'handle' and 'transport'. The extent to which they can be performed simultaneously depends on the degree of control required. A single low-controlled motion—HA or TA—is normally limited by a body motion. There are very rare occasions when an overlap occurs; for example, when an obstruction prevents the hand from starting until after the body motion is completed or partially completed, the HA or TA might overlap the body motion.

Two low-controlled motions—HA or TA—generally affect the time to the extent that one of the case A motions will overlap the body motion. The allowable distance is normally 15. HB and TB cannot be done simultaneously with

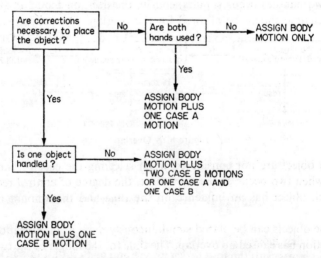

Figure 8.16 The decision model for the correct motion combination

body motions. If a single HB or TB occurs, its time is allowed as well as time for the body motion. If two HBs or TBs occur the time is allowed for both. These rules are summarized in the decision model depicted in Figure 8.16.

A procedure has been developed for analysing simultaneous 'transports' and 'handles' that reflects the different degrees of overlap that might be expected for the various degrees of control involved. In MTM-3, only motions performed with different body-members are considered. The presence of regrasp during a handle or transport would be ignored. Where simultaneous handles and transports occur it is important to know which one is time-limiting. Figure 8.17 indicates the possibility of performing two motions simultaneously. The recommendations

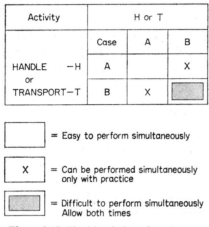

Figure 8.17 The 'simo' chart for MTM-3

given in it are normally reliable, but deviations from it may be necessary from time to time. These cases must be assessed on their own merits.

When two handles occur simultaneously, the decision model in Figure 8.18 shows what time should be allowed for the overlap. The actions required for

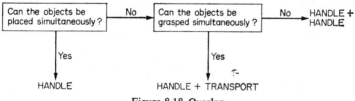

Figure 8.18 Overlap

getting the object are not considered when selecting the case of handle to be used, but when two occur at the same time, the degree of control required for getting each object has an influence on the time and the amount of overlap involved.

When the objects can be placed simultaneously there is an indication that the 'getting' action has caused no overlap. The time for the longer of the two handles is assigned. If they cannot be placed simultaneously obviously an overlap is present. The 'get' part of the handle then becomes important. If the objects can be grasped simultaneously, it indicates that the overlap is caused by precision in the 'put' part of the handle. That is, both hands grasp their objects together, but must be moved and positioned separately. A handle is allowed for the picking-up of both, and the placing of one of them. An additional transport is allowed to place the second part. If the objects can be neither grasped nor placed simultaneously, then two handles are assigned.

8.8 P.M.T.S. FOR MENTAL PROCESSES

Until recently the applications of predetermined motion–times have been confined to jobs in which manual elements predominate. Actions and movements may be assessed quite readily by the trained observer who sees, visualizes, or measures the variables which comprise the element. Mental processes cannot be appreciated in this way because the observer does not possess a sense capable of monitoring the worker's thought-process. Thus a research worker engaged in planning, or mulling over a problem, apparently will be doing no work in terms of physical action.

It is true that most systems of P.M.T. include standard elements under the heading of 'Mental Work', but generally these so-called mental elements are restricted to eye movements, which basically consist of eye-travel time and eye-focus time. Other mental elements cover reaction times and very simple decision-making. All of these, in essence, are on the *binary decision level*, that is, a simple 'yes–no' situation of the type 'do I press the foot-pedal or not?'.

Broadly speaking, mental work can be classified under two main headings of creative thinking, and action or reaction as a result of some external stimulus or predetermined plan of action. In the latter case some of this is autonomic reaction, through stimulation of the nervous system triggered off by one of the senses, and the remainder is deliberate, calculated mental activity. It is in these latter areas that most research for work measurement has been carried out.

It has already been postulated that jobs in the lower strata contain some decisions on a binary level, and attempts have been made in the systems of P.M.T. to set times for these. In Work-Factor for example, even before the introduction of the complete Work-Factor Mento-Factor System, basic Work-Factor Time Standards were available for such mental activities as 'Memorize', 'Compute', and 'Recall', and others were classified under the Work-Factor Standard Element of Mental Process.

In Methods–Time Measurement similar attempts were made to cover simple mental action, although this system largely ignores this aspect, confining itself to the eye motions. The eye-motion elements usually contain reaction-time sufficient to recognize and make binary decisions, such as whether or not a meter reading is within target areas, but not to allow the observer to recognize the actual reading.

The system of Clerical Work Improvement Programme (*see* Chapter 16) also provides mental-process times for binary decisions and eye motions.

Of course, many of the mental processes are performed simultaneously with other manual actions, and as such do not require extra time for their performance. Anticipatory perception also helps to eliminate the need for mental elements, and training programmes include instruction in the use of this concept, which demands that workers shift their field of vision in anticipation of the next element while the present actions are still being performed. In other cases, however, a delay in manual elements is inevitable during the performance of non-simultaneous mental processes.

Although it is true that in problem-solving individuals will think along different lines and arrive at different solutions, it is possible, of course, to direct their thought-sequences in certain simple jobs, through the use of decision models or *algorithms*. This approach is used in MTM-2, and examples are given in the preceding section of this chapter, which is devoted to MTM-2. In these instances, where directed thought can be applied it is possible to determine standard times for arriving at different solutions through the algorithms. A possible application of this is in television servicing and trouble-shooting, where complex problem-solving may be reduced to simple analyses using binary 'yes–no' replies to posed questions. A further example is the application to inspection, which is introduced later in this section.

In general it must be said that at present there is no way of precisely measuring or estimating times for decision-making, creativity, or jobs in which the work content cannot be predicted (as in repair maintenance work).

Reading. A typical mental task is reading. MTM reading times are suitable for normal easy-to-read prose. Reading speeds vary considerably with complexity of the material. An average reading speed is about 330 words per minute. It has been suggested that 10–12 point type is the easiest to read, and that an ideal line-length is 3·2 in. This book is printed in 10 point type.

When a passage is being read, the eyes do not scan smoothly over the words. A series of eye-focuses are made with a series of short eye-travels in between. The eye-focuses are called 'fixations' and the eye-travels are known as 'saccades' or 'jumps'.

A fixation enables the eye to read an average of 1·56 words. Being an eye-focus it takes 7·3 t.m.u. The fixation time per word is: 7·3/1·56 t.m.u. To this must be added the time for the saccades. It has been discovered that saccade time is about 8 per cent of fixation time. Therefore the total time taken to read one word is:

$$\frac{1 \cdot 08 \times 7 \cdot 3}{1 \cdot 56} = 5 \cdot 05 \text{ t.m.u.}$$

Thus the time to read a passage is given by $5 \cdot 05 \times N$ t.m.u., where N is the number of words.

Writing. MTM writing data apply to freehand characters under one inch in height. Either letters or figures can be analysed. All writing is composed of 'moves'—MfC, and 'positions'—P1SE. The moves are allowed every time the writing implement approaches the surface, either from the end of one word to the beginning of the next, or to make punctuation marks, dot 'i's, cross 't's, and so on. It is also allowed for every mark made on the paper. When there is a change in direction of more than 45°–90°, as in the forming of a loop, a new move is allowed. The initial movements of the pen to the surface, to start writing, may well be longer than one inch. These have to be analysed separately.

The positions are allowed every time the pen touches the surface to start a word or make a punctuation mark, and so on. They are also allowed where a very sharp change of direction occurs, as in the top of a 't'.

For example (Figure 8.19):

Figure 8.19 Analysis of writing

(The loops are marked with ticks to show where the moves start and finish. Positions are shown by circles.)
The initial move may be MfC or larger.
The remainder are MfC

$$\begin{aligned}\text{MfC} &= 26 \times 2\cdot0 & 52\cdot0 \\ \text{P1SE} &= 11 \times 5\cdot6 & 61\cdot6 \\ & & \overline{113\cdot6 \text{ t.m.u.}}\end{aligned}$$

If the initial move is also fractional, the total will be 115·6 t.m.u.

8.81 THE WOFAC MENTO–FACTOR SYSTEM

Some of the most extensive research into mental-process times has been carried out by the Wofac Company under a programme of research directed by Joseph H. Quick, formerly Chairman of that Company.

The research was initiated in 1949 and extended over nearly two decades. As early as 1962 a chapter dedicated to Mental Processes was included in their book *Work-Factor Time Standards* by Joseph H. Quick, James H. Duncan, and James A. Malcolm, Jr., and published by McGraw-Hill. By 1967 the system had been perfected, adequately field-tested, and made available as a volume of Rules of Application and Mental Process Times called *Detailed Work-Factor Mento Manual*.

The magnitude of the task is appreciated when one considers the very complex mechanism of the human brain and nervous system, much of which even now remains a mystery. Consider by way of example a simple case of an operator pulling a lever on receiving the stimulus of a red warning-light. A simplified analysis, taken from the above-named publication, shows that the following sequence is involved.

1. *Receptor stimulation.* Receive light signal and stimulate retina of eye.
2. *Neural condition to brain centres.* Conduct to brain by way of sensory-nerve path.
3. *Brain-centre process.* Register signal and organize motor response.
4. *Neural conduction to muscles.* Conduct impulse to muscles by way of motor-nerve path.
5. *Muscle contraction.* Tense muscles and prepare for action.
6. *Movement of body member.* Move arm to shift lever.

Studies involving the servo-mechanisms of the body and of machines are part of the discipline of *cybernetics*.

Of the fourteen elements defined for the fundamental Work-Factor Mental Process Times the first four are not strictly considered as true 'Mental Processes in the Research Programme'. The full list comprises the elements Eye Shift, Eye Focus, See, Nerve Conduct, Discriminate, Span, Identify, Decide, Memorize, Recall, Compute, Sustain, Convert, and Transfer Attention.

The mental-process times are compatible with the Detailed Work-Factor predetermined times for manual motions, and both are expressed in Work-Factor Time Units; one Work-Factor Time Unit equals 0·0001 min. The Wofac Mento-Factor System is applicable to inspection, gauging, testing, and checking operations, and to proof-reading, and jobs which involve making computations, including use of the slide-rule.

A brief description of the elements is given below.

Eye Motions. The anatomy of the eye is designed to accommodate the very complex actions demanded of it, to furnish man with the ability to retain in sharp focus objects at various distances from his eyes. To achieve this the muscles of the eye must adjust the lens and iris in sympathy with the intensity of light falling on the eye, and with the seeing distance.

Although extremely complex for the purposes of work measurement, these actions may be defined under just two Eye Motions, of Eye Focus to ensure a sharp image on the retina, and Eye Shift to change the line of vision. In analysing most work, Eye Motions are performed simultaneously with other Body-Member Motions, and are known as Concurrent Mental Processes: that is, they do not affect any part of the work-cycle time. Conversely, it may be necessary to perform the Eye Motions before executing the next Manual Motion, in which case they are referred to as Sequential Processes.

React. The two variables which determine the Work-Factor Time from the Work-Factor React Time-Tables are:

1. The extent to which the operator can anticipate the signal.
2. The number of Action Signals to which response must be made, and the number of different Appropriate Actions which can result.

The time necessary for a worker to react to stimuli depends, of course, on whether he is expecting the signal or not. Even with expectancy, the time-value is dependent on his previous knowledge of when to expect the signal, which may or may not be known to him. Thus Work-Factor provides three classes of 'React':

Class 1. React with Advance Notice; i.e. the operator knows just when to expect a specific stimulus.
Class 2. React Anticipated, but nature of stimulus and/or exact time unknown.
Class 3. React Unanticipated; i.e. when random signals occur, for which the worker has not had time to prepare.

A further subdivision of 'React' is 'Transfer Attention', for which Work-Factor Fundamental Mental Process Time-Tables are also available.

These divisions of 'React' are employed in cases where workers are interrupted from performing one task, and must break their concentration to transfer their attention to the new stimulus, and then react to it. The transfer may be part of the work-cycle, which is anticipated, such as at the end of one process element and before the next, or it may be in the form of an alarm signal, a telephone bell, or a call from a colleague for information or assistance.

Inspect. The Work-Factor system has a number of classifications of 'Inspect', thus enabling accurate times to be established for varying inspection objectives, for example, sorting, counting, identifying rejects, and so on, and for different visual-task conditions. Once these conditions are defined, both methods analysis and work measurement can be initiated.

In most work, whether it be clerical, maintenance, manufacturing, or even supervision, a great deal of checking goes on. Consequently the task of inspection is one which occupies much time. The Mento-Factor System has been successfully applied to inspection work in pioneering firms such as the Radio Corporation of America, N. V. Philips of Eindhoven, Motorola Inc., Sprague Electric, and R. L. Polk Inc., who have validated the data.

In very detailed research the Wofac Company's workers have investigated and included in their comprehensive Inspection Tables data about reflectivity factors, contrast, brightness, sizes of characters, illumination, and similar data.

The analysis of inspection is based upon the division of the area to be inspected into Work-Factor Inspection Units, which are those unit areas which can be inspected adequately by an 'Average Experienced Operator' with 'Adequate Task Vision' during an 'Inspect Interval'. A Work-Factor Inspect Interval is the time required by the average experienced operator after an Eye-Shift or -Focus to inspect the Inspection Character(s).

In the example illustrated in Figure 8.20 the printed-circuit board has been divided into Inspection Units according to the density of the points to be inspected. It will be seen that Area 'A' has a greater joint density than Area 'B'. The inspector's job is to inspect the soldered joints on the board to determine which (if any) are defective. Within each area the joints have been grouped into

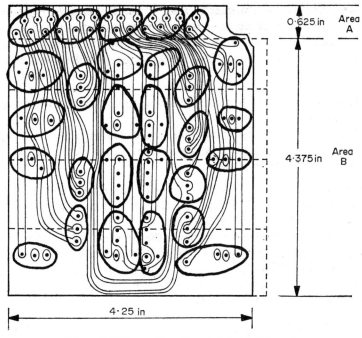

Figure 8.20 Inspection of a printed-circuit board

clusters, each cluster being an Inspect Interval, as it is possible for the eye to scan the joints within this area at one time. From Table 8.12 it will be seen that the total time for inspection of the printed-circuit board is 0·213 min.

TABLE 8.12

A MENTO-FACTOR ANALYSIS BASED ON FIGURE 8.20

OPERATION: Inspect printed circuit board

No.	Description	Work-Factor Time-Units
1.	Shift eyes 15 in. to area A on next printed circuit board	40
2.	Inspect 26 joints—area A	
	4 Inspect intervals with 6 joints	342
	1 Inspect interval with 2 joints	41
3.	Inspect 91 joints plus 6 critical-lead locations—Area B	
	9 Inspect intervals with 3 joints	556
	5 Inspect intervals with 4 joints	361
	4 Inspect intervals with 5 joints	338
	5 Inspect intervals with 6 joints	455
Totals: Inspect intervals = 28		2,133

Cycle time for inspect printed circuit board is 0·213 min.

The inspect intervals are compiled from the Work-Factor Fundamental Mental Process Times, an example of which is given below:

Mental Process	Analysis	Work-Factor Time Units
See	See 6–4	17
Conduct to brain	Con 1	1
Identify (simple)	IdA 1	4
Decide (simple)	DeA 1	4
Conduct from brain (average)	Con 1	1
	Total	27

Compute. Another common activity in indirect tasks is the activity of computation, and Work-Factor Compute includes the processes of Add, Subtract, Multiply, and Divide. The application of 'Compute' is best illustrated by way of an example. The time-values are obtained from the appropriate Work-Factor Fundamental Mental Process Times.

Description	Analysis	Work-Factor Time Units
1. See digit (first digit only; all subsequent digits are seen simultaneously with identity of previous digit).	See 1–1	6
2. Nerve conduct from eyes to brain centre (again, first digit only; note above comment).	Con 1	1
3. Identify digit (10 Action Signals).	IdA-10	41
4. Add digit.	Com-1	20
5. Carry two digits in mind from digit to digit, to next column, or to answer.	Com-2	10
6. Sustain (applied to each column separately).	Su-4	42
7. Move pencil between digits and write average digit.	Std Data*	86

Other mental-process times. As mentioned earlier, there are other Mental-Process Times which have not been dealt with here. It will be appreciated even from the foregoing description that the Wofac Mento-Factor System is a complex, but unique and comprehensive method of determining basic times for mental activities. From the complete tables, compiled with explanations in the Wofac Mento-Factor Manual of Mental Process Times, basic times for standard data compilation may be derived. Many of the other Mental-Process Times have been incorporated in the foregoing examples, which it is hoped will serve to illustrate the usefulness of the system in fields of repetitive and indirect work.

* The value is obtained from standard data compiled from Detailed Work-Factor.

8.9 HIGHER LEVELS OF DATA

So far we have been concerned with basic data systems which are generally applicable. The higher-level systems of macro-P.M.T.S. are mostly designed for specific applications. The two areas for which data are available are clerical and maintenance work. Some of these systems are described in Chapter 15.

Most systems of macro-P.M.T.S. have been based on MTM data, and rather surprisingly many still retain the original times in t.m.u.s. There are three main sources for higher-level systems: Methods–Time Measurement, Work-Factor, and Mulligan Standards (Clerical Work-Improvement Programme).

The systems are distinguished by their use of specifically named actions such as 'open file drawer', 'erect ladder', or 'search in tool-bag for tool', rather than relying on basic motions such as 'reach', 'grasp', 'get', and similar fundamental actions.

Some of the most important systems of macro-P.M.T.S. are given below.

Clerical Data
1. Universal Office Controls. This system was developed by H. B. Maynard, W. Aiken, and J. F. Lewis, for the H. B. Maynard Company.
2. Master Clerical Data. Designed by Serge A. Birn, R. M. Crossan, and H. W. Nance, for S. A. Birn Company of America, with standards in t.m.u. derived from MTM.
3. Clerical Work Data. A sister technique of B.W.D., with similar origins.
4. Clerical Work Improvement Programme (CWIP). This Australian system was developed by W. D. Scott Company from the American standards of Paul B. Mulligan, from motion-picture analyses.
5. Office Modapts. Another Australian method, this time from the Australian Association for P.M.T.S. and Research, and again based on MTM.
6. MTMA Office Data System. An official development of the MTM Association, largely the work of the U.K. MTMA, supported by the Swedish, German, and American Associations. The system was made available in 1973.

Maintenance Applications
1. Universal Maintenance Standards. A system of H. B. Maynard Company, developed from MTM and based on comparative estimating for its application.
2. Basic Work Data. The system of B.W.D. was developed from S.P.M.T.S. (itself derived from MTM) by I.C.I. and British Rail. Times are converted to milliminutes at 100 B.S. rating.
3. MTM-2 Maintenance Data. Another MTM-based technique. Uses MTM Motion Sequences and Data Blocks to synthesize standards, and widely employs Tape-Data Analysis.

9 Time Study

9.1 INTRODUCTION TO TIME STUDY

The technique of time study was the first real attempt to measure, with any reliability, the work content of a job carried out by a human being.

The theory behind the measurement of work is that the time to complete the work content of a defined job should be constant at a required level of performance. Before the introduction of work measurement, management had only the actual time taken by operatives carrying out work on which to base the standard time they needed for purposes such as planning, manning, or financial-incentive schemes. This standard time had to take into account a large number of variable factors, including the ability of the operator to carry out the work, the degree of training to which the operator had been exposed, the variability between operatives in the way in which the work was carried out, the ineffective time that operators built into their working time, variability in materials, condition of tools, plant, and equipment, and not least the ability of supervision to plan and organize the work.

Time study produces standards which are retrospective; work must first be performed and observed before standard times can be established. It is the pre-production or production-planning stages that create the first difficulty, primarily because neither operatives nor management have become accustomed to the job, but usually it is at this stage that the time for carrying out the work is established. Once management, the operatives, and unions have agreed the time for a job in this manner, by precedent, this is the time which is accepted as the standard, and subsequently it is extremely difficult to alter once it has become so established. It must also be borne in mind that in some cases management and supervision have a vested interest in excessive times for jobs, as these can be used as 'cream' to placate operatives on financial-incentive schemes, and of course easily achievable time standards remove a high proportion of the labour-relations problems that exist in industry today.

Clearly a time value that is to be used by management, whether it is to be used for financial incentives or as a time for productivity targets, must be 'right' at the outset, as subsequent revisions can only be negotiated if the methods or conditions of work change.

It was the introduction of time study that provided the relative consistency in time-values in which management could have confidence, and the means by which management could set realistic time-values. It removed the 'haggling' and bargaining aspects from the setting of time standards. It also separated the payment aspects from the production-planning aspects, in that production standards could remain consistent throughout an organization, while payment standards for financial-incentive purposes could be divorced from production standards.

Time study is slowly being replaced by predetermined motion–time systems (P.M.T.S.) in the fields of work study and industrial engineering. What must be borne in mind is that all P.M.T.S. have their bases in the theory of time study. For this reason it is essential that any work-study practitioner is trained in the technique in order that he can fully appreciate and have confidence in the work-measurement aspect of work study, and that he will fully understand labour-control data, and of course performance indices. Consequently the basis of any work-measurement training must be built on a sound understanding and practical experience of time study.

9.11 THE THEORY

Time study is based on three factors:

1. The time taken by an operative to carry out defined work.
2. A levelling factor that is used to offset variable factors affecting or effected by the operative.
3. The statistical principle of the randomness of variability.

In order to establish a time, a stop-watch is used, and the start and completion of the work-cycle accurately timed and recorded. At the same time that the work-cycle is being timed the trained practitioner assesses the tempo of working (levelling or rating factor) of the operative carrying out the work element. The work element is timed and the rating factor assessed simultaneously over a statistically calculated number of cycles. The time of the work element multiplied by the ratio of observed rating-factor to standard rating-factor produces a time that is suitable for the level of performance defined by the standard rating-factor. This calculation is called 'extension of the basic time', and gives the time required to carry out the work element on one occasion.

In assessing the rating factor for levelling purposes, the work-study practitioner recognizes the effect of three factors:

1. The speed of movement of the operative.
2. The effectiveness of the operative in carrying out the work content of the defined method.
3. The effect of effort on:
    (i) speed in carrying out the work content;
    (ii) the effectiveness of carrying out the work content.

A combined assessment of these factors is then expressed as a number on a numerical scale; the standard on the B.S.I. recommended scale being 100.

The process of rating is described fully in Chapter 5.

In general terms the actual observed time is modified by multiplying it by the ratio:

$$\frac{\text{observed rating}}{\text{standard rating}}$$

It will be seen that if a rating factor assessed during the carrying out of the work element is anything other than standard, the application of the rating ratio will effectively alter the time it actually took to carry out the work element. If the assessment of the operative is higher than standard this will have the effect of increasing the time it actually took to perform the work-cycle, but should the assessment be less than standard, the effect will be a decreasing of the time taken to carry out the work-cycle. The application of the rating ratio ensures that the resulting time will always be that at the standard rate of working, and not at the time it actually took to complete the work.

It is important to remember that the objectives of the work-study officer when carrying out the practice of time study is not to put a time to the way operatives carry out the work, but to establish a time for the total work-content of that job. In order to use time study it is necessary to have the facility of operatives actually performing the work. It is essential, even vital, that the operatives have been properly trained in the method laid down by the work-study practitioner, and that they can perform the work in a manner which is consistent from cycle to cycle. If, because of excessive cycle-to-cycle variations, the work-study practitioner is not able to compensate adequately through his rating factor, the resulting study will eventually fall into disrepute. Deviations from the defined method and other causes which make the element unrepresentative should not tempt the practitioner into faking or guessing the probable observed time or rating for the sake of producing a complete recording.

9.12 THE PRACTICE

In order that time standards set by time study shall command the respect of not only the operatives and management, but of the practitioner himself, a high degree of consistency and reliability is necessary.

As with all techniques of work study, time study is based on the analytical approach and on the statistical premiss of the randomness of errors. The analytical approach, in turn, is based upon the breaking down of a work cycle into constituent parts known as elements, which are fully discussed in Chapter 7.

In carrying out time study it is advantageous to establish the times for elements of work from which subsequent jobs may be synthesized. The objective is not necessarily to time the work cycle, although in certain cases it is economical to do just that. The importance of consistency has already been mentioned, and the aim of time study is to generate synthetic data which can be used in other situations where similar elements occur, thus providing the required consistency.

The degree to which analysis is performed depends upon the size and complexity of the job for which a time is required. If the job is complex or of long duration, the first stage in the analytical approach is to break the work down into its prime operations, each operation then being further analysed into elements of work. Having selected the elements such that the requirements of clear definition and suitability for measurement are satisfied, each element must be defined by its *break-points* (*see* page 231).

This definition is one of the most difficult aspects of the study, and provides one of time study's greatest weaknesses: the difficulty of the average person to write a description of work which is concise yet clear, which subsequently can be fully understood, and which can define clearly the variations and limitation inherent in human work. An example of the analysis of a simple job into its elements qualified by break-points is outlined in later sections of this chapter.

The practice of element breakdown has been criticized as creating an unrealistic situation; the argument being based on the fallacy of adding successive elements which are not necessarily independent. This hypothesis is pursued in depth in Chapter 7.

One of the problems of work measurement is 'drift', which occurs over a period of time; drift from the defined method and the quality standards, and drift in standards of raw materials. To safeguard the time standard the element description must take all these factors into account.

The analysis of a work-cycle into elements must be carried out without the inhibition of lengths of time taken by those elements. One of the common weaknesses in the practice of time study is that elements of work are established for their ease of timing with a stop-watch. A properly trained practitioner in good practice has the techniques available within time study to establish the time for *any* element, irrespective of its duration (*see* Section 9.2).

Because of the problems inherent both in timing elements of such a wide range, and in sample requirements to establish statistical viability of time standards, the careful planning of a time study is essential (*see* Section 9.5).

The final stage in the analysis of the work-cycle into its elements is to refer to the established library of elemental times to ensure that existing data will be used where appropriate. A good work-study practitioner will, if possible, re-adjust his element breakdown in order to incorporate such data, instead of establishing fresh time-standards.

It is, of course, essential that work-study practitioners periodically review all the elements of work in a work-cycle. This review facilitates comparison of results of their time study with existing data, thereby keeping a check on their own concept of the standard rate of working.

9.2 TIMING PROCEDURE AND PRACTICE

The description of timing procedure and practice will be separated into a review of the types of stop-watch, and the methods of timing.

9.21 THE STOP-WATCH

There are several types of stop-watch available for carrying out the various timing practices. Six types are illustrated in Figure 9.1, both manual and electronic. Watches recommended have the facility to fly back and instantly recommence the timing operation.

1. The crown of the watch has the dual function of winding the spring and when depressed, of returning the sweep-hand of the watch to zero, where it will instantly commence its rotation.

210   WORK MEASUREMENT

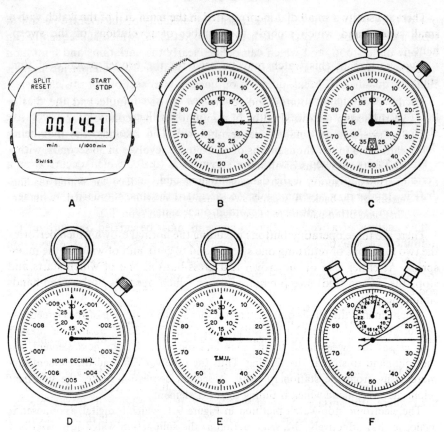

Figure 9.1 Several types of watches used in time study
    A. A digital electronic watch
    B. Decimal minute with fly-back action: side slider
    C. Decimal minute with fly-back action: side crown
    D. Decimal hour with fly-back action
    E. T.m.u. timer for use with Methods–Time Measurement
    F. Decimal minute with split-action timing

2. At the side of the watch there is a further mechanism, either:
    (a) a slide which when moved around the periphery of the watch will stop the sweep-hand instantly, and when pushed in the opposite direction will allow the sweep-hand to continue recording time. Alternatively it has:
    (b) a button which has the same action on the sweep-hand except that it is depressed once to stop the hand, and when pressed a second time will allow the sweep-hand to continue rotating.

With both types of mechanism, if the sweep-hand has been stopped by either sliding or depressing the button, if the crown knob is depressed it will zero the sweep-hand without allowing it to commence rotation. In order to start the sweep-hand the slide or button must again be operated to re-start.

There is usually a small dial incorporated in the main dial of the watch with a small sweep-hand which records the number of revolutions of the sweep-hand.

Two versions of this watch may be used for the proper practice of time study:

(i) the sweep-hand rotates one revolution in one minute and the dial is graduated in one-hundredths of a minute, or in seconds;
(ii) the sweep-hand rotates one revolution in 3·6 seconds, the dial being calibrated in ten-thousandths of an hour, or revolves in 36 seconds, with a dial in thousandths of an hour.

The split-action watch is sometimes used to reduce the watch reading errors of fly-back. A second sweep-hand is provided, concentric with the first, and often painted red (*see* Figure 9.1, watch type 'F').

There are three operating buttons including the normal crown which performs the two functions of returning one sweep-hand to zero, and of winding the main-spring. At either side of the crown is a press-button, one of which starts and stops the second (red) sweep-hand, and one which operates both sweep-hands simultaneously.

When timing, the hands are started together at the beginning of the element. At the breakpoint the press-button located at the '11 o'clock' position is pressed. This causes the sweep-hand to fly back and start retiming the next element, and the split-hand to stop at the reading. Thus the observer can read the element time more easily from the stationary hand. A second depression makes the split-hand rejoin the sweep-hand which is timing the next element.

The electronic stop-watch pictured in Figure 9.1, which is digital, as opposed to analogue, has effectively the same action as the split-action watch just described. When the button is pressed to start the watch the passing time appears on the digital readout. At the breakpoint pressing the button freezes the reading, but allows the electronics to continue timing the element. A second depression causes the current time to appear on the readout which continues to display the time until the next breakpoint when the process is repeated.

The most up-to-date equipment is an electronic study board. Typically a board is similar to the manual board, but is about 2 cm thick. Chip technology is incorporated into the board. Timing is electronic, operating via a button which is pressed at the breakpoints. Times are stored in the computer's memory. Ratings are estimated and entered into the memory using the key-pad. On completion of the study the board is plugged into a microcomputer which analyses and prints out the study completely. A full description is given in Chapter 17.

9.22 METHODS OF TIMING

Several methods of timing are available to suit the different circumstances under which the practitioner may be working:

1. Consecutive-element timing
2. Selective-element timing

3. Differential timing (consecutive)
4. Differential timing (selective)
5. Continuous (cumulative) timing.

1. *Consecutive-element timing* is the technique of rating and timing each element as it occurs in sequence in the Plan of Study, or the cycle of work, and recording these consecutively.
2. *Selective-element timing* is the technique of rating and timing a selected element, or more than one element, out of sequence in the Plan of Study or cycle of work. It is used where there is sufficient time between each selected element to change the emphasis of concentration on the part of the practitioner from rating and timing the operative, to reading the recorded time on the stop-watch and recording this accurately with the assessed rating on the Observation Record Sheet. The stop-watch is started at the beginning of the element and stopped at the break-point at the end. This technique is primarily of use when the duration of an element is extremely short and errors in the accuracy of reading and recording the observed time are likely to be significant.
3. *Differential timing (consecutive)* is a technique of consecutive-element timing where the nature of the job lends itself to planning a study programme so that the frequency of occurrence of some of the elements can be used in arriving at the times of very short elements. This is achieved by combining an adjacent short element with a longer element, the longer element having a frequency of more than one in a cycle. The time for the shorter element is found by subtracting the time for the longer element from that for the combined elements.
4. *Differential timing (selective)* is used to determine the time for an element which is too short to establish by the other techniques. As with (3) the short element is combined with an adjacent long element, both elements being rated and timed together for a sufficient number of occasions to ensure confidence in the prescribed degree of accuracy. The combined elements having been dealt with in this way, the time study is continued, rating and timing only the longer of the two elements. As before, the difference between the two standardized times is the extended time for the shorter element.
5. *Continuous timing*. In this technique the stop-watch is not touched as fly-back is not used. The actual readings are recorded at the break-points; the individual times for elements being deduced from subsequent subtraction of successive times. One advantage of this method is, of course, the elimination of fly-back errors.

9.23 THE PRACTICE OF TIMING

Accuracy and dexterity in the use and reading of a stop-watch are essential features of the overall consistency required of the technique of time study.

The time should always be recorded to the nearest centiminute, except when timing an element that is so consistent that this convention will create a bias. The recordings should then alternate, nearest up and nearest down.

Before any attempt is made to use a stopwatch for the purpose of timing, the practitioner must have achieved a high degree of dexterity and accuracy of reading, and this is possible by constant practice away from the shop-floor.

The procedure for manual or electronic timing practice is as follows:

An observation study-sheet is positioned on a normal time-study board with the watch in position on the clip. The exact time is recorded from an independent watch or clock with a second-hand, and at the same instant the stop-watch is started from zero. Having taken up a comfortable stance and conditioned his thinking to the commencement of the practice, the practitioner reads the total time on the stop-watch, including any revolutions on the small minutes dial. After he has taken the reading he zeros the stop-watch, and records the time on the sheet as quickly as possible. Initially the practice periods should be not less than 5 min, with subsequent increases to 15-min duration as competence and accuracy are achieved. At the end of each practice the stop-watch is left running until the next whole minute appears on the independent watch. When this occurs, at the same instant the stop-watch is stopped by means of the slide or button and the total stop-watch time recorded.

The sum total is found of all the recordings of the study, excluding the time spent prior to, and subsequent to the recordings of the test durations. The summation of the test recordings should be within plus or minus 2 per cent of the time on the independent watch, less the time spent prior to and subsequent to the test recordings.

A graphical record of the results of each study should be kept by the practitioner, and no attempt at time study at the workplace should be undertaken until he can prove his ability to record time consistently within the range of plus or minus 2 per cent.

If the above procedure is carried out it will become evident to the practitioner that there is more white space between the graduations of a stop-watch than there are graduations, and in order to achieve any degree of accuracy according to the above check-method it will be necessary to randomize the recording of time to the nearest graduation. To achieve the required dexterity and accuracy the practitioner must position himself comfortably and 'line up' his eyes with the column of figures he is recording and the dial of the watch, so that no head movements are required from reading the watch to recording the time. Likewise the crispness with which he operates the crown of the stop-watch is essential with those watches which have no spring-loaded mechanism for zeroing the sweep-hand.

The co-ordination necessary for reading and recording the time on a stop-watch to the high standard necessary requires a good deal of training and, just as important, continuous practice.

The practitioner who reaches such a level of competence finds that he is able to focus his attention on the process of rating and on ensuring that the worker is adhering to the specified method. Thus it becomes essential for the practitioner mentally to ignore the watch until the approach of the break-point of the element, when instantly his concentration must switch from rating to reading the stop-watch. The reading of the watch and the pressing of the crown to zero the sweep-hand, must be an instantaneous operation, and it will be found in practice that in fact one does not read a stop-watch immediately, but merely retains an image of the position of the sweep-hand, while instantly zeroing the sweep-hand. The latent image of what has been seen is interpreted in terms of actual reading, which is recorded on the form; the pencil having been previously positioned at the correct column in anticipation. This practice allows the practitioner to spend the maximum amount of the elemental time available observing the work and assessing the rating of the operative.

9.24 EXTENSION OF THE BASIC TIME

The stage of compilation is described in the example which follows, but the process of extension needs a few words of explanation.

As already stated, time study is based on the concept of modifying the time observed to be taken by the operative in performing his work to that which theoretically he would have taken had he been working at *standard rating*. This is achieved by multiplying the observed time by the ratio:

$$\frac{\text{observed rating}}{\text{standard rating}}$$

Basically there are five main methods of effecting the adjustment to the observed times; three of which are arithmetic and two graphical. In the *individual extension* method the extension formula is applied to every element of every cycle in turn, after which the basic times for each element are averaged to produce mean basic times. In the second method of *over-all extension*, the ratings for an element are first summed and averaged, the mean observed time found in a similar way, and then, using these two means, the extension formula is applied once only to each element's respective means. *Modal extension* requires the selection of the most frequently occurring rating for an element, and the most frequently occurring observed time. Again the extension formula is applied once only to each element's modal rating and modal time.

Of the foregoing methods, the first is the most accurate in mathematical terms. Consider, for example, the following ratings and observed times:

ratings:	80	85	80	90	70	85	95	90	*Total*
	90	80	90	85	90	75	90		1275
times (cm in):	18	17	19	16	22	17	15	17	
	15	17	15	17	16	21	16		258
basic time:	14·4	14·5	15·2	14·4	15·4	14·5	14·3	15·3	
	13·5	13·6	13·5	14·5	14·4	15·8	14·4		217·7

Individual extension (third row above) produces a basic average time of

$$\frac{217 \cdot 7}{15} = 14 \cdot 51 \text{ centiminutes}$$

Overall extension gives

$$\frac{\text{average rating} \times \text{average time}}{100} = \frac{85 \times 17\cdot 2}{100} = 14\cdot 62 \text{ cm}$$

Modal extension produces

$$\frac{90 \times 17}{100} = 15\cdot 30 \text{ centiminutes}$$

because the most frequently occurring (or modal) rating is 90 (occurring six times) and the modal time is 17 (occurring five times).

Comparison of the three values will show significant discrepancies; the over-all and modal methods always overestimate the basic time.

In practice the rating should not vary greatly, so the values obtained by the three methods will be closer.

9.25 EXTENSION BY GRAPHICAL METHODS

Until recently the only graphical method of determining basic times from observed times and ratings was the reciprate method. The main advantage of graphical extension is the speed with which the 'average' basic time may be derived for an element. Unfortunately the special nature of reciprocal paper has often restricted the use of this method. As a consequence of this an alternative graphical method is now presented based on easily obtained log paper.

For non-mathematicians

In determining a basic time for an element by graphical means the observed times and ratings are plotted against their respective scales. A line of best fit is then drawn through the points, and the basic time read off at the intersection of this line and the standard rating (on the particular rating scale in use).

Unfortunately there is a drawback. The line of best fit which results when ratings and observed times are plotted on linear (equally spaced) scales will not be a straight line but a curve known as a hyperbola. This is shown in Figure 9.2. Hyperbolas are very difficult to sketch, but anyone can draw a straight line. Both graphical methods referred to above rely on the ability of mathematicians to transform curves such as hyperbolas into straight lines. These transformations are discussed later in this chapter.

The reciprate graph is shown in Figure 9.3. In using this chart it is necessary merely to plot the ratings and observed times against the axes in the usual way, ignoring the fact that the ratings scale is not linear. A straight line of best fit is now plotted through the points. In practice the ratings of one operator will not vary to any great extent, so the points will lie very close together on the chart. How, then, does one determine the direction of this line of best fit through this 'plum pudding' of points? As the accuracy of the resulting basic time depends on the precise angle of the line, clearly this must be drawn accurately. Were it possible to fix a point on the chart which was some distance away from the other points, this would serve to indicate the direction the line should take. In fact such a fixed point does exist; it is the point where infinite rating and zero observed-time intersect. In using the reciprate chart this origin is joined to the centre of the mass of plotted points, and the basic time read off at the intersection of this line and the standard rating (*see* Figure 9.3 and compare results).

An alternative method of determining the direction of the line is by fixing its angle of slope (or gradient). This cannot be achieved on the reciprate chart, as its slope varies with each element. However the log log chart is unique in that *all*

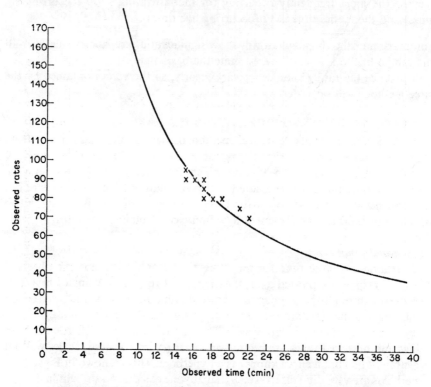

Figure 9.2 A hyperbola drawn through readings from a typical study

elements of all jobs produce the same slope when their lines of best fit are drawn.

Figure 9.4 shows that the chart is used in a similar way to the previous chart. This time the line of best fit is constructed to pass through the centre of the points but parallel to the guide-line A–B. The basic time is found as before at the intersection of the line and 100 rating. Guide-line A–B is pre-printed on the chart, but if ordinary log log paper is used the construction of A–B is a very simple matter, and is described later in this chapter.

The reader is invited to compare the graphical methods with the arithmetic methods for speed of application and accuracy using the figures given.

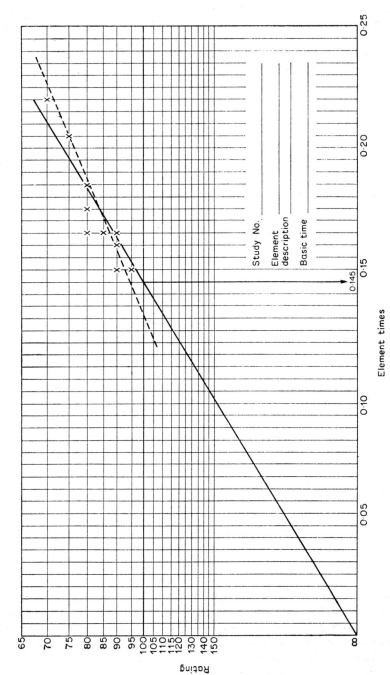

Figure 9.3 The reciprate graph for determining the basic time

How is it done?
Consider first the basic-time formula:

$$\text{basic time} = \frac{\text{observed rating} \times \text{observed time}}{\text{standard rating}}$$

In symbolic form, this is

$$b = \frac{r \times t}{s}$$

Now, if the time study is performed in a competent manner the basic times for cycles of a given element will be reasonably constant. Thus one may assume both b and s to be constant, and t and r to be variable.

Rearranged, the formula becomes $r \times t = b \times s$, or $r.t =$ a constant, which, as the mathematician (but possibly not the time-study practitioner) will know, is the formula for a hyperbola. As already explained, to be of use in graphical extension this curve must be transformed into a straight line.

Now any straight line has the form: $y = mx + c$, where y and x and the variables, and m and c, are constants. Of course, m can be negative and the formula becomes $y = -mx + c$, which may be rearranged as $y + mx = c$.

In this last equation it will be seen that the variables y and x are *added* to equal the constant c, whereas in our time-study formula the variables t and r are *multiplied* to produce the constant (bs). Clearly if the time-study formula can be manipulated so that t and r are added instead of being multiplied a straight line will result, thus achieving our objective.

This may be done in two ways, which are described below.

The equation $bs = rt$ may be written as

$$\frac{1}{r} = \frac{t}{bs} \text{ or } \frac{1}{r} = \frac{1}{bs} \times t$$

If we write R for $\frac{1}{r}$, the equation becomes:

$$R = \frac{1}{bs} t$$

which compares well with:

$$y = mx + c \text{ if } c = \text{zero}$$

The fact that there is no constant equivalent to c implies that the straight line passes through the origin (or zero) of the graph.

Thus by plotting R (which is actually $1/r$) on the y axis, and t on the x axis (Figure 9.3) a straight line will result when ratings and their equivalent observed times are plotted on a chart.

Alternatively, how can *multiplication* be performed by *adding* figures, which was our requirement for a straight line? The answer is by the use of logarithms. It will be recalled that in order to multiply numbers, their logarithms are added.

TIME STUDY

Thus, from the above formula $tr = bs$
or $\log t + \log r = \log(bs)$, which compares favourably with $y + mx = c$, i.e. a straight line.

The above comparison will show that $\log t$ is equivalent to y, and $\log r$ is equivalent to x. As there is no constant associated with $\log r$, then m the slope is equivalent to 1, and is, therefore, constant for all values of $\log t$ and $\log r$.

On the log log chart the guide-line A–B of Figure 9.5 must be constructed with a slope of unity, determined as follows:

First, any rating is chosen (say 70), and a suitable observed time which fits into the scale on the chart (say 0·4 min). These are purely arbitrary points. The point (B) corresponding to these values is plotted. When extended by the time-study formula this will produce a basic time of $(70 \times 0.4) \div 100$, or 0·28 b.m. A second rating is chosen (say 140), and the equivalent observed time calculated which will give a basic time of 0·28 b.m. This observed time must clearly be 0·2 min. The second point (A) is then plotted at the position of rating = 140 and time = 0·2 min. The two points are now joined as line $A-B$.

The construction of the reciprocal scale is shown in Figure 5.5. First the linear scale is plotted, with the reciprocals of these figures added on a second scale. It is this second scale which will be retained as the actual rating scale, while the first linear scale is discarded. The zero on the linear scale (infinity on the rating scale) indicates the origin from which the line of best fit will be drawn.

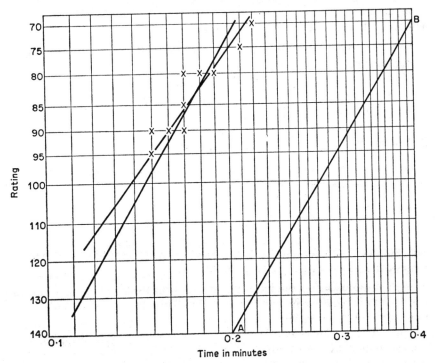

Figure 9.4 A log log graph for determining the basic time

A further use for these charts is in estimating the competence of rating. The charts depicted in Figures 9.3 and 9.4 show the theoretical lines of best fit for the data already used in the earlier examples of this chapter. The *actual* trends or lines of best fit are also plotted, and in this example indicate flat rating because the actual line is less inclined than the theoretical line. Compare this method with the chart of Figure 5.6.

9.3 TAKING A TIME STUDY

The procedure in taking a work-measurement study was outlined in Chapter 7. Naturally time study adheres to these broad principles. The following description relates to time study specifically, the procedure for which is shown diagrammatically in Figures 9.5 to 9.7.

The main steps have been described in previous chapters, but here are brought together in summary form, to be expanded in further sections of this chapter.

1. Consult with supervisor and operatives.
2. Examine the work and analyse into elements. Search for existing synthetic data.
3. Define the elements and their break-points.
4. Determine the number of cycles to be timed and decide the method of study.
5. Take a time-check (if appropriate to the technique being used).
6. Time and rate the elements.
7. Take a final time-check and calculate the timing error (again, if appropriate).
8. Calculate the basic time and hence the standard time after assessing all allowances.
9. Calculate the study error where more than two work-study practitioners have participated. If satisfied, issue the time standard.
10. Extract suitable synthetic data.
11. Monitor the progress of the reference period.
12. Up-date the standard at periodic review periods.

The complete time study will be described in more detail in the example which follows in Sections 9.4 to 9.8.

9.31 DOCUMENTS AND FORMS USED IN TIME STUDY

The diversity of work measured by time study is such that the firms employing the technique find it necessary to design their own forms and backing documents to suit their particular requirements. The practice of time study is by now reasonably standardized throughout industry, thus allowing the use of forms, which to a great extent are common in many respects. The forms indicated in Figure 9.8 are those used in the exercise which follows, and are typical of documents in general use. Terminology and names of forms may vary from firm to firm, but those depicted are considered to be the minimum necessary to allow complete documentation of the study for current and future use.

TIME STUDY PROCEDURE
ELEMENT BREAKDOWN

ESTABLISH (1)

Good relations with supervision and operatives

ENSURE (2)

A thorough appreciation of the work to be measured and the factors involved

(a) Management instructions are adhered to as to quality and technical aspects

(b) Safety regulations are conformed to

(c) Materials and equipment are to specification and in good condition

(d) The defined method is in operation and the operatives are suited and accustomed to the task

DIVIDE (3)

The job into its different operations and record operations

(4)

Each different operation into its elements and record the elements

EXAMINE (5)

(a) The elements for like elements. Adjust the breakdown as necessary to establish like elements

(b) Existing data for like elements. Adjust the breakdown as necessary to suit existing data

ESTABLISH (6)

The different elements to be measured and establish element breakpoints

PLAN THE STUDY PROGRAMME (see Figure 9·6)

RECORD (7)

In detail on the Study Observation Sheet

(a) In sequence, a description of all the operators' movements with distance involved in doing the work for each different element to be studied

(b) At the end of each element the established breakpoint

(c) Number each element

Figure 9.5 Time-study procedure: element breakdown

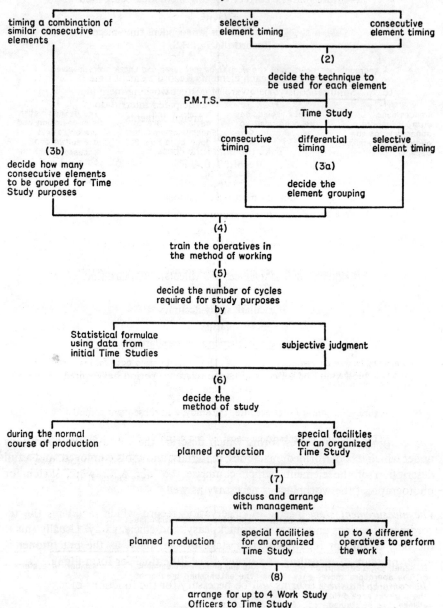

Figure 9.6 Time-study procedure: planning the study

TIME STUDY

(1)

(2)
Observe a few cycles
to ensure that the correct working method is being used

(3)
Take a check time from an independent time-piece
Record the T.E.B.S.

(4)
Rate and record each element. Record element time
Simultaneously: be aware of correct working method
anticipated interruptions
foreign elements

(5)
Take the final time check
Record T.E.A.F.

(6)
Thank operative and supervisor

(7)
Estimate timing error
Calculate: basic time, R.A., and work content

(8)
Evaluate other allowances and calculate standard time

(9)
Calculate study accuracy

(10)
Extract suitable elements as standard data

(11)
Check study parameters after the reference period has expired
Restudy if necessary

Figure 9.7 Time-study procedure: the study, and subsequent compilation

The analysis stage requires forms which describe the job and the conditions under which it is performed, machines, materials, and tools employed, and a full description of the elements which comprise the job. A drawing, sketch, or photograph of the workplace is necessary as well.

The measurement stage needs an observation record which facilitates the recordings of ratings and times in a quick, easy, and efficient way. Usually this is the only document required at this stage of the study, as the practitioner is under pressure of recording and would have little time for form-filling.

The compilation stage. Forms will be necessary for the functions of:

1. Calculation of timing error, from check times;
2. Extension of basic-element times from observed times and ratings;

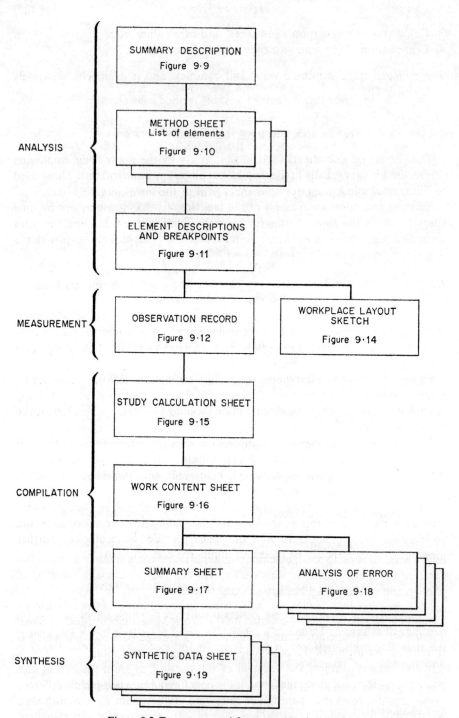

Figure 9.8 Documents and forms used in time study

TIME STUDY 225

3. Calculation of relaxation allowances, and other allowances;
4. Compilation of the standard time, and the study error.

The synthesis stage is more diverse and complex, and is dealt with separately in Chapter 11.

9.4 A PRACTICAL EXAMPLE

The forms and documents used in time study by the many firms employing this technique vary greatly in the details, but all follow basic formats. Those used to illustrate the following example are typical of the forms in general use.

This exercise offers an example of the practice of work measurement by time study to derive the allowed time for filling by hand twelve 0·285 litre tins with processed peas and packing them into a carton. The example concludes with the selection of data suitable for use as synthetics.

Job. Fill by hand twelve 0·285 litre tins with processed peas and pack into a carton.

The method used for the job was devised purely for the purpose of demonstrating this example. The number of cycles used for the study is sufficient only to depict a principle, and is not intended to represent a typical time study for a job of this nature.

Figures 9.5 to 9.7 outline the various stages required in making a time study. Some of the steps may seem very elementary to the more experienced work-study practitioner, but it is emphasized that none should be omitted, particularly those shown in Figure 9.5.

The first step in the study is to complete the relevant details on the top half of the Work Measurement Summary Sheet (Figure 9.9). Of utmost importance is the recording of the details of *Materials*, *Equipment*, and *Machines*, as those are essential to the defined method.

Where possible code numbers and machine numbers should be quoted. Where these data are not available all the facts and details should be recorded to ensure no possible doubt in the future. If the space provided is insufficient, further sheets of this form should be used and numbered consecutively.

9.41 THE OPERATION BREAKDOWN AND ELEMENT ANALYSIS

The operation breakdown is the first stage in the analysis of the job into its component elements. This is recorded on the Work Measurement Method Sheet (Figure 9.10). An operation can be defined as 'a specific part of a job which is different by nature from the preceding and subsequent operations'.

In the present example the removal of the twelve tins from the carton and placing them in a row along the front of the bench requires a completely different motion-pattern from the operation of removing the lids from the tins, although in both cases the operations are for handling tins. It is necessary to separate these consecutive operations for the purposes of data and the assessment of rating. The

WORK MEASUREMENT—SUMMARY SHEET		Sheet 1 of 1	Reference P.P.E.
Operation or job Fill twelve 0·285 litre tins with processed peas and pack into cartons		Manning 1 female	**Standard time**
			Work content minutes
			Unoccupied time minutes
Materials (a) Expendable 0·285 litre tins to code 1835 (85 o.d. × 105 high) Cardboard cartons code E.19 Cardboard box 460 × 460 × 230		Machines Nil	Contingency allowance minutes
			Interference allowance minutes
			Standard time minutes
(b) Worked on Processed peas to code 'B'			Allowed time minutes
			Issued allowed time minutes for work minutes for delay
Tools and equipment Lid opener lever type 100 × 19 × 3 Aluminium scoop 76 × 13 Work bench 1370 × 760 × 760 high Storage bench 1370 × 760 × 760 high		Access Easy ✓ No restriction at workplace Normal Difficult	Per unit of production
Protective clothing/ equipment Company issue of white cap and overalls		Safety arrangements As laid down in department training instructions	Governing time
Range of working conditions Lighting good no glare Dry – carpeted floor Temperature –18°C		Relaxation allowance special conditions Nil	Plant 'A'
			Department Packing
Quality standards In accordance with quality control criteria published May 1967			Branch —
Variable factors affecting data			Location Ground floor
Note: all lengths are in millimetres			Operating instructions reference
Unoccupied time reference	Contingency allowance reference	Interference allowance reference	Date 10 July	Date 11 July
			Derivation by G.S.	Checked by A.K.

Figure 9.9 Time-study summary sheet, completed as far as possible

WORK MEASUREMENT – Method sheet			Sheet No. 1 of 1	Ref. P.P.E.	
FILL 12 – 0.285 litre TINS WITH PROCESSED PEAS, AND PACK			Classification	Date 10 July	
Element Number	Ref.	ELEMENT DESCRIPTION AND BREAK POINT		Origin	% confidence limits	Basic time
		Operation breakdown				
		Operation 1. Change cartons				
		2. Lay out tins				
		3. Remove lids				
		4. Fill tins				
		5. Replace lids				
		6. Replace tins in carton.				
		Element breakdown				
		Operation 1. Element No. 1. Pick up carton of full tins and place aside to storage. Pick up carton of empties and position to bench.				
		" 2. Remove 2 tins from carton and place in row of 12 along front of bench.				
		" 3. Pick up lid-opener and position to tin.				
		" 4. Release lid from one tin and place aside to bench. 5. Aside lid-opener to bench.				
		" 6. Pick up scoop from pea stock, and reach for next tin. 7. Pick up tin from bench, and fill with 4 scoops of peas. Replace full tin to bench.				
		" 8. Replace and secure one lid to tin.				
		" 9. Replace 2 tins in position in carton.				

Figure 9.10 Time-study sheet showing element breakdown

job is, therefore, broken down into consecutive operations which are listed and numbered as shown.

The purpose of the element breakdown is to establish all the different elements in the job for which the basic times are required; it is not necessarily a list of the consecutive elements in the job. From the subsequent basic times for these elements it must be possible to synthesize the time for the job according to the defined method (not necessarily the method used during the study).

The work content for each operation is examined and analysed, and broken down into elements. An operation may itself be classified as an element, or it may contain two or more elements. In the final selection consideration must be given to its suitability for yielding either company-wide or local data. As far as possible an element should contain one type of work only. This will make rating easier than if the work in a given element were mixed, and will make for clarity in the rest-allowance assessment.

9.5 PLANNING THE STUDY

In order to decide which means of work measurement is to be used to establish the basic times for the elements selected and listed in the element breakdown it is necessary to know roughly the duration of each element. This is achieved by taking an observed time for each element. The observed time will give an indication which should be sufficient for the purpose.

The number of cycles to be studied may now be calculated or assessed, as described in Section 7.5. Alternatively, subjective judgment which comes with experience and practice through the initial use of the nomogram or formulae may be used by the seasoned practitioner.

It is considered inadvisable to rate and time more than ten different elements in any one study. In planning the study programme it is necessary to take this into account and arrange facilities accordingly. Advantage should be taken of repeating elements in order to reduce the sample size and to select the elements for study appropriately. Reference should also be made to existing elemental data. Elements for which there is available data should be eliminated from the study programme, provided the accuracy of the data satisfies the needs of the proposed study.

Having now planned the study to give times for all the different elements in the job, it is necessary to organize the study conditions for the fulfilment of the study plan. The following are some of the factors to be considered:

1. Arrangements, if necessary with management, for production to:
    (i) give a sample adequate to the required accuracy of the results, and to cover all the acceptable variables likely to be met with in the materials and equipment used,
    (ii) have a sufficient number of cycles to cater for the study plan,
    (iii) enable more than one experienced operative, preferably three or four, to be studied.

2. The operatives should have had sufficient practice to have become accustomed to the defined method. Minor differences in operator methods of working should not be greater than that which may be allowed for in the *effectiveness* aspect of rating.
3. Arrangements should be made for more than one practitioner to participate in the study.
4. A sufficient number of Observation Record Sheets should be prepared to cover the period of the study.

The following are some considerations affecting the elements in the example:

Element 1. This element has a duration of approximately 8 cm, which should not lead to any difficulty in rating and timing.

Element 2. This element has a duration of approximately 3 cm, but as the distance varies with each pair of tins an average time covering the variable distance is required. Consequently the time for laying out twelve tins will be determined, from which an average time for two tins will be found.

Element 3. Element 3 has a duration of about 2 cm, which is too short to time using conventional consecutive-element timing. In this case either selective-element timing or differential-element timing must be considered as alternatives. Selective-element timing is rejected in the present circumstances because it demands an increase in the sample required, and hence lengthens the study. Using the latter method this element may be combined conveniently with Element 4 for subsequent evaluation.

Element 4. Having a duration of approximately 3 cm, this element again is too short for consecutive timing by itself. However, this element repeats twelve times consecutively, so some of the repeats may be grouped together to form a study element of a duration that can be easily rated and timed by the consecutive-element timing method. It must be borne in mind that the longer the duration of the element the less confidence there will be in rating. An element duration of approximately 8 cm will be required for three tins. This will give four study elements for summary, two of which will be required for differential timing of adjacent elements, leaving two from which to arrive at the basic time for this element. Of course, this study element must be divided by three to determine the basic time for Element 4.

Element 5. Element 5 has a duration of about 2 cm and must be dealt with by differential timing. Thus, Element 5 combined with the last of study element 4 of each cycle will extend the duration of the combination to one which is reasonable for the purposes of rating and timing. Subsequently the time for Element 5 may be extracted during the compilation stage.

Element 6. Again recourse must be had to differential timing, on this occasion combining Elements 6 and 7, because of the extremely short duration of Element 6 (approximately 1 cm). In this case the next element has a duration of about 10 cm, but there is an additional problem to be dealt with because the element will be an average of the variable factor of distance between the work area and the different tins in the row. To eliminate the effect of the variable distance on the extended time (for summary purposes only) the basic times for the second, third, and fourth tins will be averaged, then subtracted from the combined Elements 6 and 7 to leave the basic time for Element 6.

Element 7. In this case there is no difficulty in rating and timing each element individually because of the relatively long duration (approximately 11 cm). The summary of the times will average the variable factor of distance of each tin in relation to the working area. When summarizing, eleven of the elements of each cycle will be used. This includes the second, third, and fourth tins, which were used in the summary for determination of the basic time for Element 6.

Element 8. This is an extremely short element, having a duration of approximately 1 cm, but as twelve elements are repeated consecutively with a total duration of about 19 cm the study element consisting of twelve of these elements will be used. The elements follow in a rhythmic manner, so no difficulty should be experienced in rating and timing.

Element 9. This element is dealt with in exactly the same way as Element 2.

9.51 DEFINITION OF ELEMENTS

The following description refers to Figure 9.10. Each different element in the proposed study must be clearly defined so as to leave no doubt about the agreed method of working. The detailed element-description must be completed immediately before the study so that the time-study practitioner has a clear and fresh picture in his mind of the method of working, to facilitate rating for effectiveness.

Each element must be qualified with a clearly defined break-point to facilitate timing and the subsequent establishment of data for synthesis. If the resulting basic time is within acceptable limits of accuracy and confidence (*see* Section 9.71) this detailed element-description, with the basic time, should be recorded for future use as elemental data.

9.6 AT THE WORKPLACE

Armed with his prepared Observation Record Sheet the practitioner makes his way to the department. After establishing communication with the supervisor (with whom the job should have already been discussed during element breakdown) the supervisor conducts the practitioner to the workplace and introduces him to the operatives. The practitioner stations himself so that he can see the operation, but so that he does not distract the operative. It is necessary for him to be in the view of the worker because an invisible pair of eyes 'boring into one's neck' is as distracting as a visible onlooker.

WORK MEASUREMENT – Method sheet		Sheet No. 1 of 3	Ref. P.P.E		
Fill 0.285 litre tins with processed peas, and pack into cartons (12 tins per carton)		Classification	Date 10 July		
Element Number	Ref.	ELEMENT DESCRIPTION AND BREAK POINT	Origin	% confidence limits	Basic time
1		Simultaneously with both hands reach for and grasp opposite sides of full carton and lift clear of bench – turn 180° – place full carton aside to storage bench and release in position. With both hands simultaneously reach for and grasp opposite sides of next carton of empties (distance approximately 600mm) lift clear of bench – turn 180° – place carton in position at left-hand of work bench and release (distance between benches = 760mm). BREAKPOINT :– As both hands release carton to bench.			
2.		Simultaneously with both hands reach for next 2 tins in carton and grasp with thumbs and fingers the top rims of the next 2 tins, one in each hand. Simultaneously lift 2 tins vertically clear of carton. Turn body and move as necessary to place tins in position in single row of 12 tins along front edge of bench (approximately 40mm apart). Release tins to bench. BREAKPOINT :– As hands release 2 tins to bench.			
3.		With right hand reach for and pick up lid-opener from right-hand side of bench, and move to position to insert under rim of next tin – left hand moves simultaneously to grasp tin. BREAKPOINT :– As lid-opener is in position to insert under rim of lid.			

Figure 9.11 Full description of elements and break-points

WORK MEASUREMENT – Method sheet

Sheet No. 2 of 3
Ref. P.P.E
Classification
Date 10 July

Element Number	Ref.	ELEMENT DESCRIPTION AND BREAK POINT	Origin	% confidence limits	Basic time
4.		With fingers of left hand reach for and encircle top of next tin, with lid-opener continuously held in right hand. Simultaneously insert end of lid-opener under rim of lid. Firmly grasp tin with left hand, and lever off lid of tin with lid-opener. Release tin from left hand and reach for and grasp loosened lid with thumb and first finger of left hand. Lift lid clear and place aside, top upward, on bench immediately behind tin. Release lid to bench and reposition both hands for next tin, simultaneously moving to next tin as necessary. BREAKPOINT :– As lid-opener is in position for next tin.			
5.		Take one step and place lid-opener aside at right-hand side of bench; release lid-opener to bench. BREAKPOINT :– As right hand releases lid-opener to bench.			
6.		With right hand reach for scoop in pea stock and grasp scoop by handle. Simultaneously with left hand reach for next tin. BREAKPOINT :– As scoop is in position in right hand.			
7.		With scoop continuously held in right hand, with left hand reach to next empty tin in row of 12 and grasp with thumb and first finger round side of tin, lift clear and move to position over pea stock. Simultaneously with right hand take first scoop of peas and with tin in position empty scoopful of peas into tin. With three further scoopsful complete the filling of the tin to within 5mm of the top. With left hand replace filled tin to bench in original position, and release. BREAKPOINT :– As left hand releases filled tin to bench.			

WORK MEASUREMENT – Method sheet

		Sheet No. 3 of 3	Ref. P.P.E.		
		Classification	Date 10 July		
Element Number	Ref.	ELEMENT DESCRIPTION AND BREAK POINT	Origin	% confidence limits	Basic time
8.		With left hand pick up next lid from bench and place on top of tin. Release lid in position on top of tin and with left hand reach for next tin lid on bench. Simultaneously with palm of right hand move to position on tin lid and press down to secure. Release pressure from lid with right hand and move to next lid. BREAKPOINT :– As right hand releases pressure from tin lid			
9.		Simultaneously with both hands reach for and grasp tops of 2 tins (adjacent tins), lift clear and move to place in position in carton, in 2 layers of 6 tins. Release tins in position. BREAKPOINT :– As hands release tins in position			

WORK MEASUREMENT   OBSERVATION RECORD (TIME STUDY)

SHEET REF. 1 of 2
STUDY NO. P.P.E.
DATE 10 July

ELEMENT	RATING	OBSERVED TIME	INEFF. OR CHECK TIME	BASIC TIME	ELEMENT	RATING	OBSERVED TIME	INEFF. OR CHECK TIME	BASIC TIME
Brought forward					Brought forward		213	1·37	227·7
Check time			1·37						
Change cartons	1	100	8	8·0		1	105	7	7·4
Lay out 12 tins	2		18	18·0		2	100	18	18·0
P.U. open and remove 3 lids	3+4		9	9·0		3+4	95	10	9·5
Remove 3 lids	4	110	5	5·5		4	105	7	7·4
Remove 3 lids	4		6	6·6		4		7	7·4
Remove and put aside opener	4+5	95	8	7·6		4+5	120	8	9·6
P.U. scoop fill 1st tin	6+7	110	11	12·1		6+7	105	12	12·6
Fill 2nd tin	7		10	11·0		7	110	10	11·0
Fill 3rd tin X	7		10	11·0	X	7		11	12·1
Fill 4th tin	7		9	9·9		7	100	12	12·0
Fill 5th tin	7		9	9·9		7		11	11·0
Fill 6th tin	7		10	11·0	Use handkerchief	7		11	29	11·0
Fill 7th tin	7		11	12·1		7		12	12·0
Fill 8th tin	7		10	11·0		7	110	10	11·0
Fill 9th tin	7		10	11·0		7		12	13·2
Fill 10th tin	7		10	11·0		7		13	14·3
Fill 11th tin	7		11	12·1		7	90	14	12·6
Fill 12th tin	7		12	13·2		7	105	13	13·7
Replace 12 lids	8	115	17	19·6		8	115	16	18·4
Replace 12 tins	9	95	19	18·1		9	100	17	17·0
WM 3 Totals		213	1·37	227·7	Totals		444	1·66	468·9

Figure 9.12 A time-study observation record sheet, completed

TIME STUDY

WORK MEASUREMENT

OBSERVATION RECORD (TIME STUDY)

SHEET REF.	2 of 2
STUDY NO.	P.P.E.
DATE	July 19XX

ELEMENT	RATING	OBSERVED TIME	INEFF. OR CHECK TIME	BASIC TIME	ELEMENT	RATING	OBSERVED TIME	INEFF. OR CHECK TIME	BASIC TIME
Brought Forward		444	166	468·9	Brought Forward		675	166	711·8
1	100	7		7·0	1	105	7		7·4
2	90	20		18·0	2		17		17·9
3 + 4	95	9		8·6	3 + 4	110	7		7·7
4	105	6		6·3	4		6		6·6
4	105	7		7·5	4		6		6·6
4 + 5	105	8		8·4	4 + 5		7		7·7
6 + 7	100	12		12·0	6 + 7	110	11		12·1
⎧ 7	110	10		11·0	⎧ 7		10		11·0
X ⎨ 7	110	11		12·1	X ⎨ 7		10		11·0
⎩ 7		10		11·0	⎩ 7		10		11·0
7		10		11·0	7		10		11·0
7		10		11·0	7		10		11·0
7		11		12·1	7		11		12·1
7		12		13·2	7		11		12·1
7		12		13·2	7		11		12·1
7		11		12·1	Talk to supervisor 7		12	50	13·2
7		13		14·3	7	95	12		11·4
7		15		16·5	7		12		11·4
8	115	17		19·6	8	120	17		20·4
9	90	20		18·0	9	85	20		17·0
					Check Time			77	
TOTALS		675		711·8	TOTALS		892	293	942·5

Before he starts timing, the practitioner must be prepared to note on his study sheet any foreign elements or interruptions which may occur during the study, so that all times, whether productive or non-productive, may be recorded for subsequent timing-error check.

On completion of these preliminaries the time study proper may start.

The B.S.I. definition for *check time* is:

The time intervals between the start of a time study and the start of the first element observed (known as T.E.B.S. or time elapsed before starting) and between the finish of the last element observed and the finish of the study (known as T.E.A.F. or time elapsed after finishing).

A time check should be included whenever a time study is made using the techniques of consecutive-element timing or consecutive-element differential timing. When selective-element timing is used a time check is superfluous, but it should clearly be understood that a practitioner should have the ability to read and record time within the prescribed limits (i.e. ± 2 per cent) before dispensing with the time check. Reading and recording the observed time accurately should be regarded as more important than reading and recording ineffective time. Wherever possible, provided that the quality of the study be preserved, a time check should be made for the satisfaction of the practitioner concerned. This may be plotted so that a constant control is always available. The time of starting and of completion of the study should be recorded under the description 'time check' on the Observation Record Sheet (Figure 9.12).

The first check time is obtained in the following way. The practitioner waits for the sweep-hand of the check-time watch or clock to reach the 12 when he flicks his stop-watch to zero and commences timing. The actual clock time is recorded as the Starting Time. The time between this action and the start of the first element timed is the check time (also known as Time Elapsed Before Starting).

The practitioner should be conversant with the study elements and the break-points, making sure that he knows the work content of each as described in the detailed element description, so that he may rate for effectiveness. During the study he should be able to record the figures without needing to look at what he is writing.

Rating should be made just prior to the break-point, and should only be recorded when the value differs from the preceding one. If the performance of an element is beyond the rating range of an individual time-study practitioner, or he cannot rate with confidence, then a dash should be recorded in the column to denote this. An observed time, which is not qualified either by a value from the rating scale or a dash, is considered to carry the figure of the last recorded rating, and should be extended and summarized accordingly.

Rating, timing, and recording continues until the requisite number of cycles

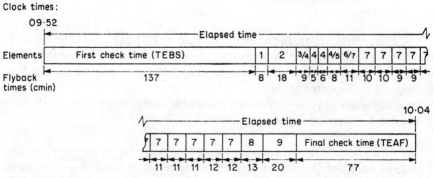

Figure 9.13 The time study in pictorial form (values are from the observation sheet of Figure 9.12)

have been observed. Finally the second check time is taken. On completion of the very last element timed, the observer allows the stop-watch to continue timing. As soon as the sweep-hand on the check-time clock indicates '12' both the stop-watch reading and the actual clock time are recorded; the former as the final check time (otherwise Time Elapsed After Finishing—T.E.A.F.), and the latter as the Finishing Time. The starting, finishing, and check times are used later in the calculation of the timing error where this is necessary.

All the data have now been collected for the completion of the standard time.

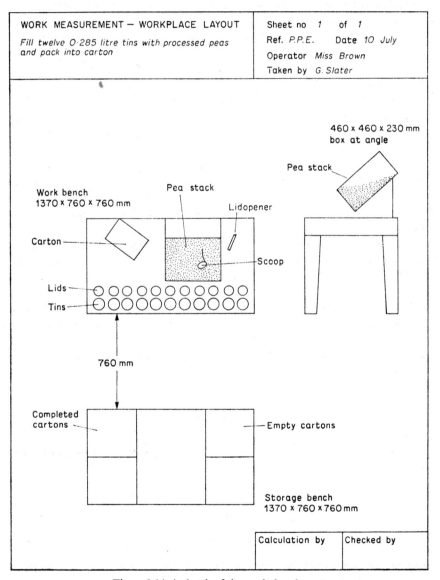

Figure 9.14 A sketch of the workplace layout

The practitioner's last act at the workplace is to acknowledge the co-operation of the operatives and supervisor.

To complete data-recording the practitioner should make a sketch of the workplace layout. This sketch (Figure 9.14) should be complementary to the detailed element description, and should show all the relevant dimensions.

Before starting any subsequent time studies in connection with the job, reference should always be made to the original sketch of the layout to ensure

WORK MEASUREMENT – STUDY CALCULATION SHEET	Sheet No 1 of 1
Fill twelve 0·285 litre tins with processed peas and pack into carton	Ref. P.P.E. Date 10 July Operator Miss Brown Taken by G. Slater

Element numbers

1	2	3+4	4	4+5	6+7	7x	7	7	8	9
8·0	18·0	9·0	5·5	7·6	12·1	11·0	11·0	11·0	19·6	17·0
7·4	18·0	9·5	6·6	9·6	12·6	11·0	11·0	12·1	18·4	18·1
7·0	18·0	8·6	7·4	8·4	12·0	9·9	9·9	11·0	19·6	18·0
7·4	17·9	7·7	7·4	7·7	12·1	11·0	9·9	11·0	20·4	17·0
4⌈29·8	4⌈71·9	4⌈34·8	6·3	4⌈33·3	4⌈48·8	12·1	11·0	11·0	4⌈78·0	4⌈70·1
7·5	18·0	8·7	7·5	8·3	12·2	12·0	12·1	12·1	19·5	17·5
			6·6			11·0	11·0	13·2		
			6·6			12·1	11·0	13·2		
			8⌈53·9			11·0	11·0	12·1		
			6·7			11·0	12·1	14·3		
						11·0	13·2	16·5		
						11·0	11·0	11·0		
						12⌈134·1	12·1	11·0		
						11·1	12·0	11·0		
							11·0	11·0		
							11·0	11·0		
							12·0	12·1		
							11·0	12·1		
							13·2	12·1		
							14·3	13·2		
							12·6	11·4		
							13·7	11·4		
							257·1	264·8		

Element 1 = No.1 = 7·5 = 7·5
Element 2 = No.2 ÷ 6 = 18·0 ÷ 6 = 3·0
Element 3 = No.3+4 − No.4 = 8·7 − 6·7 = 2·0
Element 4 = No.4 ÷ 3 = 6·7 ÷ 3 = 2·2
Element 5 = No. 4+5 − No.4 = 8·3 − 6·7 = 1·6
Element 6 = No.6 + 7 − No.7x = 12·2 − 11·1 = 1·1
Element 7 = No.7 = 11·9 = 11·9
Element 8 = No.8 ÷ 12 = 19·5 ÷ 12 = 1·6
Element 9 = No.9 ÷ 6 = 17·5 ÷ 6 = 2·9

Total = 521·9
521·9 ÷ 44 = 11·9

Timing error

Elapsed time = 12·00 minutes
Observed time = 8·92 minutes
Ineffective time = 2·93 minutes
Residual elapsed time =
12·00 − 2·93 = 9·07 minutes
Timing error
9·07 − 8·92 = 0·15 minutes
∴ $\frac{0·15}{9·07}$ × 100 = 1·7 %

Calculation by	Checked by
G.S.	A.K.

Figure 9.15 Analysis of element times

that the conditions have not changed. This is, of course, when both studies are to be used in arriving at the standard time for the same job, and is not part of a study for a new method or layout.

WORK MEASUREMENT – Work content sheet

Sheet No. 1 of 1 Ref. P.P.E.

Element number	Ref.	ELEMENT	Basic time minutes	\multicolumn{5}{c	}{RELAXATION ALLOWANCE %}	Total % R.A.	Work content of element	Frequency per cycle	Work cycle content
				A	B	C	D	E	F	G	H				
1.		Pick up carton of full tins and place aside to storage. Pick up carton of empties and position on work bench.	0.075	7½	1	0	0	4	0	0	0	12½	0.084	1/1	0.084
2.		Remove 2 tins from carton, and place in row of 12 along front of bench.	0.030	6	1	0	0	4	0	0	0	11	0.033	6/1	0.198
3.		Pick up lid-opener, and position to tin	0.020	6	1	0	0	4	0	0	0	11	0.022	1/1	0.022
4.		Release lid from one tin, and place aside to bench.	0.022	6	1	0	0	4	0	0	0	11	0.024	12/1	0.288
5.		Aside lid-opener to bench.	0.016	6	1	0	0	4	0	0	0	11	0.018	1/1	0.018
6.		Pick up scoop from pea stock and reach for next tin.	0.011	6	1	0	0	4	0	0	0	11	0.012	1/1	0.012
7.		Pick up from bench and fill with four scoops of peas. Replace full tin to bench.	0.119	6	1	0	0	4	0	0	0	11	0.131	12/1	1.572
8.		Replace and secure one lid to tin	0.016	6	1	0	0	4	0	0	0	11	0.018	12/1	0.216
9.		Replace in position 2 tins in carton	0.029	6	1	0	0	4	0	0	0	11	0.032	6/1	0.192
													TOTAL MINUTES		2.602
										Less 6% for tea breaks					2.446

Derived by | Check by E.J. | UNIT OF PRODUCTION per carton of 12 tins

Figure 9.16 Time-study calculation sheet for relaxation allowances

9.61 RELAXATION ALLOWANCE ASSESSMENT

Immediately on completion of the time study, reference should be made to the Study Summary Sheet and a check made of the lists of materials and equipment used during the study.

The Guide to Relaxation Allowances (Table 12.1) should always accompany the practitioner when taking a time study; this piece of equipment is just as important as the watch. The relaxation-allowance assessment should be made before leaving the scene of the study.

Figure 9.16 shows the R.A. Assessment for this study.

9.7 THE STUDY EXTENSION

When a time check has been made, the first operation before extending the observed times is to calculate the timing error. This should be calculated on the observed times only and should not include ineffective time. Reference is made to Figure 9.15, which shows how the error is determined, and to Figure 9.13, which illustrates the time study in pictorial form (values are taken from the Observation Record Sheet of Figure 9.12).

It will be seen that the total observed time, including effective, ineffective, and check times, should be equal to the total elapsed time as registered by the check-time clock; any inequality indicating the presence of error. Several formulae are in use to calculate either accuracy or error. The following formula determines the error:

$$\text{timing error} = \frac{\text{(elapsed time less ineffective and check times)} - \text{(observed times)}}{\text{(elapsed time less ineffective and check times)}}$$

In the example of Figure 9.13 the timing error as a percentage is:

$$\frac{(12 \cdot 00 - 2 \cdot 93) - 8 \cdot 92}{(12 \cdot 00 - 2 \cdot 93)} \times 100 = \frac{9 \cdot 07 - 8 \cdot 92}{9 \cdot 07} \times 100$$
$$= 1 \cdot 7\%$$

Studies having a timing error greater than ± 2 per cent are not considered valid and should not be used.

Each observed time qualified by a rating should be extended to three significant figures and recorded in the Basic Time column in centiminutes. Each of the columns, i.e. observed time, ineffective or check time, and basic time, should be totalled, and the total recorded at the bottom of each column. This sum should be carried forward to the head of the next appropriate column for inclusion in that total. The last column in the study should have the grand total for each column, from which can be calculated the timing error and average rating for the study.

The numbers of each study element are listed across the top of the Study Calculation Sheet (Figure 9.15). The basic time for each study element is listed under the appropriate study-element number. Observed time not rated should

WORK MEASUREMENT—SUMMARY SHEET		Sheet 1 of 1	Reference P.P.E.
Operation or job Fill twelve 0.285 litre tins with processed peas and pack into cartons		**Manning** 1 female	**Standard time** Work content 2.446 minutes Unoccupied time Nil minutes
Materials (a) Expendable 0.285 litre tins to code 1835 (85 o.d. x 105 high) Cardboard cartons code E.19 Cardboard box 460 x 460 x 230 (b) Worked on Processed peas to code 'B'		**Machines** Nil	Contingency allowance Nil minutes Interference allowance Nil minutes Standard time 2.446 minutes **Allowed time** 3.261 minutes
			Issued allowed time minutes for work minutes for delay
Tools and equipment Lid opener lever type 100 x 19 x 3 Aluminium scoop 76 x 13 Work bench 1370 x 760 x 760 high Storage bench 1370 x 760 x 760 high		**Access** Easy ✓ No restriction at workplace Normal Difficult	**Per unit of production** 1 carton of 12 tins
Protective clothing/ equipment Company issue of white cap and overalls		**Safety arrangements** As laid down in department training instructions	**Governing time** —
Range of working conditions Lighting good no glare Dry - carpeted floor Temperature -18°C		**Relaxation allowance special conditions** Nil	**Plant** 'A' **Department** Packing
Quality standards In accordance with quality control criteria published May 1967			**Branch** —
Variable factors affecting data Nil			**Location** Ground floor
Note: all lengths are in millimetre.			**Operating instructions reference**
Unoccupied time reference —	**Contingency allowance reference** —	**Interference allowance reference** —	**Date** 10 July **Date** 11 July
			Derivation by G.S. **Checked by** A.K.

Figure 9.17 The time-study summary sheet of Figure 9.9 now completed

also be included in the list, but ringed. The total basic time for each study-element is calculated, excluding any of the ringed unrated times. The total is then divided by the number of basic times used in arriving at the total. The reason for listing unrated but ringed observed times is to give a pictorial indication of the quality of the study. Each of the study elements is then dealt with in accordance with the study plan to deduce the basic time for each different element originally selected in the element breakdown.

The abbreviated element-descriptions are listed and numbered as on the Work Content Sheet, and against each is entered the basic time derived, as shown in Figure 9.17.

The relaxation allowance is now agreed. Any significant differences between R.A. assessments made by the time-study officers concerned in taking the studies should be reconciled by discussions, and finally confirmed. Because no tables of R.A. are in any sense accurate it pays for individual practitioners to be consistent in the allowances they allocate. The agreed R.A. for each of the factors A to H is then entered for each element. The calculation of the Work Content of Element column is made by the addition of the total percentage R.A. to the basic minutes for each element.

The frequency with which each element occurs in a job can often be calculated, but sometimes a study has to be made to establish this frequency. This can take the form of a production study, or an activity-sampling study. The work content of the element is then multiplied by the frequency of occurrence to obtain the work content per element.

The introduction of fixed rest-periods or tea-breaks goes a long way to reducing fatigue and should, therefore, be considered as part of the relaxation allowance and deducted from the total R.A. agreed. Generally these rest periods are in mid-morning and mid-afternoon. In this example a 15-min break in the morning and 15 min in the afternoon are allowed. This 30-min allowance is equivalent to 6 per cent, which is deducted from the total work-content minutes. Where applicable, to the work content is added the unoccupied time, contingency, and/or interference allowances to obtain the *standard time*. In this example there are no additional allowances, so the total work content is equal to the standard time.

9.71 ASSESSMENT OF ACCURACY

Where three or more practitioners have studied a job it is possible to assess the error to which the mean standard time obtained by averaging their results is subject (see Chapter 6 for a description of the statistical theory). Using a form such as that shown in Figure 9.18, the individual basic time from each study is listed in column (a) against the appropriate practitioner's initials. Calculation is facilitated by converting minutes into milliminutes, thus providing whole numbers from which to evaluate the square roots. Column (a) is then totalled, and the mean of all the studies established as (b). The next column, 'Deviation from Mean Squared' is completed by squaring the deviation from the mean of each practitioner's basic time. This column is then summed to give the variance (c). The

TIME STUDY

standard deviation (d) is found by dividing (c) by the number of time-study officers taking part less one, and then taking the square root. The standard error (e) is calculated by dividing (d) by the square root of the number of practitioners participating. Line (f) is calculated by multiplying the result (e) by the value of X obtained from the small table contained in the calculation sheet.

ANALYSIS OF TIME STUDIES FOR % ERROR OF MEAN

This Analysis should not be used for studies of less than three W.S.O.s

Element 1. Pick up carton of full tins and place aside to storage. Pick up carton of empties and position on work bench

Sheet.No. 1 of 9
Ref. P.P.E.

W.S.O Initials	(a) Mean of W.S.O. Total Basic Time (Milliminutes)	(a − b) Deviation From Mean	(a − b)² Deviation of Mean Squared	(n) Number of W.S.O.s	Value of X
G.S.	74	1	1	3	4·3
T.C.T.	77	2	4	4	3·2
M.G.L.	79	4	16	5	2·8
A.A.	71	4	16	6	2·6
TOTAL	301		(c) = 37		
MEAN	(b) = 75				

DATE | CALCULATED BY | CHECKED BY

(d) = Standard Deviation = $\sqrt{\frac{c}{n-1}}$ = $\sqrt{\frac{37}{3}}$ = $\sqrt{12\cdot 3}$ = 3·5

(e) = Standard Error = $\frac{d}{\sqrt{n}}$ = $\frac{3\cdot 5}{\sqrt{4}}$ = $\frac{3\cdot 5}{2}$ = 1·75

(f) = (c) with 95% confidence = e × X
1·75 × 3·2 = 5·6 = 0·0056 min

(g) = Error of study with 95% confidence =
$\frac{f}{b}$ × 100 = $\frac{5\cdot 6}{75}$ × 100 = 7½%

Mean Basic Time with 95% confidence
= b ± g %
= 0·075 ± 7½%
= 0·075 ± 0·0056 b.m.

Figure 9.18 Calculation of statistical error when three or more work-study observers participate

Finally the percentage error of the study (at the 95 per cent confidence level) is found by dividing the result of (f) by the mean basic time (b), expressing this as a percentage.

This gives the basic time for the element qualified by the error with 95 per cent confidence. A separate calculation is required for each element in turn. It is important to remember that this analysis should *not* be used for studies involving less than three time-study practitioners' results.

WORK MEASUREMENT – Method sheet			Sheet No. 1 of 1	Ref. P.P.E		
			Classification C	Date 10 July		
Element Number	Ref.	ELEMENT DESCRIPTION AND BREAK POINT		Origin	% confidence limits	Basic time
1	PPE/C	Remove 2 tins from carton and place in row of 12 along front of bench. Simultaneously with both hands reach for next two tins in carton and grasp with thumbs and fingers the top rims of the next two tins, one in each hand. Simultaneously lift two tins vertically clear of carton. Turn the body and move as necessary to place two in position in a single row of 12 tins along front edge of bench (approximately 40mm apart). Release the tins to the bench. BREAKPOINT :– As the hands release two tins to the bench.		Time study	7	0.030

Figure 9.19 A synthetic-data form for Element 1

9.8 USE OF DATA RECORDED

On pages 276 to 286 is a suggested layout for the filing and issue of elemental data.

Elemental data are listed according to their categories, and each category is given a classification letter. This letter codifies the data for subsequent use as synthetics, thereby eliminating the necessity for further element description. Each element is given a number, reference letter, and classification letter. In the present example, since all these examples are derived from the pea-packing exercise, the reference is P.P.E., and each element is classified accordingly; either, in this case, 'C' for cans, or 'B' for tool handling.

Against each element number and reference is recorded a brief element-description, followed by a detailed element-description with break-points.

In the next column, headed 'Origin', is recorded the technique used, in this case 'time study'. The percentage error calculated on the form (Figure 9.15) is recorded in the percentage confidence-limits column. Then the basic time is recorded in the last column. Use of these data at subsequent times for establishing standard times requires only the quotation of the reference and element number for the data on the set-up sheet.

All elemental data published on the form will be accompanied by supporting information giving details of the conditions under which the basic times were obtained, as shown on the Summary Sheet (Figure 9.17).

9.81 APPLYING SYNTHETIC DATA

The following illustrates the use of the data from the foregoing time study as synthetic data. The basic time for packing twenty-four half-pint tins into a carton will be deduced.

The basic time in minutes is required for the packing operation. Of the different elements in the job the only one which is likely to be different is Element 1: Pick up cartons of full tins and place aside to storage. Pick up carton of empties and position to work bench. For other elements there are data available: check-study of Element 1 confirms this. Owing to increased size and weight of the F24 carton used, the time for this element will be significantly different from the time for the existing Element 1, which applies to the E19 carton. A time study is arranged and completed to establish the time for this element only, the results being given on the Work Measurement Method Sheet illustrated (Figure 9.19). Reference to the 'Job Specification' will show the method to be used.

In these examples the job specification for carton E19 would be the element breakdown for the job, with the elements repeated as appropriate according to the frequency-per-cycle columns.

The job specification for the carton F24 would be a little more complex, owing to the fact that it is most economical to deal with twelve tins at a time while filling. The method would comprise the following sequence of data:

C7 × 1, C1 × 6, B1 × 1, C3 × 12, B2 × 1, B3 × 1, C5 × 12, C4 × 12, C1 × 6, C2 × 6, B1 × 1, C3 × 12, B2 × 1, B3 × 1, C5 × 12, C4 × 12, C2 × 6.

10 Rated Activity Sampling

10.1 BACKGROUND

A somewhat controversial technique of work measurement, but one which is rapidly gaining credibility and acceptance, is an extension of time study known generally as rated activity sampling. The technique has several variants which are refinements of the basic system. These include systematic rated activity sampling, and high-frequency work sampling, which incorporates systematic sampling with very short time-intervals of the order of 0.25 min.

In spite of its name, the technique may be regarded as an extension of time study rather than of activity sampling, because it bears a greater resemblance to the former; activity sampling is concerned with the *proportion* of time spent on a particular activity, whereas the objectives of rated activity sampling are identical to those of time study, the common purpose being to determine a *standard time* at a defined level of performance.

Before the 1960s little research into the method had been carried out, but around the middle of the decade several workers in America and the United Kingdom produced papers describing viable methods in this field. As early as 1952, however, tentative experiments were being conducted by Dr Ralph Barnes, one of which is described in his major text on work sampling.* A notable contribution to the literature on the subject was made by the National Coal Board.† A comprehensive bibliography is given at the end of this book.

10.11 COMPARISONS WITH TIME STUDY

The objective of rated activity sampling is to determine a basic time for a task. The B.S.I. definition is: 'A work measurement technique which may take the form of either: (a) an extension of activity sampling which establishes the percentage of time during which activities occur, and develops these by the use of rating the number of units produced, to give a basic time per unit, or (b) an extension of time study which assesses ratings of elements at random or fixed time intervals, to produce a basic time per unit' (Term number 41011).

In contrast with time study, although the elements are defined, no attempt is made to observe the break-points. The criterion which determines the recorded time is the end of the time-interval rather than the break-point between elements. Furthermore, the observation times are pre-recorded on an observation sheet, these times being at random, or regularly spaced, intervals, whereas those ob-

---

* R. M. Barnes, *Work Sampling* (New York: Wiley, 1957).
† A. Flowerdew and P. Malin, 'Systematic Activity Sampling', *Work Study and Management Services*, December 1963, p. 542.

tained during a time study must be recorded from the stop-watch as the elements occur. Because the times are pre-recorded they may be anticipated, but with time study, observation must be made continuously to detect the possibility of delayed break-points.

Some similarities between the two techniques are summarized below.

(i) The techniques have a common objective in establishing basic times for jobs.
(ii) Rating is employed as a means of extending observed times to a common basis for pace of working.
(iii) Both techniques are suitable for short-cycle, or highly repetitive, as well as long-cycle jobs.
(iv) Rated activity sampling is carried out on a sampling basis in the same way as activity sampling, but it must be remembered that the cycles studied by time study are also only a sample of the total population.
(v) The assessment of error may be carried out using similar reasoning to that used in time study. The activity-sampling formula is not applicable because the situation covered by rated activity sampling is no longer a purely binomial situation.

Advantages of the method over conventional time study include:

(i) Observed times are pre-recorded, thus obviating the necessity for writing down the stop-watch readings: the observer merely records the ratings.
(ii) The observer is not required to watch for break-points between elements.
(iii) It is not necessary for him to be ready for unexpected occurrences because he has plenty of time before the occurring of the time-interval to observe these.
(iv) The observer has the whole of the time-interval between observations to observe the job, rate the work, and watch for the unexpected.
(v) The observer can concentrate on looking for correct working method, and for any deviations from this, or foreign elements.
(vi) One observer is able, in suitable circumstances, to study teams of two or more workers.
(vii) On long-cycle jobs the study may be extended over the whole shift or day, using long time-intervals.

10.2 PROCEDURE FOR MAKING A STUDY

Each application is unique, and thus requires individual planning. However, certain common steps may be identified. A study falls roughly into seven parts, the general pattern following that of other work-measurement techniques. It is assumed that the reader is conversant with the principles of time study in order that he may understand the description which follows. The procedure will be illustrated with the aid of two examples, but here will be summarized.

1. Consultation

In common with other techniques already described, the first essential step in any application is that of consultation with all interested parties. This is fully described in Chapter 4.

2. Definition of elements

The second step requires a preliminary study of the work, to obtain a factual, accurate, and detailed job-description. This may be achieved with the assistance of the supervisor, and operatives engaged on the work. The rules for element breakdown as described in Chapter 7 (Section 7.4) are valid for rated-activity sampling. It is essential to specify break-points, even though these will not be used for the purpose of timing: they may be of use for subsequent checking and maintenance of the standard time, or in the event that the need for validation by time study arises.

One of the requirements for a successful study on short-cycle work is the need for the time-interval to be less than the shortest element in the study. To satisfy this requirement it is necessary to make an assessment of the duration of each element, which may be accomplished by timing a few cycles using a stop-watch.

3. Choice of the time-interval

The third step requires two decisions: the choice of sampling method to be used, i.e. random or systematic, and the selection of the time-interval. In the case of random sampling the *average* time-interval is chosen. Using this figure and the number of observations (when this value is known later in the study) the duration of the study may be calculated. The random times may then be arranged to fit into this duration, thus providing the required observations. A worked example of this procedure is given in the second case-study (Section 10.5).

In certain cases, with very short elements it will not be possible to satisfy the requirement regarding the time-interval, in which event either elements must be combined or the study must be extended to ensure adequate coverage of all elements. This and other requirements for the time-interval are described in Section 10.32.

4. Number of observations

For a given accuracy the number of observations required is calculated from the formula derived in Section 10.33. Accuracy, of course, must be paid for in terms of the number of observations made. In practice an acceptable value for error is ±5 per cent, but it must be remembered that this figure is based on the inherent *statistical* error and does not take into account the human errors of, say, rating and incorrect recording. All errors are additive, but this does not necessarily mean that the actual total error for the study will be excessive, because of the occurrence of compensating errors.

In the case of random observations it will be necessary to generate random times in the manner described in Chapter 7, of the required number and extending over an appropriate duration, which will produce the required average time-interval. The random times, in chronological order, must be entered on a suitable observation record sheet, such as that shown in Figure 10.4.

When using systematic sampling it is usual to employ an observation sheet upon which are pre-printed the times of the observations. Several versions of this observation sheet are required, one for each different time-interval. Suitable time-intervals, for example, are: 0·2, 0·25, 0·33, 0·40, 0·50, 1·0, 2·0, 3·0, 5·0 minutes. Blank sheets must be available to cater for time-intervals not covered by the standard sheets.

5. Study check

Immediately prior to the study the observer should check that he is about to observe a representative period of working. His observation record-sheet should carry the times of observation, the worker(s) names, and a coding of the elements to be observed. He should be sure that he can identify the workers and is also conversant with the elements.

6. The study

Only complete cycles should be observed to preserve the statistical accuracy and eliminate bias. This is an important point to observe at the start and finish of the study.

Suppose the observer is using ½-min intervals, and is taking his observation on the minute and on the half-minute (i.e. 100 and 50 respectively on the decimal-minute stop-watch), he should wait until Element 1 commences, and then record the rating the next time the minute-hand of the watch indicates either the 100 or the 50, whichever occurs first. Similarly at the end he should record his last rating for the last element of the final cycle he is to observe.

Clearly absolutely 'snap-rating' is impossible, because an appreciable time is required for the observer to assess the rate of working. The practitioner must be consistent in his method of making this assessment; for instance, he may decide to start rating each time-interval by observing the work for four or five seconds prior to the observation time and using this time to make his assessment, recording the rating exactly on the occurrence of the time prescribed for the observation.

A suitable code should be devised for recording ineffective time, interruptions, and other occurrences which have not been previously noted down as elements.

7. Analysis and extension of the study

The analysis of the study carried out using this technique is similar to that used for time study, and falls conveniently into three parts:

(i) Extension of the study to determine the basic elemental times;
(ii) Application of the frequencies of occurrence;
(iii) Addition of relevant allowances for relaxation and contingencies, together with applicable process and other allowances.

The formula used for extending the study times uses the familiar time-study equation, suitably adjusted to the requirements of rated activity sampling.

These formulae are derived in Section 10.34. Elements are extended separately.

Frequencies of occurrence are applied to the elements to reduce occasional-element times to basic times per cycle. These frequencies are found by the usual methods of activity sampling, production studies, or from past records or diary sheets kept by the operatives themselves. The question of frequencies is discussed in Chapter 7.

Finally allowances are assessed and added to the basic-element times, or to the basic work-content of the job, as appropriate. Allowances are described in full in Chapter 12.

10.3 MATHEMATICAL ASPECTS

In common with other sampling methods of work measurement, rated activity sampling has a statistical basis. This section is devoted to the mathematics of the method.

10.31 SAMPLING METHODS

The usual sampling techniques of activity sampling described in the previous chapter are available to the practitioner of rated activity sampling. The most favoured method, and one used widely, particularly by the work-study sections of local authorities, is systematic sampling.

One drawback to systematic sampling, usually quoted, but often over-emphasized, is the danger of cyclic coincidence. It is agreed that should the time-interval exactly synchronize with the cycle time the result will be biased, because exactly the same part of the cycle will be observed every time. Thus something —length of cycle, or time-interval between observations—must vary randomly.

In practice it is unlikely that there will be perfect synchronization because of the usual occurrences which prevent this, such as variations in rating, foreign elements, fumbles, and other interruptions to the rhythm. However, to ensure that such a situation is completely obviated the following precautions should be taken.

1. The time-interval should be chosen so that it is not a sub-multiple of the cycle-time, nor, where possible, a sub-multiple of any elemental time. The cycle and element times to which reference is made are observed, actual times, and not basic times. Where the rating varies appreciably it will be very difficult to avoid some synchronization, but again, this does not present such a problem as is often alleged.
2. With long-cycle work there should be a relatively large number of observations per element.
3. If all else fails, the time-intervals should be of random durations.

It has been found that systematic sampling provides a much more reliable result than does an equivalent study based on random sampling. Furthermore it is the simpler of the two methods to implement, because the observation times may be memorized, thus avoiding the need to consult the list of times on the observation sheet before each reading is made.

When random times are required in circumstances where several workers are being studied, the times appearing on the observation record-sheet may be regularly spaced, but the order in which the workers are being observed may be randomized to produce a similar effect to random times. This method has the advantages of systematic sampling, but introduces the disadvantage of requiring the practitioner to note the order of observation each time he observes.

10.32 TIME INTERVALS

The choice of the times at which observations will be made will depend on several factors, not the least of which is method of sampling, that is, whether random or systematic sampling is employed. In either case there are certain common rules which must be observed. The determination of the statistical error inherent in the result depends on the fact that no complete element occurs between consecutive observations. This ensures that every element is observed at least once in each cycle. In order to satisfy this requirement it is essential that the time-interval be less than the shortest of the observed times for the elements. This presents difficulties, because even elements of relatively long duration (in short-cycle, repetitive work) of, say, 0·15 min require a time-interval of the order of 6 to 8 seconds, which demands continuous attention to rating and recording, and mitigates some of the advantages of this technique.

In certain cases elements may be combined to effectively increase the elemental times. Otherwise it will be necessary to make a much larger number of observations, which will ensure that even the short elements receive their share of the study.

The minimum time-interval recommended is 0·2 min, but longer intervals are to be preferred. Where element times are of the order of 1 min, studies based on around $\frac{1}{2}$ min or more, avoiding synchronization, are reasonable.

10.33 INHERENT ERROR, AND NUMBER OF OBSERVATIONS

Various formulae for determining the error inherent in the sampling may be derived, each depending on the type of error to be considered.

It is not possible to use the formula deduced for application, to time study in Section 9.5, because this equation relates to the variance of individual recurrences of elements. Unfortunately rated activity sampling assesses *total* and *average* times for each element, without the facility of recourse to individual element-times. Another formula based on the variation of individual elements must be found. In the case of systematic sampling, differences in the durations of individual recurrences of one particular element may be shown to be confined to the ends of elements. The reader is referred to Figure 10.1. The time intervals between observations must be constant by definition, leaving the leading and trailing portions (x and y respectively) which vary from element to element. One of the conditions for this form of sampling already stated in the previous section was that the time-interval must not exceed the duration of the shortest element.

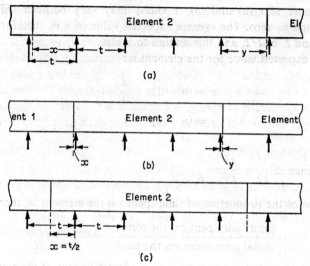

Figure 10.1 Diagrammatic representation of sampling errors

This ensures that no element may occur unobserved between two consecutive observations. This being so, Figure 10.1(a) must depict the maximum case, and Figure 10.1(b) the minimum length of the element for a given number of observations.

Flowerdew and Malin,* writing on the subject for the National Coal Board, have produced an interesting formula for error, based on the above reasoning.

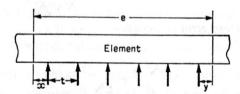

Figure 10.2 Diagram used in the proof of the formula for number of observations

Figure 10.2 shows the essential symbols used in the following proof. From this figure it is evident that there are five time-intervals (t) and six observations (n). Therefore the number of time-intervals is always one less than the number of observations.

Thus $$e = (n - 1)t + x + y$$

where e is the length of the element in centiminute
n is the number of observations on this element
t is the time-interval for the study in centiminutes

* Flowerdew and Malin, op. cit.

From Figure 10.1 (a) and (b), x and y may vary between values almost equal to t, and to zero. The average expected value of x is, therefore, mid-way between 0 and t, or $t/2$, and this applies to y also.

Thus the expected value for the element is:

$$e = (n-1)t + \frac{t}{2} + \frac{t}{2}$$
$$= (n-1)t + t$$
$$= nt$$

The variance $= \dfrac{t^2}{12} + \dfrac{t^2}{12} = \dfrac{t^2}{6}$

The estimate of the proportion of time spent on the element is, therefore,

$$\frac{\text{total time spent on the element}}{\text{total time spent on the study}} = \frac{\Sigma nt}{Nt} = \frac{E}{N}$$

where E is the number of observations on *all* recurrences of the element
N is the total number of observations made during the study

The absolute standard error of this estimate $= \sqrt{(e/Nt)}$, which if it is assumed that successive values of e are independent, since there are

R occurrences of the element,

$$\text{absolute standard error} = \sqrt{\frac{Rt^2}{6N^2t^2}}$$
$$= \frac{1}{N}\sqrt{\frac{R}{6}}$$

The relative standard error is therefore

$$\frac{\frac{1}{N}\sqrt{\frac{R}{6}}}{\frac{E}{N}} = \frac{1}{E}\sqrt{\frac{R}{N}}$$

For 95 per cent confidence, two standard errors must be considered, which, together with the factor of 100 to bring the values to percentages, convert the formulae into:

$$\text{absolute error} = \frac{200}{N}\sqrt{\frac{R}{6}}$$

$$\text{relative error} = \frac{200}{E}\sqrt{\frac{R}{6}}$$

The number of observations may be extracted from the first formula:

$$N = \frac{200}{\text{error}}\sqrt{\frac{R}{6}}$$

10.34 EXTENSION

Extension of individual occurrences of elements as practised in time study, although possible with rated activity sampling, is not practicable owing to the gross errors inherent in the method of sampling: as much as twice the time-interval as shown in the previous section. It is usual, therefore, to extend the *total* time for each element, using the *average* rating for that element.

Now if the rating for a given element j is R_j, and the time-interval is t, and the total number of observations on the particular element is n, then the basic time per cycle is:

$$\text{total time} \times \frac{\text{average rating}}{\text{standard rating}} \div \text{number of cycles}$$

$$= \frac{T \times \dfrac{\Sigma R_j}{n}}{100 \times c}$$

where ΣR_j is the sum of all ratings for the element
c is the number of cycles or occurrences of the element

This formula may be used when the observations are random.
In the event of systematic sampling a much simpler formula is possible:

$$\text{total time} = \text{time interval} \times \text{number of observations}$$
$$= nt$$

$$\text{therefore basic time per cycle} = \frac{nt \times \dfrac{\Sigma R_j}{n}}{100c}$$

$$= \frac{t \Sigma R_j}{100c}$$

It will be noticed that with systematic sampling, the basic time is completely independent of both the number of observations and the overall time of the study.

10.4 A CASE STUDY—SHORT-CYCLE WORK

The practical aspects of rated-activity sampling will now be illustrated with the aid of simulated examples; one for short-cycle and one for long-cycle work. The first example is described in the form of a continuous log. It is emphasized that the log (Table 10.1) is *not part of the study*, but is merely to simulate the actual events as they occur, to afford a means by which the reader may follow the reasoning behind the sampling and, if he so wishes, to take his own validating study using different time-intervals (Section 10.32).

The first example concerns a study made on a repetitive, short-cycle job at the firm of a greetings-card manufacturer. Cards are printed on thin, flat sheets of

TABLE 10.1

A LOG OF THE PERIOD OVER WHICH THE STUDY WAS TAKEN

Basic times have been included to enable the reader to check the accuracy of the technique of rated activity sampling

Elmt	Rating	Ob. time	Cumul. time	Basic time	Elmt	Rating	Ob. time	Cumul. time	Basic time
5	80	0·35	−06·97		1	85	28	97	23·8
1	80	31	07·28	24·8	2	85	21	19·18	17·8
2	90	20	48	18·0	3	85	13	31	11·1
3	80	12	60	10·8	4	85	11	42	9·4
4	90	10	70	9·0	1	85	28	70	23·8
1	80	32	08·02	25·6	2	85	20	90	17·0
2	90	20	22	18·0	3	90	12	20·02	10·8
3	90	13	35	11·7	4	90	10	12	9·0
4	90	11	46	9·9	Ineffective		65	77	65·0
1	85	28	74	23·8	1	85	27	21·04	23·0
2	85	21	95	17·8	2	90	20	24	18·9
3	90	12	09·07	10·8	3	85	12	36	10·2
4	90	10	17	9·0	4	85	12	48	10·2
1	85	29	46	24·6	1	85	29	77	24·6
2	90	19	65	17·1	2	85	21	98	17·8
3	85	13	78	11·1	3	90	12	22·10	10·8
4	85	12	90	10·2	4	95	9	19	8·6
5	100	33	10·23	33·0	5	100	33	52	33·0
1	80	31	54	24·8	1	90	27	79	24·3
2	85	22	76	18·7	2	90	20	99	18·0
3	90	11	87	9·9	3	85	13	23·12	11·1
4	95	9	96	8·5	4	85	11	23	9·4
Ineffective		86	11·82	86·0	1	90	26	49	23·4
1	85	29	12·11	24·6	Ineffective		85	24·34	85·0
2	85	21	32	17·8	2	90	21	55	18·9
3	90	12	44	10·8	3	90	12	67	10·8
4	90	10	54	9·0	4	90	10	77	9·0
1	80	30	84	24·0	1	90	27	25·04	24·3
2	85	21	13·05	17·8	2	90	20	24	18·0
3	90	12	17	10·8	3	90	14	38	12·6
4	85	12	29	10·2	4	90	11	49	9·9
1	85	28	57	23·8	1	85	28	77	23·8
2	90	19	76	17·1	2	85	19	96	16·2
3	90	13	89	11·7	3	90	11	26·07	9·9
4	90	11	14·00	9·9	4	90	10	17	9·0
5	95	35	14·35	33·3	5	100	34	51	34·0
1	85	27	62	23·0	1	85	28	79	23·8
2	85	22	84	18·7	2	90	19	98	17·1
3	90	12	96	10·8	3	95	11	27·09	10·5
4	90	10	15·06	9·0	4	90	10	19	9·0
Ineffective		118	16·24	120·0	1	85	27	46	23·0
1	90	27	51	24·3	2	90	19	65	17·1
2	90	20	71	18·0	3	90	12	77	10·8
3	95	11	82	10·5	4	90	10	87	9·0
4	95	9	91	8·6	1	90	27	28·14	24·3
1	90	28	17·19	25·2	2	90	20	34	18·0
2	95	19	38	18·0	3	95	11	45	10·5
3	80	14	52	11·2	4	90	10	55	9·0
4	80	12	64	9·6	1	90	27	82	24·3
1	90	27	17·91	24·3	2	85	21	29·03	17·8
2	85	21	18·12	17·8	3	85	13	16	11·1
3	80	13	25	10·4	4	90	9	25	8·1
4	85	12	37	10·2	5	100	32	29·57	32·0
5	100	32	69	32·0					

24 cycles, 112 observations

board which are subsequently cut to size. The operation studied is the folding, decorating, and boxing of the '21st Birthday' card, type B21F. The card carries a gold-coloured cardboard key on the front, with an insert of red foil which shows through a cut-out portion of the card.

At the request of management an observer was assigned to establish a standard time for the job as it was then being performed. After due consultation with the supervisor he was introduced to the section and to the operative engaged on the work. The observer watched as many cycles of the job as was necessary to produce a detailed description of the job, from which an element breakdown was prepared. The complete analysis, together with the appropriate break-points, was subsequently entered on the appropriate form (Figure 10.3). The job fell naturally into five elements.

RATED ACTIVITY SAMPLING
ELEMENT DESCRIPTION

Job: Folding, decorating, and boxing *File No.:* EX 327
'21st Birthday' cards, Type B21F

Unit of production: per card.

El. No.   ELEMENT DESCRIPTION

1. Take up card with L.H. and move to work area, simultaneously taking up bone with R.H. Make first fold with both hands, carefully aligning bottom edge to top edge. Make crease with bone held in R.H. Turn card through 90° and make second fold, aligning bottom edge to top edge. Make crease with bone in R.H. Place aside bone.
Break-point: As bone touches bench (audible break-point). Freq: 1/1

2. Take up gold key from tote-bin, picking up adhesive with R.H. Apply spot of adhesive to key back and replace tube of adhesive. Transfer key to R.H. and holding card with L.H. position key to front of card. Press.
Break-point: On release of pressure (visual break-point). Freq: 1/1

3. Lift corner of card with L.H., taking up adhesive with R.H. Squeeze tube and apply a spot of adhesive to each corner of the top sheet of the pile of coloured foils. Replace adhesive, take up foil and position under top fold of card. Place aside adhesive.
Break-point: As tube touches the bench (audible break-point). Freq: 1/1

4. Press card down to stick foil to card. Place aside card to bench.
Break-point: As card touches bench (visual break-point). Freq: 1/1

5. When four cards have accumulated on the bench move four boxes from pile to work area. Insert the four cards one at a time to the boxes. Take up two lids, one in each hand from their respective piles, and place these on boxes. Repeat for the other two boxes.
Collect boxes up into a pile of four and place to pile of finished boxes.
Break-point: As hands release boxes (visual break-point). Freq: 1/4

NOTE: Materials are replenished, and full boxes removed periodically by a separate servicing operative, so these elements are not included.

Figure 10.3 The element-description form for rated activity sampling

RATED ACTIVITY SAMPLING

A few stop-watch readings taken simultaneously with the element recording revealed the shortest element to be of the order of 10 centiminutes. Approximate times for the five elements were 30, 20, 12, 10, and 35 cm respectively. Practically, it is extremely difficult to employ a time-interval shorter than the shortest element in this case, so the practitioner decided that nothing would be lost by combining Elements 3 and 4, thus facilitating the use of 20 centiminutes as a suitable and practical time-interval for this study. (Systematic sampling was to be used.)

RATED ACTIVITY SAMPLING STUDY

Observer: EKJ
Time: to:
from: 10·07·00
Elapsed:

Job: Fold, decorate, and box 21st Birthday, Type B21F

Date of study: 23 May
Time interval: 0·2 min
No. of operators: 1 man

Time	1	2	3	4	5	6	Time	1	2	3	4	5	6	Time	1	2	3	4	5	6
07·00							60							20						
20							80							40						
40							00							60						
60							20							80						
80							40							00						
00							60							20						
20							80							40						
40							00							60						
60							20							80						
80							40							00						
00							60							20						
20							80							40						
40							00							60						
60							20							80						
80							40							00						
00							60							20						
20							80							40						
40							00							60						
60							20							80						
80							40													
00																					
20																					
40																					
60																					
80																					

Figure 10.4 A systematic rated activity sampling observation record sheet

However, for the purpose of demonstrating the effects of such an amalgamation he decided to effect the combination *after* the study had been completed, a move which is quite valid with elements which are consecutive.

For the purposes of this study the observer calculated that 110 observations covering about twenty-four cycles of the work) would suffice.

The observer's next step was to select the appropriate Observation Record Sheet for this study, which was printed with time-intervals of 0·2 min (Figure 10.4). After completing as many of the headings as possible at this stage, he

RATED ACTIVITY SAMPLING STUDY

Observer: E.K.J
Time: to: 11-29-40
from: 11-07-00
Elapsed: 22·40 min

Job: Fold, decorate, and box 21st Birthday, Type B21F

Date of study: 23 May
Time interval: 0·2 min
No. of operators: 1 man

Time	1	2	3	4	5	6	Time	1	2	3	4	5	6	Time	1	2	3	4	5	6
07·00	80						14·60	85						22·20					100	
20	80						80		85					40					100	
40		90					15·00					90		60		90				
60			90				20						✓	80		90				
80	80						40						✓	23·00				85		
08·00	80						60						✓	20				85		
20		90					80						✓	40		90				
40			90				16·00						✓	60						✓
60	85						20						✓	80						✓
80		85					40		90					24·00						✓
09·00			90				60		90					20						✓
20	85						80			95				40		90				
40	85						17·00	90						60			90			
60		90					20		95					80	90					
80			85				40		80					25·00	90					
10·00				100			60			80				20		90				
20				100			80	90						40			90			
40	80						18·00		85					60	85					
60		85					20		80					80		85				
80		90					40				100			26·00			90			
11·00						✓	60				100			20					100	
20						✓	80	85						40					100	
40						✓	19·00		85					60		85				
60						✓	20			85				80		90				
80						✓	40				85			27·00				95		
12·00	85						60	85						20	85					
20		85					80		85					40	85					
40			90				20·00			90				60		90				
60	80						20						✓	80				90		
80	80						40						✓	28·00	90					
13·00		85					60						✓	20		90				
20			85				80	85						40				95		
40	85						21·00	85						60		90				
60		90					20		90					80		90				
80			90				40				85			29·00		85				
14·00				90			60	85						20				90		
20					95		80		85					40					100	
40	85						22·00			90				60						
Total	1070	700	450	350	295	5	Total	1850	1400	970	690	495	14	Total	2820	2110	1425	1045	995	18

Figure 10.5 The observation record sheet of Figure 10.4 completed

RATED SAMPLING ACTIVITY

coded the elements, using columns 1 to 5 to denote the respective elements, and column 6 to record observations made during ineffective time.

Having satisfied himself that the present situation was as near 'normal' as possible the observer had a few words with the operative, recorded his name, and proceeded to observe a few cycles of the job.

As one cycle was nearing its conclusion the observer recorded the time he anticipated the first recording would occur, in this case 10·07 exactly. Consequently he added the '07' to the first '00' on the Observation Record Sheet. In fact Element 5 ended at 0·03 min before 10·07, which gave the observer a couple of seconds before and after 10·07 in which to assess the rating, which was duly recorded for Element 1 (*see* Figure 10·5). The second observation was timed for 0·2 min after 10·07, and this, too, occurred during Element 1. The observer proceeded to rate and record continuously throughout the twenty-two minutes at 0·2-minute intervals. The final Observation Record Sheet for the study is depicted in Figure 10.5.

After duly thanking the operative and taking his leave of the supervisor the observer carried out his analysis of the study.

10.41 ANALYSIS

The study was extended using the formula from Section 10.34 on each element separately. The elements numbered 3 and 4 were combined, as explained earlier, to provide a greater accuracy, and justify the use of a time-interval of 20 cm. The formula used was:

$$\text{basic time} = \frac{\Sigma R.t}{100c}$$

the time-interval, t, in this case being 0·2 min and the number of cycles, c, being 24.

The full analysis for the study is given in Figure 10.6.

To each element's basic time the appropriate relation allowance was added, in this case $12\frac{1}{2}$ per cent, which was assessed as explained in Chapter 12.

10.42 ASSESSMENT OF ACCURACY

In practice, of course, it would not be possible to test the accuracy of such a study unless a check were made using some other 'perfect' means of alternative measurement which could be used as a yardstick against which results could be compared. In the case of this simulated exercise it is possible to carry out this check using the basic times calculated from the observed times and ratings. These basic times are shown in the last column in Table 10.1.

Some very interesting results accrue from the analysis (*see* Table 10.2). It will be seen that Elements 3 and 4, when separated, are relatively inaccurate; errors being 9·2 and 6·5 per cent respectively. It may be shown that a difference of just one observation made 0·01 min later in the very first cycle would have been sufficient to alter the errors from +9·2 to a mere +1·8 per cent for Element

STUDY ANALYSIS SHEET

1. Job Description: **Job No :** 3271   **File No :** Ex 327
Make and Pack '21st Birthday' cards, Type B21F

Observer(s) 1. E K Johnson   Operative(s) 1. E P Hulbert   6.
                2.   2.   7.
                3.   3.   8.
                                                      4.   9.
                                                      5.   10.

2. Study Error Check

         Time to: 11-29·40   Time interval: 0·20
            From: 11-07·00   No. of observations: 112
          Elapsed: 22·40 minutes   22·40 minutes

                              Difference : 0 minutes
                              % difference : 0 %

                      Number of cycles studied = 24

3. Extension

     Element   1. $\frac{2820 \times 0·2}{24 \times 100}$ = 0·235   bm per   cycle

                 2. $\frac{2110 \times 0·2}{24 \times 100}$ = 0·175   bm per   cycle

                 3. $\frac{1425 \times 0·2}{24 \times 100}$ = 0·119   bm per   cycle   ⎫

                                                                                                ⎬ combine 0·206 bm

                 4. $\frac{1045 \times 0·2}{24 \times 100}$ = 0·087   bm per   cycle   ⎭

                 5. $\frac{995 \times 0·2}{6 \times 100}$ = 0·332   bm per   4 cycles

                 6. $\overline{\text{INEFFECTIVE}}$ = 18 × 0·2   = 3·6 minutes

                 7.   ———— =   bm per

4. Summary

El. code	basic time (bm)	freq.	basic time (bm)	R.A. %	R.A. min	std. time (sm)
1	0·235	1/1	0·235	12½	0·029	0·264
2	0·175	1/1	0·175	12½	0·022	0·197
3	⎫ 0·206	1/1	0·206	12½	0·026	0·232
4	⎭					
5	0·332	1/4	0·083	12½	0·010	0·093
6						

                                                              std. time = $\overline{0·786}$ sm per card

Figure 10.6 Analysis of the rated activity sampling study

TABLE 10.2
A COMPARISON OF THE RESULTS OF THE STUDY

Element:	1	2	3	4	5	Ineffective
Study result:	0·235	0·175	0·119	0·087	0·083	3·6 min
Actual result:	0·241	0·178	0·109	0·093	0·082	3·55 min
Difference:	−0·006	−0·003	+0·010	−0·006	+0·001	+0·05 min
% difference:	−2·5	−1·7	+9·2	−6·5	+1·2	+1·4

Combined 3 + 4:
  Study result: 0·206
  Actual result: 0·202
  Difference: +0·004
  % difference: +2·0%

Over-all study result:
  Study result: 0·699 b.m.
  Actual result: 0·703 b.m.
  Difference: 0·004 b.m.
  % difference: −0·6%

3, and from −6·5 to just +2·2 per cent for Element 4: a great improvement. The existing gross errors are due to statistical error, or in lay language, 'the luck of the draw'. However, the requirement to justify the use of a time-interval of 0·2 min was not satisfied until Elements 3 and 4 were combined. The effect of this combination resulted in an error of just +2·0 per cent.

Although the new results for Elements 1, 2. 3/4, and 5 are −2·5, −1·7, +2·0, and =1·2 per cent respectively, the *overall* result for the complete study must be more accurate than any of these, owing to compensating errors. The reason for the inevitability of compensation is given in the last paragraph.

The final study-error is seen to be −0·6 per cent; an extremely accurate value to have been produced by this method of measurement. It is apparent that, provided the principles of rated activity sampling are closely followed, there is no reason why a reliable result should not be obtained from this method of work measurement. It should be remembered that the error shown here is the *statistical* error, and does not take into account rating error, human errors and mistakes, or any other form of error.

10.5 A CASE-STUDY—LONG-CYCLE WORK

The second case-study, which illustrates the use of random sampling and the application to more than one operative, is described here in very much abridged form. The study as a whole would follow the principles laid down in the last section, and in previous chapters of this book.

The two parts which will be described here are: (i) the generation of the required number of random-observation times, and (ii) analysis of the results of the study to produce a basic time per unit of production.

10.51 OUTLINE OF THE JOB

The job which is the subject of this case-study is that of digging a trench 29 inches deep in earth which consists of 15 inches of top-soil and 14 inches of sub-soil. Three men are engaged on the digging, and the requirement is to determine a basic time per unit of production for the work. The study is one of many being taken for the purpose of building up a set of synthetic times for this type of work. Travelling to and from the site, and preparation and clearing up, are dealt with in other studies to be performed by the work-study team involved. The present study is simply concerned with the digging operation.

10.52 UNIT OF PRODUCTION

One important decision to be made is on the unit of production on which to base the basic time. In nearly all work of a so-called 'long'-cycle nature there are repetitive elements, but these are not always suitable as units of production for the purposes of control, nor for synthetic data. With short-cycle work the cycle often provides a natural unit of production. In the present example the closest resemblance to a repetitive cycle is a shovelful of earth. This would not be a practical unit for control purposes because (*a*) it would not be possible to monitor the number of shovels of earth removed, and (*b*) it is extremely difficult to estimate the number of shovels of earth in a given volume for the purpose of setting a standard time for a projected job. In such circumstances the unit of production must be based on an easily measured unit, such as volume of earth to be removed, which is readily obtained from the dimensions of the trench. Such a unit would be 'minutes per cubic metre'.

10.53 RANDOM TIMES

The first consideration when generating random times is that of the number of observations required; in the present example it was estimated to be 360. Three workers were engaged on the task; thus 120 random times would suffice to provide the required number of observations. Further, it was decided to spread the study over a period of two hours. The average time-interval, therefore, was to be:

$$\frac{120 \text{ minutes}}{120 \text{ observations}} = 1 \text{ min per observation}$$

The next problem was to spread the 120 observations over the two-hour period, using the fact that the *average* time-interval was 1 min. This was accomplished as follows.

It is not possible, when generating random times, to regard the distribution of the population of times as continuously variable; the population must be divided into discrete quantities. (Such a situation already pertains, of course, the divisions being seconds and minutes, etc.) The size of the discrete unit must be selected so that the time-interval is of a practical size for activity-sampling purposes, such a value in this case being, say, $\frac{1}{4}$ min. Thus the random observations will be set to the nearest $\frac{1}{4}$ min. Clearly a discrete unit of one second would be far too small, for when using random numbers it is possible by chance to obtain consecutive numbers, which would require the observer to make two separate rating-assessments within as many seconds.

It is necessary to code the times in some way in order to relate them to the random numbers. In the present example the random number '1' was used to represent the time-interval '$\frac{1}{4}$ min', while '2' represented $\frac{1}{2}$ min, 3 represented $\frac{3}{4}$ min, 4, 1 min, and so on.

To achieve the desired result two assumptions were to be made:

1. That the distribution of random numbers around the average required time-interval is rectangular (i.e. each number has an equal chance of being picked).
2. That the distribution extends from zero to a value equal to twice the average number required. For example, if the average time-interval is to be 1 min, it is assumed that the distribution ranges from zero to 2 mins, so that when many values are taken between zero and 2 mins the average should then be 1 min.

In view of the second assumption it will only be necessary to use random numbers 1 to 8 to represent time-intervals of $\frac{1}{4}$ min to 2 min respectively.

One other problem arises from this. Suppose each value does, in fact, occur an equal number of times. Taking the most simple case, assume each time-interval occurs once only, i.e.: $\frac{1}{4}$, $\frac{1}{2}$, $\frac{3}{4}$, $1\frac{3}{4}$, 2, $1\frac{1}{2}$, 1, $1\frac{1}{4}$ min, the total is 9 min, and the average time-interval is $\frac{9}{8} = 1\cdot 125$ min, whereas we need an average of 1·0 min. In order to rectify the situation it is necessary to include the time-interval of zero minutes, in which case the random time-intervals may appear thus: $\frac{1}{4}$, $\frac{1}{2}$, $\frac{3}{4}$, $1\frac{3}{4}$, 0, 2, $1\frac{1}{2}$, 1, and $1\frac{1}{4}$ min; the total is again 9 min, but the average is now $\frac{9}{9} = 1\cdot 0$ min. The implication of this in practice is that the observer must record the rating of the previous time-interval once again when he sees the zero time-interval indicated, as shown in the recording in Figure 10.8.

The next step is to prepare a table of random numbers from which random times may be obtained. From a set of random-number tables 120 random numbers are recorded, using only those which are 8 and below and including the zero. Adjacent to these the corresponding time-intervals are written according to the following coding:

Random number:	0	1	2	3	4	5	6	7	8	
Time-interval:	0	$\frac{1}{4}$	$\frac{1}{2}$	$\frac{3}{4}$	1	$1\frac{1}{4}$	$1\frac{1}{2}$	$1\frac{3}{4}$	2	min

Starting from 0·00, which is the first observation, the time-intervals are added cumulatively to form random observation-times (see Figure 10.7) at which the observer is to make his rating assessments.

The foregoing is the preparatory work which, necessarily, is carried out before the actual study. On arriving at the site of the study the observer may quickly convert the cumulative times to actual times, or alternatively he may use a continuously running stop-watch which he zeros on taking the first reading, thus facilitating the use of the cumulative times as they stand.

Random number	Random interval	Cumulative time	Random number	Random interval	Cumulative time	Random number	Random interval	Cumulative time
		00	5	1¼	42¼	7	1¾	1·31
1	¼	00¼	7	1¾	44	6	1½	1·32½
6	1½	01¾	7	1¾	45¾	5	1¼	1·33¾
2	½	02¼	0	0	45¾	7	1¾	1·35½
7	1¾	04	3	¾	46½	2	½	1·36
4	1	05	8	2	48½	6	1½	1·37½
6	1½	06½	1	¼	48¾	7	1¾	1·39¼
2	½	07	4	1	49¾	3	¾	1·40
7	1¾	08¾	1	¼	50	2	½	1·40½
4	1	09¾	8	2	52	6	1½	1·42
1	¼	10	8	2	54	7	1¾	1·43¾
0	0	10	7	1¾	55¾	0	0	1·43¾
4	1	11	0	0	55¾	8	2	1·45¾
2	½	11½	8	2	57¾	1	¼	1·46
7	1¾	13¼	3	¾	58½	1	¼	1·46¼
3	¾	14	7	1¾	1·00¼	7	1¾	1·48
2	½	14½	8	2	1·02¼	4	1	1·49
7	1¾	16¼	5	1¼	1·03½	8	2	1·51
3	¾	17	2	½	1·04	0	0	1·51
3	¾	17¾	7	1¾	1·05¾	2	½	1·51½
6	1½	19¼	3	¾	1·06½	5	1¼	1·52¾
5	1¼	20½	1	¼	1·06¾	3	¾	1·53½
4	1	21½	2	½	1·07¼	1	¼	1·53¾
0	0	21½	7	1¾	1·09	0	0	1·53¾
6	1½	23	4	1	1·10	2	½	1·54¼
4	1	24	8	2	1·12	8	2	1·56¼
7	1¾	25¾	8	2	1·14	6	1½	1·57¾
5	1¼	27	5	1¼	1·15¼	3	¾	1·58½
1	¼	27¼	4	1	1·16¼	1	¼	1·58¾
2	½	27¾	4	1	1·17¼	6	1½	2·00¼
7	1¾	29½	1	¼	1·17½	1	¼	2·00½
4	1	30½	8	2	1·19½	2	½	2·00¾
3	¾	31¼	4	1	1·20½	2	½	2·01¼
3	¾	32	1	¼	1·20¾	1	¼	2·01½
7	1¾	33¾	3	¾	1·21½	2	½	2·02
5	1¼	35	8	2	1·23½	7	1¾	2·03¾
8	2	37	8	2	1·25½	1	¼	2·04
7	1¾	38¾	2	½	1·26	5	1¼	2·05¼
4	1	39¾	8	2	1·28	1	¼	2·05½
5	1¼	41	2	½	1·28½			
			3	¾	1·29¼			

Figure 10.7 Generation of random times using random numbers

In the present study the observer took his first observation at 9·10 a.m., which caused him to add 9·10 to all his cumulative readings in order to convert them to clock times. This conversion was performed on the Observation Record Sheet itself to save time (see Figure 10.8).

10.54 ANALYSIS OF THE STUDY

The mechanics of the study resembled the previous example in nearly every way. The operatives were rated in turn; the first rating being made at the

Job: *Excavating trenches by hand*
Condition of ground: *damp but not clogged*
Depth of first stratum: *0·4 metres* Type: *top-soil*
Depth of second stratum: *0·35 metres* Type: *sub-soil*
Depth of third stratum: — Type: —
Volume of earth removed: *4·2 cubic metres*
Study begins: *as first worker begins to use his pick*

Cumul time	Clock time	1	2	3	4	Cumul time	Clock time	1	2	3	4	Cumul time	Clock time	1	2	3	4
00·0	9·10	60	U	U		42¼	9·52¼	75	R	90		10·29¼	39¼	75	75	80	
00¼	10¼	65	U	U		44	54	80	R	90		31	41	75	80	80	
01¾	11¾	60	55	80		45¾	55¾	80	60	85		32½	42½	75	85	90	
02¼	12¼	65	65	80		45¾	55¾	80	60	85		33¾	43¾	75	85	90	
04	14	65	60	85		46½	56½	80	65	R		35½	45½	70	85	90	
05	15	65	60	85		48½	58½	85	60	80		36	46	70	85	90	
06½	16½	65	65	90		48¾	58¾	80	60	80		37½	47½	70	80	90	
07	17	75	70	85		49¾	59¾	80	60	85		39¼	49¼	70	80	90	
08¾	18¾	75	70	85		50	10·00	U	U	U		40	50	60	R	60	
09¾	19¾	70	70	80		52	02	U	U	U		40½	50½	R	R	R	
10	20	70	70	85		54	04	60	60	65		42	52	R	R	R	
10	20	70	70	85		55¾	05¾	65	65	80		43¾	53¾	R	R	R	
11	21	75	75	80		55¾	05¾	65	65	80		43¾	53¾	R	R	R	
11½	21½	70	70	80		57¾	05¾	75	60	90		45¾	55¾	R	R	55	
13¼	23¼	70	60	80		58½	08½	70	65	90		46	56	R	60	90	
14	24	75	55	65		1·00¼	10¼	70	65	90		46¼	56¼	R	70	90	
14½	24½	65	R	60		02¼	12¼	65	60	90		48	58	70	85	90	
16¼	26¼	60	R	R		03½	13½	70	60	85		49	59	75	80	90	
17	27	60	65	60		04	14	75	60	85		51	11·01	70	80	90	
17¾	27¾	70	60	70		05¾	15¾	75	60	85		51	01	70	80	90	
19¼	29¼	75	60	80		06½	16½	60	65	90		51½	01½	75	80	90	
20½	30½	70	60	85		06¾	16¾	60	65	R		52¾	02¾	75	80	90	
21½	31½	70	60	85		07¼	17¼	60	60	R		53½	03½	75	85	95	
21½	31½	70	60	85		09	19	R	R	R		53¼	03¼	70	85	95	
23	33	U	65	U		10	20	R	R	R		53¾	03¾	70	85	95	
24	34	U	60	U		12	22	R	R	70		54¼	04¼	85	85	95	
25¾	35¾	60	60	U		14	24	R	70	80		56¼	06¼	80	85	95	
27	37	70	65	60		15¼	25¼	65	75	80		57¾	07¾	80	80	95	
27¼	37¼	75	60	75		16¼	26¼	70	70	85		58½	08½	75	80	90	
27¾	37¾	75	60	75		17¼	27¼	70	70	90		58¾	08¾	75	R	90	
29½	39½	75	65	75		17½	27½	70	75	90		2·00¼	10¼	75	R	90	
30½	40½	60	60	75		19½	29½	75	80	90		00½	10½	70	80	90	
31¼	41¼	R	65	R		20½	30½	70	80	90		00¾	10¾	70	80	U	
32	42	R	50	R		20¾	30¾	75	80	90		01¼	11¼	70	85	U	
33¾	43¾	60	R	80		21½	31½	75	85	95		01½	11½	75	80	90	
35	45	70	60	80		23½	33½	75	80	90		02	12	70	80	90	
37	47	70	60	85		25½	35½	U	R	90		03¾	13¾	70	80	90	
38¾	48¾	75	65	85		26	36	U	R	85		04	14	75	80	90	
39¾	49¾	75	60	85		28	38	70	70	90		05¼	15¼	70	R	85	
41	51	75	50	85		28½	38½	70	75	80		05½	15½	70	R	85	

Figure 10.8 A random rated activity sampling observation record sheet

observation time, with the other two following as soon after as was practicable. The columns for the three workers on the Observation Record Sheet were used for the three types of observation, suitably coded: either the actual rating, or 'resting' (R), or 'unavoidable delay' (U).

The full record is shown in Figure 10.8.

The data required for analysis was extracted from the Observation Record Sheet, and include:

1. The grand total of all ratings for the three operatives, a sum of 22,060;
2. The number of ratings in the study—296;
3. The duration of working for the three operatives.

Other information recorded at the time was:

4. The number of cubic metres of soil removed, and the depth of soil removed;
5. Other dimensions of the trench;
6. Number of square metres of turf removed.

In addition checks were made on:

7. Relaxation taken (to cross-check the relaxation allowance);
8. Unavoidable delays.

The ratings were analysed as follows:

	Remove turf	Excavate trench
Total ratings for Man 1	950	6,075
Total ratings for Man 2	250	6,440
Total ratings for Man 3	—	8,345
Grand totals:	1,200	20,860

The total time taken for rest and unavoidable delays = $69\frac{1}{4}$ min. Study times:

$$\text{duration of study, } 125\frac{1}{2} \times 3 \text{ men} = 376\frac{1}{2} \text{ min}$$
$$\text{rest and unavoidable delays} = 69\frac{1}{4}$$
$$\text{time of actual working: } 307\frac{1}{4}$$
$$\text{time removing turf} = 16\frac{1}{4}$$

therefore, time spent on excating trench = 291 min

$$\text{Basic time} = \frac{T \times \frac{\Sigma R}{n}}{100c} \text{ basic min}$$

$$\text{Removing turf, b.t.} = \frac{16\frac{1}{4} \times \frac{1,200}{16}}{100 \times 5\cdot 2} \text{ b.m.}$$

$$= 2\cdot 35 \text{ b.m. per square metre}$$

$$\text{Excavating, b.t.} = \frac{291 \times \dfrac{20{,}860}{280}}{100 \times 4\cdot 2} \text{ b.m.}$$

$$= \frac{21{,}680}{420}$$

$$= 51\cdot 6 \text{ b.m. per cubic metre}$$

11 Synthesis

11.1 DEFINING SYNTHESIS

Chapter 5 outlined the three basic ways of measuring work, and the three stages of implementing these procedures. The third stage, compilation, is sometimes known as 'synthesis', but as already stated, this term should be reserved for describing the method of applying existing elemental basic times as standard data, as defined by B.S.I. Term number 31002 (Glossary of Terms Used in Work Study):

> A work measurement technique for building up the time for a task or part of a task at a defined rate of working, from previously established elemental times.

The elemental times, referred to in the definition may be obtained from tasks which contain similar elements, and have previously been work measured. The elemental times are known as *synthetic data*, a term which is synonymous with the term *standard data*, one used by some practitioners in the United Kingdom and almost exclusively in the United States. Synthetic data is not defined by B.S.I. but a definition which may be used is:

> Tables and formulae derived from the analysis of accumulated work measurement data, arranged in a form suitable for building up standard time, machine process times, etc., by synthesis.

In many ways synthesis may be regarded as a form of macro-P.M.T.S. because the basic times originate from previous studies employing the basic work-measurement techniques of estimating, timing, and P.M.T.S., and, therefore, are predetermined. However it is more convenient to consider synthetic data as derived or *second-level data*.

There are several advantages in using synthesis which may not be apparent in the basic work-measurement techniques:

1. In many cases basic or standard times synthesized from standard data may be more consistent than those for similar jobs obtained by several observers at different periods of time.
2. Synthetic data can be used in pre-production planning for jobs which at the time do not physically exist and cannot be demonstrated, hence cannot be timed.
3. Jobs which are still in the development stage may have the labour content costed, for the purposes of cost-estimating.

4. Similarly synthetic data may be used for predicting the manning required on projected jobs.
5. Synthesis may be used to set a standard time in jobbing and small batch-work sections, where timing or P.M.T.S. may be impractical or uneconomical.
6. Synthesis is quicker to apply than time study and in many cases some P.M.T. systems.
7. Many of the work-measurement techniques for indirect and non-repetitive work are based on synthesis because of its relative ease of application.

11.11 PROCEDURE

There is nothing new in the concept of standard or synthetic data for work measurement. In 1911 Frederick W. Taylor presented his paper 'Shop Management', to the American Society of Mechanical Engineers, in which he gave details of standard-data formulae for determining standard times for barrowing earth. The formulae included allowances for relaxation.

One general formula for the standard time for shovelling one cubic yard was:

$$T = \left([s + t + (w + w') \text{ (distance covered)}]\frac{27}{L}\right)(1 + p)$$

where s = time for filling the shovel and straightening up ready to throw
t = time for throwing one shovelful
w = time for walking one foot with loaded shovel
w' = time for returning one foot with empty shovel
L = the load of a shovel in cubic feet
p = percentage of the day required for rest and necessary delays.

Simplified formulae are also given, such as that for moving sand: Time (B) per cu.yd for picking, loading, and wheeling sand a given distance when wheeler loads his own barrow:

$$B = 25\cdot86 + 0.071 \text{ (distance hauled in feet)}$$

The general procedure for applying synthesis follows that for work measurement outlined in an earlier chapter. There are certain guide-lines which should be considered when generating synthetic times.

1. The method of working should be agreed and standardized before the measurement starts.
2. The full range of work covered by the standard should be clearly defined.
3. The most suitable method or methods of work measurement should be selected for generating the data.
4. The job must be broken down into elements suitable for subsequent use as synthetic data. It may be necessary to analyse the job or elements further to suit the technique selected (for example P.M.T.S.).
5. Many jobs consist of constant and variable elements. The variable elements must be separated from the constants and the form of variability determined.

Sufficient studies must be made to identify the shape or law of the variable, or at least to cover the range of variability as far as possible. Even where the data are assumed to be linear, more than two studies should be made to verify the linearity.

6. The data should be analysed and presented in the form most convenient for subsequent use. The various methods of presentation are described in Section 11.3.
7. The data, graphs, and formulae should be validated during the reference period, applying the data to jobs for which the standard times are known, or which may be measured by another method such as the production study.
8. Synthetic times must be filed in a way which facilitates data-retrieval, and for this many cross-references may be required.

11.2 COLLECTION AND DOCUMENTATION OF DATA

Synthetic-data banks may be compiled from former work-measurement studies, or from studies specifically designed to generate such data.

All work is considered as a source for the collection of data. The main purpose of analysis is to obtain elements which have the maximum potential use within the field of synthetic data.

All elemental times derived from P.M.T. systems and time study are forwarded to the data section of the Central Work Study Department for inclusion in the synthetic-data files for use in the company. Data should be fully documented on forms such as those depicted in Figures 11.1, 11.2, and 11.3.

Work Measurement—Summary Sheet (Figure 11.1)

This sheet contains all relevant details and conditions of the data under the following headings. Each heading is carefully considered and any relevant facts recorded under it.

(i) operation or job
(ii) manning
(iii) materials—(a) expendable, (b) worked on
(iv) machines
(v) tools and equipment
(vi) access
(vii) protective clothing and equipment
(viii) safety arrangements
(ix) range of working conditions
(x) relaxation-allowance special conditions
(xi) quality standards
(xii) variable factors affecting the data
(xiii) plant, department, and/or branch
(xiv) location

WORK MEASUREMENT—SUMMARY SHEET		Sheet 1 of 1	Reference CC/1
Operation or job	Hand wrap 250g 'sticks' of Canadian Conventional Cheese	Manning 1 Female	**Standard time** Work content _____ minutes Unoccupied time _____ minutes
Materials (a) Expendable	Propafilm 'C' 70/375 cut in sheets 250 × 250mm with one corner cropped 25mm	Machines Nil	Contingency allowance _____ minutes Interference allowance _____ minutes Standard time _____ minutes
(b) Worked on	250g 'sticks' of Canadian Conventional Cheese		**Allowed time** _____ minutes Issued allowed time _____ minutes for work _____ minutes for delay
Tools and equipment	Heat sealing plate to company specification operating at 170° centigrade. Wrapping bench to drawing P23/40	Access Easy No restrictions at work place Normal _____ Difficult _____	Per unit of production 250g 'stick' portion
Protective clothing/ equipment	Company issue of cap and overalls	Safety arrangements As laid down in depot training instructions	Governing time
Range of working conditions	Lighting in accordance with I.S.E code. Dry:- Temprature 18°C	Relaxation allowance special conditions Nil	Plant 'B'
Quality standards	In accordance with quality standard criteria published March 1968		Department Pre-pack cheese Branch ——
Variable factors affecting data	Nil		Location Wrapping room on first froor
			Operating instructions reference
Unoccupied time reference	Contingency allowance reference	Interference allowance reference	Date 3 Sept	Date 5 Sept
			Derivation by A.K.	Checked by G.S.

Figure 11.1 A summary sheet for synthetic data

WORK MEASUREMENT – Method sheet.		Ref.		
		Sheet No. 1 of 1		
		Classification WRAPPING		
		Date 3 September		
		Origin	% confidence limits	Basic time
Element Number	Ref.	ELEMENT DESCRIPTION AND BREAK POINT		
1	CC	Hand wrap 250 g Stick of Canadian Cheese With the left hand reach to stack of film-wrappers on operator's left-hand side, and grasp one sheet of film-wrapper. Position film-wrapper diagonally on wrapping bench with the cropped corner nearest to the heat-sealing plate. Simultaneously with the right hand reach 300 mm to stack of cheese at operator's right-hand side, grasp one piece of cheese, and position cheese on film-wrapper with one long edge lying parallel to the conveyor belt. With the thumb and forefinger of each hand grasp the side corners of the wrapper (i.e. corners "A" and "C" as shown in the sketch on form WM5) and fold over the cheese with corner "C" over corner "A". Smooth down the folds and hold in position with the forefinger of the other hand. With the thumb of the left hand and thumb and first finger of the right hand, turn in the gussett at each side of the corner nearest to the operator (i.e. corner "D" as shown on form WM5) and fold over on top of the cheese. Smooth down the folds and hold folds in position with thumb of right hand. Using thumb and forefinger of left hand and first finger of the right hand turn in the gussett at each side of the corner furthest from the operator (i.e. corner "B" as shown on form WM5) and fold over on top of the cheese. Holding the folds firmly in position with the thumb and first finger of the left hand, turn the cheese over with the right hand, left hand assisting, and slide the cheese across the table to the heat sealing-plate to seal the wrapper. Turn the cheese on to the narrow edge and seal the tail of the wrapper. Move the cheese to the conveyor beyond the heat-sealer, and release. BREAKPOINT: As the right hand releases the wrapped 250 g portion of Canadian Conventional cheese to the conveyor belt.		

Figure 11.2 A synthetic-data element breakdown sheet with break-points

SYNTHESIS

Work Measurement—Method Sheet (Figure 11.2)

This sheet carries the study reference of the element, and for quick recognition a brief element-description followed by a detailed element-description with element break-points. The basic times in minutes and the technique of measurement used are stated, and in the case of time study, the basic time will be qualified by the percentage confidence limits.

The *Work Measurement Study Calculation Sheet* (Figure 11.3) is used for calculating the basic time from the Study Observation Record. In this particular

WORK MEASUREMENT – Study Calculation Sheet	
Wrap 250g 'Stick' of Canadian Conventional Cheese	SHEET NO. 1 OF 4 REF. CC/1 DATE September 19XX OPERATOR K.S. TAKEN BY M.J.G.
	T.M.U.
1 Cheese to film and wrap over 2 sides	65·8
2 Push in front gussetts and fold front	31·8
3 Push in back gussetts – fold over back – Seal (2 seals) and aside to conveyor belt	60·5
	158·1
Basic Time = 0·0791 minutes	
	Calculation By Checked By M.J.G A.K.

Figure 11.3 A synthetic-data study calculation sheet

case the technique employed was MTM, the analysis for which is shown in Figure 11.4. The sheet is used additionally for recording a sketch of the workplace layout. Photographs and film loops, when used, bearing their correct reference number, will be attached to this sheet.

After completion the forms are photocopied and the originals, together with

MTM ANALYSIS

MTM-1 ✓
MTM-2
MTM-3

Job description: Wrap 250g sticks of Canadian Conventional cheese

Sheet 1 of 2
Analyst D. J.
Date 3 September

El.		LH	tmu	RH	
1.	To near film pile	(R-B)	6.9	M4B	Push cheese aside
	½x	(G1B)	2.0	RL1	Let go of wrapped cheese
	Get hold of film.	{G1B	12.9	R12B	To cheese on tray
		{G2	1.0	G1A	½x) cheese on tray
	Film to table	M10B	3.7	G1C1	½x)
	Open fingers	(M1B)	—		
	Let go of film	(RL2)	13.4	M12B	Cheese to film
	Hand away	(R-E)	2.0	RL1	
	To edge of film	(R2B)	6.4	R4B	To edge of film
	Film (thumb + forefinger)	G1A	2.0	G1A	Film (thumb + forefinger)
	Fold over	M7B}	9.7	(M5Bm)	Film to cheese
		G2	—		
	Reposition 1st. finger	(RL2)	2.9	mM2B}	Complete fold to cheese
	and hold film	(R2B)	—	G2	
		M1B	2.9		
	1st. finger	G5	—	RL2	
			65.8		
2.	Thumb only	R2B	4.0	R2B	Thumb + 1st. finger to gusset
		G5	—	G5	Thumb
		M1B	2.9	M1B	Push in gusset
			—	RL2	
			6.4	R4B	To front tail of film
		(RL2)	2.0	G1A	
		(R2B) G5	8.0 —	M5B} G2	Wrap over cheese
	To hold film tight	G2	5.6	(RL2)	
	Push down tail 1st finger	M1B	2.9		
			31.8		

Figure 11.4 The MTM analysis

MTM ANALYSIS

MTM-1 ✓
MTM-2
MTM-3

Job description: 250g "sticks"

Sheet 2 of 2
Analyst
Date

El.		LH	tmu	RH	
3.					Leave thumb to hold film
	To front gusset	R2B	4.0	R2B	To front gusset
		G5	—	—	
	Push in gusset	M1B ⎤	2.9	M1B	Push in gusset
		~~RL2~~ ⎦	—		Hold
		~~R-Bm~~	—		"
	To front tail of film	mR2B	2.7		"
		G1A	2.0		"
		M5C ⎤	9.2	RL2	Finger (thumb holds)
		~~G2~~ ⎦	—		
			4.0	R2B	⎤ Regrasp to turn
			2.0	G1A	⎦
		(T90S	6.9	M4B	Lift up, turn put down
		RL1)	—		
			9.7	M7B	To hot plate
			5.6	Process	
			5.7	M3B	Turn to edge
			5.6	Process	
			60.3		

for the study in the text

any film loops and photographs, will be forwarded to the Data Section of the Department.

All issues of data are fully documented with the relevant information from the forms. The data section of the Central Work Study Department examine, verify, and classify the information, reissuing it in data format to the sections. Copies of data issued are forwarded to a nominated work-study officer in each of the sections. It is usual to restrict the issue, and justification for more copies must be obtained from the data section.

Nominated work-study officers are made responsible for the correct filing of the data and current maintenance of data files. All data sheets replaced or cancelled must be returned to the data section of the department to ensure that only current data are in use.

When completing forms shown in Figures 11.2 and 11.4 it will be necessary to record only the appropriate reference number where the element-description for the study is identical with the element-description of the data. Where the study element-description differs slightly, minor modifications can be made to the data elements by the use of P.M.T.S. or other techniques. These minor modifications are included in the forms (Figures 11.2 and 11.4) with the correct data-reference, and then forwarded to the data section of the Department for further assessment, for inclusion in the data files.

In manual systems data-retrieval presents a problem, and careful consideration must be given to the indexing and coding of data sheets.

11.3 PRESENTATION OF DATA

To be of practical use, synthetic data must be presented in a way which facilitates data-retrieval. Depending on the size of the data-bank, storage may be as simple as a card-index, or so comprehensive as to warrant the use of a computer.

In general terms there are five main methods of presenting synthetic data:

1. In tabular form;
2. In graphical form;
3. As a formula;
4. In nomogram form;
5. As a computer print-out of automatic data-processing.

11.31 TABULATION

The method of setting out data in tabular form is useful where a 'ready reckoner' type of reference is required, but its use may be restricted by its physical size when the more compact formula or nomogram may be preferred. Furthermore it must be confined to data which are in the form of *discrete variables* (i.e. those which occur in clearly defined steps or finite intervals). Such data include twist-drill sizes, number of holes drilled, and so on.

For data which are *continuous* the practitioner must resort to a graphical method such as the nomogram, graph, or formula. If the data are linear, certain sizes or dimensions may be listed in the table and values between these found by interpolation.

Although in two dimensions, the table may be designed so that the captions (column headings) and stub (row headings) are subdivided to allow more than two variables. A page extracted from a file of standard data-sheets is shown in Figure 11.5. It will be seen that this particular table for *turning* accommodates three independent variables of Diameter, Diameter reduction, and length turned, with subdivision of the study values for *diameter* into three *diameter reduction* sizes each.

11.32 GRAPHICAL PRESENTATION

The line graph offers flexibility and simplifies the process of interpolation, especially where the data follow a non-linear law. The graph is a valuable aid for presenting data which are in the form of continuous variables.

The guide-lines in Section 11.11 serve as an indication of the procedure to be followed. When compiling graphs for synthetic data it is not sufficient to make two or three studies on the assumption that the graph will be linear. Drilling ten holes in a piece of metal may take ten times as long as drilling just one hole (excluding loading and unloading to the jig). However, there could be other factors, such as accumulation of swarf, which would extend the time by a greater factor than 10 and cause the graph to be non-linear. Sufficient studies should be taken to confirm the actual slope of the graph when plotted.

A typical linear graph for duplicating sheets on a wax-stencil machine is illustrated in Figure 11.6. It is quite reasonable to suppose that when sheets are duplicated on a machine which runs at a selected constant speed the time for running off will be directly proportional to the number of sheets run off. Nevertheless in this case four studies were made (for 50, 100, 150, and 200 copies) at each set speed to confirm this assumption.

The line graph accommodates one independent variable (number of copies) and one dependent variable (standard time). In this case a graph for two independent variables (number of copies and machine running-speed) were required. This was realized by constructing several lines on the same graph field, each line representing one particular speed. Standard times for other speeds may be obtained by interpolation.

The original studies, of course, produced basic times, but as the relaxation allowance will be constant in fixed conditions there is no reason why the synthetic data should not show standard time to obviate the additional task of applying R.A. on each occasion. This graph (Figure 11.6) in fact shows standard times.

It will be seen that the graph indicates a standard time of 1·62 s.m. even when no copies are produced. This, clearly, is the handling time, which is independent of the number of copies duplicated. One advantage of the graphical method is that it automatically separates constant elements from variable elements if

SCREWCUTTING
STANDARD MINUTES PER THREAD

MATERIAL GROUP I

H.S.S. TOOLS
SOURCE OF DATA: SC5/ /75

Diameter of Thread (mm)	Machine (min per cm)	Manual work (min)	Length of Thread (mm)											
			10	20	30	40	50	60	70	80	90	100	110	120
10	0·30	6·93	7·23	7·53	7·83	8·13	8·43	8·73	9·03	9·33	—	—	—	—
11	0·305	7·92	8·23	8·53	8·84	9·14	9·45	9·75	10·1	10·4	—	—	—	—
13	0·27	7·92	8·19	8·46	8·73	9·00	9·27	9·54	9·81	10·1	—	—	—	—
16	0·31	8·91	9·22	9·53	9·84	10·1	10·5	10·8	11·1	11·4	11·7	—	—	—
19	0·32	8·91	9·23	9·55	9·87	10·2	10·5	10·8	11·1	11·5	11·8	12·1	12·4	12·7
22	0·40	9·90	10·3	10·7	11·1	11·5	11·9	12·3	12·7	13·1	13·5	14·0	14·4	14·7
25	0·42	9·90	10·3	10·7	11·2	11·6	12·0	12·4	12·9	13·3	13·7	14·1	14·6	15·0
31	0·52	10·89	11·4	11·9	12·4	13·0	13·4	14·0	14·5	15·0	15·6	16·1	16·6	17·1
38	0·54	10·89	11·4	12·0	12·5	13·0	13·6	14·1	14·7	15·2	15·7	16·3	16·8	17·4
44	0·62	11·33	12·5	13·1	13·7	14·4	15·0	15·6	16·2	16·8	17·5	18·1	18·7	19·3
50	0·69	11·88	12·6	13·3	13·9	14·6	15·3	16·0	16·7	17·4	18·1	18·8	19·5	20·2
65	0·82	12·87	13·7	14·5	15·3	16·2	17·0	17·8	18·6	19·5	20·3	21·1	21·9	22·8
75	0·92	14·85	15·8	16·7	17·6	18·5	19·5	20·4	21·3	22·2	23·2	24·1	25·0	25·9

SCREWCUTTING: METRIC—EXTERNAL—UNRESTRICTED
TURNING
STANDARD MINUTES PER OPERATION

MATERIAL GROUP I

H.S.S. TOOLS
SOURCE OF DATA: DF2/E/75

From diameter up to (mm)	Diameter reduced by (mm)	M/C (min per cm)	Manual work (min)	Length Turned (mm)											
				5	10	15	20	25	30	35	40	45	50	55	60
200	10	1·11	1·98	2·54	3·09	3·65	4·20	4·76	5·31	5·87	6·42	6·98	7·53	8·09	8·65
	20	1·50	2·97	3·72	4·47	5·22	5·97	6·72	7·47	8·22	8·97	9·72	10·5	11·2	12·0
	30	1·88	3·96	4·90	5·84	6·78	7·72	8·66	9·60	10·5	11·5	12·4	13·4	14·3	15·3
	40	2·26	4·95	6·08	7·21	8·34	9·47	10·6	11·7	12·9	14·0	15·1	16·3	17·4	18·5
225	10	1·52	1·98	2·74	3·50	4·26	5·02	5·78	6·54	7·30	8·06	8·72	9·58	10·3	11·1
	20	2·00	2·97	3·97	4·97	5·97	6·97	7·97	8·97	9·97	11·0	12·0	13·0	14·0	15·0
	30	2·48	3·96	5·20	6·44	7·68	8·92	10·1	11·4	12·6	13·9	15·1	16·4	17·6	18·9
	40	2·96	4·95	6·43	7·91	9·39	10·9	12·4	13·9	15·4	16·8	18·3	19·8	21·3	22·8
250	10	1·70	1·98	2·83	3·68	4·53	5·38	6·23	7·08	7·93	8·78	9·63	10·5	11·3	12·2
	20	2·26	2·97	3·10	5·23	6·36	7·49	8·62	9·75	10·9	12·0	13·1	14·1	15·2	16·2
	30	2·82	3·96	5·37	6·78	8·19	9·60	11·0	12·4	13·8	15·2	16·6	18·1	19·5	20·9
	40	3·38	4·95	6·64	8·33	10·0	11·7	13·4	15·1	16·8	18·5	20·2	21·8	23·5	25·2
350	10	2·10	1·98	3·03	4·08	5·13	6·18	7·23	8·28	9·33	10·4	11·5	12·5	13·6	14·6
	20	2·74	2·97	4·34	5·71	7·08	8·45	9·82	11·2	12·6	13·9	15·3	16·7	18·1	19·4
	30	3·38	3·96	5·65	7·34	9·03	10·7	12·4	14·1	15·8	17·5	19·2	20·9	22·6	24·3
	40	4·02	4·95	6·96	8·97	11·0	13·0	15·0	17·0	19·0	21·1	23·1	25·1	27·1	29·1
425	10	2·50	1·98	3·23	4·48	5·73	6·98	8·23	9·48	10·7	12·0	13·3	14·5	15·8	17·1
	20	3·36	2·97	4·65	6·33	8·01	9·69	11·4	13·0	14·7	16·4	18·1	19·8	21·5	23·2
	30	4·22	3·96	6·07	8·18	10·3	12·4	14·6	16·7	18·7	20·8	23·0	25·1	27·2	29·3
	40	5·08	4·95	7·49	10·0	12·5	15·2	17·7	20·2	22·7	25·3	27·8	30·4	32·9	35·5

Turning dia. Range 165–425 mm; 0–60 mm Pass

Figure 11.5 Examples of tabulation of synthetic data

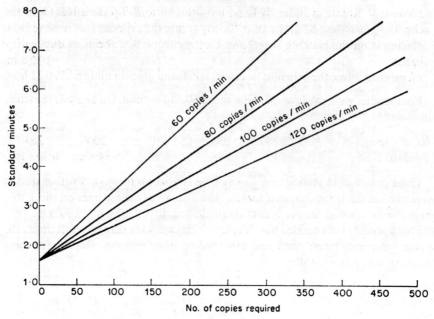

Figure 11.6 An example of graphical presentation of synthetic data

overall cycle-times are plotted. This is particularly useful when the elements are short. The method compares with *differential timing*, described in Chapter 9.

Example

The following is the work specification for duplicating wax-stencil masters.

Element 1. Take stencil to machine. Open clip, offer stencil skin to machine and close clip. 0·12 b.m.

Element 2. Turning the handle with the left hand, carefully wrap the skin around the cylinder as it revolves. Stop when the end clip appears and secure the end of the skin in the clip. Tear off the backing sheet and discard. 0·15 b.m.

Element 3. Rotate the cylinder two or three times to allow the ink to penetrate the impressions on the skin. Move the tray carrying the copy paper up the 'operate' position. 0·15 b.m.

Element 4. Try two or three copies, operating the machine by hand. Inspect for quality. 0·10 b.m.

Element 5. Adjust the settings, paper, etc. as necessary and if necessary. Frequency of occurrence 1/4. (0·40 b.m.) 0·10 b.m.

(This time includes running two or three trial copies and repeating as required.)

Element 6. Select the appropriate speed and switch on the machine. 0·04 b.m.

Element 7. Run off the required number of copies.

(Standard times obtained from graph)

Element 8. Switch off machine. Remove finished copies. Joggle to straighten, and place aside. 0·15 b.m.

Element 9. Rotate cylinder by hand until the bottom clip is visible. Open the clip and grasp the end of the skin. Slowly rotate the cylinder and remove skin, placing it on the backing sheet, and blotting using two sheets of duplicating paper. 0·42 b.m.

Element 10. Place the skin into its cover, label, and place in filing rack. 0·21 b.m.

Four studies were made at machine speed 60 copies/min. On analysis the study times were:

No. of copies	50	100	150	200	250
Standard time	2·50	3·48	4·28	5·34 s.m.	6·30 s.m.

These points were plotted and a line of best fit drawn through. Further studies were carried out using different speeds. The intercept of these lines on the Y axis gives the total value for the constants (including R.A.), which is 1·62 s.m.

Check studies were carried out during normal working using overall times, the manual elements being rated and extended to basic minutes. Results for these check studies were as follows:

Study	1	2	3	4	5	6
No. of copies	80	60	100	100	240	30
Speed	80	80	60	100	100	60
Standard time s.m. (actual)	2·60	2·52	3·37	2·60	4·72	2·00
Standard time s.m. (from graph)	2·70	2·45	3·45	2·70	4·25	2·15

Study	7	8	9	10	11	12
No. of copies	380	50	50	100	150	125
Speed	120	100	100	100	120	60
Standard time s.m. (actual)	6·60	2·05	2·25	2·60	3·04	4·23
Standard time s.m. (from graph)	5·05	2·15	2·15	2·70	2·95	3·90

The check studies confirm the validity of the graph, with the exception of Study No. 7, and to some extent Study No. 5. This may be due to extrapolation beyond 200 copies, the limit of the original set-up studies. Further checks may be made to investigate the linearity beyond 200 copies.

The degree to which individual studies are representative of the situation may be checked using the control chart described in Chapter 7 (Figure 7.18).

11.33 NON-LINEAR RELATIONSHIPS

In many other jobs the data will be non-linear, the variables not being in direct proportion: it should not be assumed that *all* data are linear. The resulting curve may follow one of the recognized equations such as parabola, square law, hyperbola, or exponential. In some cases the shape may not be recognized, but if sufficient studies are taken and plotted a reasonably smooth curve may be superimposed using a flexible ruler. The equation of the curve may be determined or estimated either by trial and error, that is, by plotting several curves to find the one which fits the best, or by using a suitable computer curve-fitting program.

In cases of recognizable curves it is often possible to transform the equation into the form of a straight line, an example of which is given in Section 9.25, dealing with the hyperbolic curve resulting from plotting ratings against observed times.

The example illustrated by Figure 11.7 is for hand-spraying metal panels for cupboards, filing cabinets, bench tops, etc. When the graph was originally compiled it was found that the relationship was not linear. The graph shown, of course, is for a fixed spray-gun setting, predetermined pressure, and fixed-size nozzle. If data for variations of these parameters are required these may be dealt with as described above.

The graph may be used as it stands, but it may be more convenient to trans-

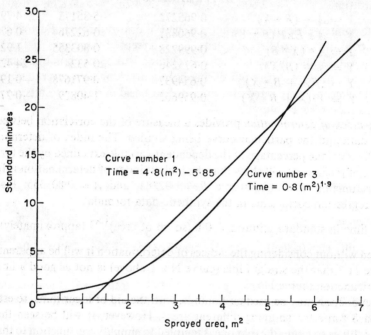

Figure 11.7 An example of a non-linear graph, from the data in the text

form this into a straight-line graph, or into a formula which may represent the situation.

The curve appears to follow a power law, i.e. $Y = aX^b$, even perhaps forming a parabola in the square-law formula $Y = aX^2$. Using a standard curve-fitting package programme the regression coefficients a and b may be determined. A programme of this type usually offers a limited range of standard formulae, such as:

$$Y = a + bX$$
$$Y = ae^{bX}$$
$$Y = aX^b$$
$$Y = a + \frac{b}{X}$$

SYNTHESIS

$$Y = \frac{1}{a + bX}$$

$$Y = \frac{X}{a + bX}$$

When the data are input in the above programme the following output, typical for such a programme, results from the input data:

	CURVE TYPE	INDEX OF DETERMINATION	A	B
1.	$Y = A + (B * X)$	0·965272	−5·85173	4·79394
2.	$Y = A * EXP(B * Y)$	0·960837	0·622764	0·675049
3.	$Y = A * (X \wedge B)$	0·999228	0·803355	1·92282
4.	$Y = A + (B/X)$	0·675246	20·5324	−24·4263
5.	$Y = 1/(A + B * X)$	0·654937	0·971678	−0·190975
6.	$Y = X/(A + B * X)$	0·959662	1·40829	−0·271885

The *index of determination* provides a measure of the correlation between the input data and the particular curve being studied. The index of determination actually gives the percentage of the dependent variable explained by the formula. A scan of the indices will show that the highest index of determination is for the power function $Y = AX^B$ where $B = 1·92282$, and $A = 0·803355$. Inserting these regression coefficients in the synthetic-data formula:

time in standard minutes = 0·8 (sq. m of area)$^{1·92}$ (approximately)

Even without considering the indices of determination it will be apparent from Figure 11.7 that the straight line (curve No. 1 above) is not as good a fit as the power function (curve No. 3).

The function may be used as it stands, with the aid of logarithms, to estimate standard times for spraying different areas. However, it will be seen that the power, 1·9, is so close to 2 that one is tempted to simplify the function to that of a parabola, i.e.

time in standard minutes = 0·8 (sq.m)^2

This approximation produces very large errors in the final standard time; the only sure way is to adhere to the original form.

Thus the time to spray 6 sq.m of area is: $0·8(6)^{1·92}$

i.e. log (time to spray) = log (0·8) + 1·92 log (6)
= $\bar{1}$·9031 + 1·92 × 0·7782
= 1·3972

thus time to spray = 25 s.m. (approximately)

(This may be verified from Figure 11.7)

The output may be arranged to include the computed values of *time* for the input values of square metres.

$Y = A * (X \uparrow B)$ IS A POWER FUNCTION. THE RESULTS OF A LEAST SQUARES FIT OF ITS LINEAR TRANSFORM (SORTED IN ORDER OF ASCENDING VALUES OF X) ARE AS FOLLOWS:

X-ACTUAL	Y-ACTUAL	Y-CALC	DIFFERENCE	PCT DIFFER
1	0·8	0·803355	−3·35501E−03	−0·4
1·5	1·8	1·75186	4·81425E−02	2·7
2	3·1	3·04602	5·39776E−02	1·7
3	6·2	6·64239	−0·44239	−6·6
4·5	14·2	14·4849	−0·284904	−1·9
5	18	17·7378	0·262235	1·4
5·75	24	23·2065	0·793496	3·4
		STD ERROR OF ESTIMATE	0·43691	

From this table an accurate line of best fit may be superimposed on the plotted points (Figure 11.7). Alternatively, where facilities are available an automatic curve-plotter linked to the computer may be used to plot this line.

The statistical analyses have been simplified, and it must be appreciated that other, more subtle methods of eliminating the unwanted variables and optimizing the objective function are usually employed. Simplification is necessary to illustrate the use of M.R.A. in work study, and perhaps stimulate the interest of the practitioner who may wish to pursue the subject in more detail.

11.34 USE OF THE FORMULA

Curves which are continuously smooth without steps or discontinuities may be assigned an accurate or, in some cases, approximate formulae which describe their forms in mathematical terms. One such formula (for the straight line) has already been introduced in Chapter 6. The science of curve fitting is well known to mathematicians and, except for simple examples, is outside the scope of this book.

As explained in Chapter 6, the straight line has the form:

$$y = mx + c,$$

where x is the independent variable

y is the dependent variable (usually time)
m is the slope or gradient of the line
c is the intercept of the line on the y axis.

In the example used for Figure 11.6 a formula may be derived for the machine speed of 60 copies per minute. Any point may be chosen which lies on this line. For example 200 copies gives a time on the vertical axis of 5·28 s.m. Thus when $x = 200$, $y = 5·28$. The value of c is the point where the line crosses the vertical axis and is 1·62. Substituting these values in the formula:

$$5·28 = 200m + 1·62$$

or
$$3·66 = 200m$$

or
$$m = \frac{3·66}{200} = 0·0183 \text{ s.m./copy}$$

Inserting this value back into the formula:

$$y = 0.0183x + 1.62,$$

or better,

$$\text{standard time} = 0.018 \text{ (number of copies)} + 1.62$$

Thus the time for running off 90 copies of a master stencil is:

$$\begin{align}\text{standard time} &= (0.018 \times 90) + 1.62 \\ &= 1.62 + 1.62 \\ &= 3.24 \text{ s.m.}\end{align}$$

This result agrees with the value when read off the graph.

The above formula is for 60 copies/min. The formula may be extended to accommodate all speeds by multiplying the m value by 60 and dividing by S, the speed in copies/min, or:

$$\begin{align}\text{standard time} &= \frac{60}{S} \times 0.018 \text{ (number of copies)} + 1.62 \\ &= 1.08 \frac{N}{S} + 1.62\end{align}$$

where $N =$ number of copies.

Thus the time for 120 copies at a machine speed of 110 copies/min

$$\begin{align}&= \frac{1.08 \times 120}{100} + 1.62 \\ &= 1.178 + 1.62 \\ &= 2.798 \text{ s.m.}\end{align}$$

which may be verified from the graph, using interpolation between the 100 and 120 lines.

It will be seen that the formula covers not only those lines on the graph but also an infinite number of speeds, i.e. continuously variable speeds. This demonstrates the extreme flexibility of the formula over the graph. Provided that the particular formula for curves can be deduced, the above process may be followed to derive a suitable formula.

11.35 NOMOGRAMS

From the foregoing description it is clear that the line graph is restricted to one independent variable, which may be increased to two if a family of curves is plotted on the graph field, or three if several graphs are constructed: one graph for each value of the third variable. Increasing the number of independent variables above three presents a problem which is generally not soluble owing to the physical size of the material.

The problem of presenting two or more variables on one chart may be overcome by using a nomogram. The one depicted in Figure 11.8 is a very simple type, with three independent variables.

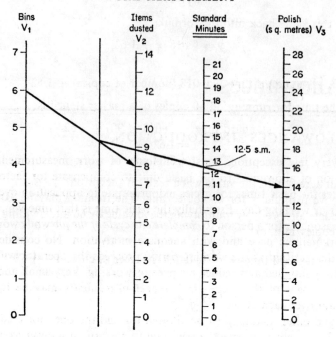

Figure 11.8 An example of a nomogram for synthetic data

Nomograms are relatively difficult to construct, often requiring non-linear scales, a common one being logarithmic. The main advantage is in the simplicity of use. In the nomogram shown in the figure, the standard time is read off the appropriate scale at the point indicated by the following method. The value of the first variable ($V1$) is joined by a straight line to the value of the second variable ($V2$), when the line will cross the dummy scale (D). A straight line is drawn from the value of variable 3 ($V3$) to the point on D, and produced to the Standard Time Scale indicating the standard time appropriate. In this particular example shown in Figure 11.8, the nomogram is for cleaning individual offices in an office block. The standard time for an office of polished area 14 sq.m, with eight pieces of furniture, and six waste bins to empty, is 12·5 s.m.

Nomograms have other uses such as in establishing allowances, and also for estimating the number of cycles to be studied in time study (Figure 7.6) and for the number of observations to make in activity sampling (Figure 7.12).

12 Allowances

12.1 ALLOWANCES: INTRODUCTION

With very few exceptions most techniques of work measurement require the addition of allowances to the basic time to compensate for factors which may render the basic times for cycles inappropriate to application over a complete shift or working day. Essentially the basic time is that time which is considered reasonable for a person to *complete one cycle of the job* while working at a brisk workmanlike pace and with adequate motivation. No consideration is given to the fact that as the working period proceeds the operator will become progressively fatigued and reduce his pace of working. Very small additions to the work content of the cycle which repeat infrequently may be ignored as insignificant over each cycle.

These and other factors must be allowed for and the operator credited with additional time which, in effect, allows him to perform at a lower average pace over the work period, yet still achieve a high relative performance.

The most usual way of applying these allowances is to apportion them, in percentage form, over each cycle, regardless of how the worker actually uses his allowances. Alternatively the *relaxation allowance* may be awarded as a fixed amount in minutes to the final total of earned minutes. The latter method is almost exclusively confined to the proprietary techniques discussed in Chapter 15.

12.2 RELAXATION ALLOWANCES

Work measurement recognizes the fact that the worker cannot sustain a high level of performance without a break in the routine to overcome the psychological effects of boredom and concentration, and without resting periodically to recover from the effects of physical exhaustion. Some jobs are less demanding of the worker and do not tax him to the extent of requiring long periods of rest. However, other biological needs must invariably be satisfied, as a consequence of which all workers are awarded an allowance for 'personal needs'.

Relaxation allowance (R.A.) is defined as 'an addition to the basic time to provide a qualified worker with a general opportunity to: (a) recover from the effort of carrying out specified work under specified conditions (fatigue allowance), (b) allow attention to personal needs, and (c) (rarely) recover from adverse environmental conditions. The amount of allowance will depend on the nature of the work and may be taken away from the place of work under management direction. *Note:* Health and Safety legislation and codes are relevant' (Term number 43025).

Of the two major components of R.A., personal needs allowance is always awarded, as previously stated, whereas fatigue allowance may or may not be applied in every case. The ergonomist and method-study engineer may be employed to carry out a prior study to improve conditions and eliminate or at least mitigate the need for fatigue allowance.

Examination of the standard R.A. tables published by the various firms for their own use will show a great disparity between values. Personal allowance ranges between 2 and 10 per cent. In most cases the P.A. for women exceeds that for men by up to 3 per cent. Table 12.1 is one example, and indicates that a personal allowance for men of $2\frac{1}{2}$ per cent and 4 per cent for women is appropriate.

The personal allowance awarded is for basic needs; extra sanitary needs such as washing hands after handling excessively dirty materials, or hygiene requirements before packing food, will require an additional personal allowance.

In addition to a fixed personal allowance, in most systems all workers receive a *fixed* fatigue allowance, to which may be added variable fatigue allowances depending on the conditions pertaining which exceed those considered to be normal for a working human being.

Fatigue is extremely difficult to define in a meaningful way. Practitioners understand in general what is meant by the term, explaining it in such words as 'tiredness', 'exhaustion', and the like. The sensations of fatigue are natural phenomena provided as a warning to the subject that he must take rest in order that he may recover from the stress to which he has been subjected.

Grandjean* quotes the work of several experts in support of the theory that general fatigue is caused by reduction of stimuli to the activating system of the brain. Indeed it has been shown that an inhibitory system exists which reduces the general metabolism and increases the activity of the digestive organs, which together aid recovery and build-up of new energy.

The term 'work measurement' itself may be regarded as a misnomer, because in fact it is time which is measured rather than work in the mechanical or physiological sense. It may be better to regard work measurement as determining the time required to perform a task; the time being modified by the energy expenditure, which is compensated for by the addition of fatigue allowance. Unfortunately the measurement of energy expenditure is complicated by certain factors:

1. The practical difficulties of evaluating the energy expended;
2. The presence of static work, which is not easily recognized;
3. The effects of environmental conditions, which may be extremely difficult to assess;
4. The presence of mental stress, which again is difficult to measure.

Work involves the use of muscles which are activated as a result of chemical reaction taking place within the muscle. During this reaction heat is liberated and oxygen is consumed to regenerate the chemicals of the muscle ready for further

* E. Grandjean, *Fitting the Task to the Man* (London: Taylor & Francis, 1969).

work. During the past sixty years many attempts have been made to measure the energy expenditure in different jobs by using these effects of muscular activity. The methods include metering of air inhaled or exhaled, measurement of pulse rate and heart beats, and measurement of heat liberated.

In spite of the volume of data collected no entirely satisfactory way of calculating appropriate allowance for recovery from fatigue have been developed, although numerous tables of values have been compiled by various firms based on empirical data. The values tabulated are essentially for the 'average worker', but Murrell* has demonstrated the wide variation in energy expended by different workers performing the same task. He gives an example of a man weighing 144 lb walking at 75 B.S. rating who expends 4 Kcal/min, whereas a man of 200 lb walking with him expends 33 per cent more energy. From this it is evident that the effects of individual differences in the workers may swamp the effects of environment and other factors of the job. In practice the variation is reduced by the personnel-selection procedure of the firm.

Adequate rest-pauses or rest-allowance are necessary to allow time for the necessary chemical changes to take effect. Generally the amount of rest required increases with the physical effort expended, but this is not necessarily a linear relationship. The presence of static work will increase the need for rest. The oxygen necessary for chemical regeneration is supplied by way of the blood stream. Static work is occasioned by contraction of the muscle, which reduces the blood flow and hence the oxygen supply. The muscle need not be performing dynamic work to cause this restriction. Working in an abnormal position, say with the arms above the head, causes muscle contraction in the operator's limbs; the action of holding up the arms being an example of static work.

From the foregoing account it is evident that much is known about the causes and effects of fatigue. However, in summary it must be emphasized that the published tables are largely empirical and that no entirely satisfactory way has yet been found to assess the amount of allowance which should be awarded in given circumstances. The intensity of fatigue experienced may vary with individual workers, and with such factors as the motivation of the person, sex, nationality, and with external environmental conditions and distractions which may include ambient temperature, humidity, noise, lighting, tedium, and monotony. In spite of this, relaxation allowances are always assessed for the job conditions and not for the particular worker. Furthermore the tables do not attempt to specify when and how the allowed rest should be taken. The frequency and the lengths of rest pauses are significant factors in the recovery from the effects of fatigue.

Published tables of relaxation allowances should be regarded as guides to the levels of allowances which *could* be awarded for a job to be performed under its usual conditions by the average worker. Individual companies must select the allowances most appropriate to their local conditions, situation, and types of work, and adjust the values as appropriate.

* K. F. H. Murrell, *Ergonomics: Man in his Working Environment* (London: Chapman & Hall, 1965).

The factors which contribute to fatigue will now be discussed as a guide to the practitioner who wishes to apply such allowances more appropriately.

12.21 FACTORS WHICH CONTRIBUTE TO FATIGUE

Fatigue allowance may be subdivided into two main groups. Those factors which depend on the conditions and type of work being performed are defined as *variable allowances*, while those which apply to all jobs irrespective of conditions are grouped under the heading of *fixed allowances*.

In the current literature on the subject Barnes* publishes a list of bench-mark jobs with appropriate allowances, against which the job being studied may be compared and the corresponding allowances extracted. Murrell† observes interestingly that true fatigue rarely exists except in heavy work, and that the effects felt by the worker (which the worker attributes to fatigue) and his reduced performance, are due to diminution in motivation. If this be the case fatigue allowances and rest pauses should be designed to regenerate the drive or motivation to do work, rather than just to recover from fatigue.

In recent years many tables of allowances have been published by individual concerns and management consultants whose values are not comparable, and indeed sometimes conflict. Most of the tables are empirical, often being based on estimated values without the backing of adequate research or experience. Furthermore, when applying the data values extracted from the various sections of allowance tables are summed to produce a comprehensive fatigue allowance. Summation is valid where the factors in the sections are independent of each other. Unfortunately in some cases there is a high degree of intercorrelation, which tends to invalidate the calculation.

The value of using tables of allowances is not so much in their accuracy (and the validity of using the word 'accuracy' in this context may be questioned) but because of the consistency any one system provides, and a consistent guide is better than no guide at all.

One important factor considered is the *physical effort* involved in the work. All work, of course, demands some degree of physical effort in moving weights or working against resistance. Fatigue allowance is credited only for that effort required in excess of what is considered 'normal'.

As previously explained, *abnormal working posture* will cause a build-up of lactic acid in the muscles, oxydation of which is retarded by the muscle's contraction. This static work causes fatigue and must be compensated for by a measure of allowance. Examples of abnormal posture are working with the arms held above the head, encountered in car maintenance or interior decorating, or long periods of work in a stooped position, as when planting bulbs in parks and nurseries. Prolonged standing may be included under this classification.

Restricted working causes frustration as well as physical fatigue. Anyone

---

* R. F. Barnes, *Motion and Time Study* (New York: Wiley, 1962).

† K. F. H. Murrell, *Ergonomics—Seminar Papers* (London: British Productivity Council, 1961).

concerned with plumbing or car maintenance has experienced the disproportionate rise in body temperature (and hence energy expenditure) which accompanies work done in restricted conditions. The description 'restricted work' may apply also to tasks which require the wearing of protective clothing (in heavy rubber or asbestos).

Concentration is affected by boredom or prolonged periods of vigilance. Very little is known about so-called mental fatigue, but it is a fact that deterioration in performance and concentration results from monotony. In allowing for this factor, consideration is given only to excessive monotony as it is realized that all jobs are more or less monotonous. This is particularly true of jobs which do not provide diversionary activities at frequent intervals which break the monotony. Machine-minders without the opportunity of chatting to neighbouring workers, or who are denied the chance of doing ancillary work, will find repetitive work monotonous. This is recognized in the award of an allowance of up to 2 per cent for working in complete absence of company, in Table 12.1, Section H4.

It has been suggested that mental fatigue may be due to a feeling of tiredness in the eyes. From the foregoing it would appear that fatigue allowance for monotony should be taken in the form of frequent short rest-pauses, which will not so much compensate for fatigue, but regenerate motivation to work and renew vigilance.

The distractive effect of noise depends on individual attitudes to the noise, and on the characteristics of the noise to which the workers are subjected. Thus the tables of allowances usually specify whether the noise is high- or low-pitched (although mostly in the working situation it is 'white noise' made up of many different notes on the musical scale), and tables differentiate between intermittent and continuous noise, and repetitive or irregular sounds. A steady, low-amplitude noise is merely background hum to which people very quickly become accustomed. Furthermore, background noise tends to mask unexpected bangs which could destroy concentration where they are to be heard in a quiet situation. The allowance table depicted in Figure 12.1 indicates that up to 5 per cent may be allowed for intermittently loud noises.

A related effect, *vibration*, may cause discomfort, or in the extreme, physical damage to the worker, but it is the detrimental effect on effective working which concerns the practitioner when awarding allowances for fatigue.

Visual fatigue may be due to eye-attention or to the effects of lighting. Once again research into this aspect of fatigue has been found wanting. Fatigue of the muscles of the eye may result from continuous eye-attention and the need to refocus at frequent intervals. This is aggravated by lighting which may be unsuitable both in intensity and colour, as well as by the effects of contrast between the objects being observed and the background against which they are being viewed. The association between the so-called mental fatigue and visual fatigue has been mentioned earlier in this chapter.

In compensating for the effects of *heat* on the ability of a man to work effectively due allowance must be made for *humidity*, which must form an inseparable part of any table of allowances for heat. Body heat is produced by

TABLE 12.1

A GUIDE TO RELAXATION ALLOWANCES

(1) These tables are intended as a guide in assessing the relaxation allowances for individual elements.
(2) Although allowances are usually additive for individual elements, assessments for elements incurring high allowances may require reduction when followed by elements incurring less.
(3) The appropriate allowance for any factor present in an element should not be given until the influence of other factors upon it has been considered.
(4) Only in A, E, and H has a separate allowance been indicated for women. Discretion should be used when assessing allowances for women under other factors.
(5) When special protective clothing such as gloves, footwear, suits or goggles have been worn, additional fatigue may arise. Care should therefore be taken in making allowances under factors A, C, D, F and G.

FACTORS	TYPICAL EXAMPLES		ALLOWANCES		REMARKS
		Equivalent to handling	Men	Women	
A. *Energy Output* (affecting muscular recovery)			per cent	per cent	
1. Negligible.	Light bench-work—seated.	No Load	0 – 6	0 – 6	When selecting the appropriate allowance for an element, the influence of the energy output in adjacent elements should be considered.
2. Very light.	Light bench-work—standing.	0 – 5 lb.	6 – 7½	6 – 7½	
3. Light	Light shovelling.	5 – 20 lb.	7½ – 12	7½ – 16	
4. Medium.	Hacksawing or filing.	20 – 40 lb.	12 – 19	16 – 30	
5. Heavy.	Swinging heavy hammer 7–28 lb.	40 – 60 lb.	19 – 30	—	
6. Very heavy.	Loading weights.	60 – 112 lb.	30 – 50	—	
7. Exceptional.	Loading heavy sacks.	above 112 lb.	requires special consideration		
B. *Posture*			per cent		
1. Sit.	Normal sedentary work.		0 – 1		When selecting the appropriate allowance for an element, the influence of the posture in adjacent elements should be considered.
2. Stand (both feet).	Whenever body is erect and support on feet only.		1 – 2½		
3. Stand (one foot).	Standing on one leg (using a foot control).		2½ – 4		
4. Lying down.	On side, face or back.		2½ – 4		
5. Crouch.	When body is bent, but supported on feet or knees.		4 – 10		
C. *Motions*					
1. Normal.	Free swing of hammer.		0		When selecting the appropriate allowance for an element, the influence of the restricted motions in adjacent elements should be considered.
2. Limited.	Limited swing of hammer.		0 – 5		
3. Awkward.	Carrying heavy load in one hand.		0 – 5		
4. Confined (limbs only).	Working with arms above head.		5 – 10		
5. Confined (whole body).	Working at thin coal seam.		10 – 15		

			per cent	
D. *Visual Fatigue*				
1. Intermittent eye-attention.	Reading meters or gauges.		0	
2. Nearly continuous eye-attention.	Precision machine work.		2	
3. Continuous eye-attention—varying focus.	Inspecting moving or stationary cloth for faults.		2	All colour contrasts wherever occurring must be considered in addition to light intensity.
4. Continuous eye-attention—fixed focus.	Inspecting minute and/or moving objects.		4	
E. *Personal Needs*			Men per cent 2½	Women per cent 4
F. *Thermal Conditions*		Humidity		
	Temperature	Normal	Excessive	The selection of the allowance between the ranges given must be related to the type of work done which may offset the temperature effects, and to the type of ventilation.
		per cent	per cent	
1. Freezing.	below 30°F	over 10	over 12	
2. Low.	32° – 55°F	10 – 0	12 – 5	
3. Normal.	55° – 75°F	0 – 40	5 – 100	
4. High.	75° – 100°F	over 40	over 100	
5. Excessive.	above 100°F			
G. *Atmospheric Conditions*			per cent	Additional allowance will be necessary for special conditions of altitude and climate.
1. Good. 2. Fair.	Well ventilated rooms or fresh air.		0	
3. Poor.	Badly ventilated air, presence of non-toxic but fetid odours or non-injurious fumes.		0 – 5	
	Presence of toxic dusts or heavy concentration of non-toxic dusts involving use of breathing filters.		5 – 10	
4. Bad.	Presence of toxic fumes or dusts involving use of respirator.		10 – 20	
H. *Other Influences of Environment*			per cent	
1. Clean, healthy, dry and bright surroundings, low noise-level. Influences without effect on work.			0	
2. Where work cycle is continuously repetitive and between 5 and 10 seconds.			0 – 1	
3. Where work cycle is continuously repetitive and less than 5 seconds.			1 – 3	
4. Where there is a complete absence of company Day – Men. Day – Women. Night – Men.			2 – 5	
5. Excessive noise, *e.g.* riveting (allowance related to continuity of noise).			0 – 5	
6. Where effect of such disturbing influence might be detrimental to quality of output.			0 – 5	
7. Vibration of floors or machines, *e.g.* pneumatic drilling (allowance related to continuity of vibration).			5 – 10	
8. Extreme conditions, *e.g.* dirt, noise, etc.			5 – 15	

metabolism and, in very hot work-areas, absorbed from the environment by radiation or convection. At normal working temperatures of, say, 15 to 25°, if the subject is suitably clothed this heat is lost by radiation and convection, but as the temperature increases he relies to an increasingly greater extent on perspiration to remove excess heat. A damp atmosphere will retard this evaporation process, causing the body to retain and increase its heat. Evaporation is assisted by air currents, so in awarding allowances for heat and humidity the practitioner must take into account the amount of ventilation and cooling through movement of air. As with other allowances, the awards for heat and humidity should not be made on the temperature and hygrometer readings alone. Due consideration should be given to the type of work being performed, as this will clearly have a profound effect on the rate of sweating, and hence the fatigue.

12.22 COMPUTATION OF R.A.

After calculating the basic time for a job the practitioner computes the R.A. applicable to the work. As already stated, all jobs receive the personal allowance and the constant part of the fatigue allowance. The general allowance is added to the basic time. In circumstances where the conditions deviate significantly from 'normal' the excesses are identified and the extent of the deviation estimated to enable the appropriate percentages to be extracted from the relevant tables. The sum of these percentages and those of the fixed allowances gives the total relaxation allowance.

R.A. may be added as a blanket allowance to the job basic-time in cases where the conditions under which all elements of the job performed are similar. However, it is recommended that individual relaxation allowances be computed for each element in turn to produce element standard-times from which the job standard-time is obtained by summation.

An example of the application of the R.A. chart in Table 12.1 is given in the chapter dealing with time study.

12.3 PROCESS ALLOWANCES

Work which consists entirely of manual elements and whose cycle-time is completely under the control of the operator requires only the addition of R.A. and work contingency-allowance to convert the basic time to standard time.

Many jobs involve the use of machines or automatic processes. Much of the work in industry is performed by teams of workers, and assembly lines on which the job cycle-times may or may not be balanced (i.e. all being equal in duration). In those cases where line balancing is not ideal and where machine-cycle and manual-work times are unequal, the worker or one or more of the machines in his charge may be idle for portions of the work-cycle. In general, if a machine operator working automatic machines has too few to supervise he will experience *unoccupied time*, whereas if he has so many in his care that he cannot keep them all running *machine interference* will result. In most jobs of this kind either one (or even both) of these situations will be present.

ALLOWANCES

Before proceeding with the theory of unoccupied time and interference it is necessary to appreciate the terminology used.

The following definitions apply to machine work. Most of the terms are illustrated in Figure 12.1.

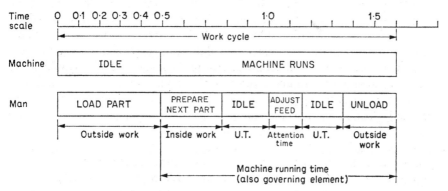

Figure 12.1 Illustration of the terms used in machine process

work-cycle: The sequence of elements which are required to perform a task or to yield a unit of production.

machine-running time: The time during which a machine is actually operating, i.e. the machine available-time less any machine down-time, machine idle-time and machine ancillary-time.

governing element: An element occupying a longer time than that of any other element which is being performed concurrently.

inside work: Elements which can be performed by a worker within the machine-controlled time.

outside work: Elements which must necessarily be performed by a worker outside the machine-controlled time.

attention time: The time during which the presence of a worker is necessary to ensure the safe and proper functioning of a machine (or process) although the worker is not carrying out physical work. (May include some R.A.)

12.31 UNOCCUPIED-TIME ALLOWANCE

The enforced idleness of unoccupied time will be due to the provision of insufficient work for the operator to perform while the machine is running (inside work) causing the worker to wait for the machine-cycle to finish. Regardless of the speed with which he carries out the inside work he cannot produce more parts. Consequently he must be compensated for this enforced idleness by way of an unoccupied-time allowance (U.T.A.).

To find the unoccupied time the inside and outside work is measured at a defined level of performance using one of the techniques of work measurement. The difference between the basic time for inside work and the machine-running time is the unoccupied time. The practitioner is now faced with certain problems the answers to which are usually decided by the company as part of general policy. These decisions which must be made and defined before commencement of measurement are three in number:

1. The rating at which unoccupied time is to be allowed (because, of course, unoccupied time cannot itself be rated).
2. The rating at which attention time is to be allowed.
3. The R.A. applicable to unoccupied time.

Many firms determine the basic time for a machine-controlled job simply by adding the basic time for outside work to the machine-running time, which, in the case of Figure 12.1, would be 1·6 min. This is, in effect, allowing both unoccupied time (U.T.) and attention time (A.T.) at the same performance as the work elements (i.e. standard time or equivalent). In other cases it is argued that the worker is doing no work during the U.T. period of the work-cycle, and therefore cannot be assessed at standard performance. Moreover a worker performing outside and inside work at a relatively low rating will benefit through having U.T. assessed at a higher level, which will credit him with working faster during unoccupied time when in fact he is doing nothing. Similarly he would have an advantage over other employees engaged entirely on manual elements who were working all the time to achieve their standard performance.

On the other hand he is not to blame for his enforced idleness, so some allowance should be made. Thus U.T. may be extended to a lower rating (say 75 B.S.) or even to an average of the performance usually achieved by operators engaged on this work. The calculation of the standard time is shown in the example below.

Attention time, of course, cannot be rated or measured by P.M.T.S. because it is composed mainly of mental work and vigilance. The time spent on ensuring that the machine is processing correctly must be assessed by measuring the observed time and extending this by a predetermined rating. The allowed rating will be decided by company policy or by agreement, and often this is standard rating.

R.A. is applied in the usual way for the manual elements of the work, but one problem of U.T. is to determine the amount of R.A. attributable to the operator while he is not working. During the U.T. periods he will not be using his energy; in fact he will be relaxing. Because of this it is not unusual to insist that he take some of his R.A. during his U.T. However it is essential that he be allowed to quit his workplace for some of his R.A.

Although he may not be increasing his fatigue level during U.T. periods his body is still performing its biological functions, and he will require at least a personal needs allowance. Therefore it is customary for firms to credit the unoccupied time with anything between full R.A. and just personal needs allowance.

Example

A standard time for the job shown in Figure 12.1 will be calculated.

The firm has decided to allow attention time at 100 B.S. rating U.T.A. to be allowed at 75 B.S. performance. R.A. is 15 per cent on outside work and $12\frac{1}{2}$ per cent on inside work and attention time. Personal-needs allowance is $2\frac{1}{2}$ per cent.

Outside work	0·750 b.m.
R.A. at 15 per cent	0·113
Inside work	0·300 b.m.
R.A. at $12\frac{1}{2}$ per cent	0·038
Standard time for work	1·201 s.m.
Attention time allowed at 100 B.S.	0·150
R.A. on A.T. at $12\frac{1}{2}$ per cent	0·020
Total	1·371 s.m.
Machine-running time	0·850 min
less inside work and A.T.	0·508
U.T.	0·242 min
U.T.A. allowed at 75 B.S.	0·182
R.A. at $2\frac{1}{2}$ per cent	0·005
	0·187

Thus standard time is 1·371 s.m. + 0·187 min U.T.A.

12.32 TEAM WORK

In flow-line work it is unusual for workers to pass partly completed work from hand to hand without intermediate buffer stocks of at least two or three items in which case U.T. will be calculated over several cycles. For example, suppose four operators are working in such a team, the production will depend on the worker with the longest work-cycle, assuming that all are performing at similar ratings. U.T. for the other three workers will be the difference between their cycle-times and the lead operator's cycle-time.

Thus if the governing element is 36 min per 100 parts and the cycle-times for the other three members of the team are 20 b.m., 32 b.m., and 18 b.m. respectively, the unoccupied times will be 16, 4, and 18 minutes respectively, giving a total of 38 minutes. The policy of the firm may be to allow U.T.A. at, say, 80 performance. The U.T.A. in this case will be 80 per cent of 38, or 30·4 min, and thus the cycle-time for 100 parts will be (36 + 20 + 32 + 18 b.m.) + 30·4 min U.T.A., or 136·4 min.

Of course the job would be made more effective by re-scheduling the work, reducing the effect of the bottleneck operation with the 36-min work-cycle. This would be achieved by doubling the staff engaged on the 36- and 32-min work-cycles, effectively reducing their cycle times to 18 and 16 respectively. The jobs in the line as a consequence would be 18, 20, 16, and 18 min respectively. This re-arrangement would make the line more 'balanced'. Alternatively it might be possible for the operator engaged on the 20-min cycle to perform the 18-min job; the flow-line would now comprise the jobs of 36, 38, and 32 min, which again achieves better balance.

12.33 MULTI-MACHINE OPERATION

Where several machines are operated the unoccupied time may be determined by graphical means using a man/machine chart. Where the machines have different running times or the work-content differs it is almost inevitable that the situation will be complicated by the occurrence of interference as well as U.T.

The problem is simplified in the case of one operator attending more than one machine if the machines and the manual elements involved are identical. The cycle of operations quickly settles down to a repetitive uniform pattern. This is illustrated in Figure 12.2. The amount of U.T. may be determined in two ways,

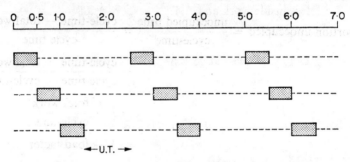

Figure 12.2 Unoccupied time for three identical machines

either by plotting the situation on a man/machine chart, or by using a ratio known as the *load factor*. From Figure 12.2 it is apparent that the amount of unoccupied time is 1·0 min.

12.34 LOAD FACTOR

From Figure 12.2 can be deduced the unoccupied time were the operator to attend to just one machine. The cycle-time is 2·5 min and the outside plus inside work 0·5 b.m. Thus the unoccupied time is $2·5 - 0·5 = 2·0$ min. The addition of Machine 2 to his work-load has the effect of reducing this U.T. by the amount of the total work (0·5 b.m.) to 1·5 min. Attending to Machine 3 again reduces this remaining U.T. of 1·5 min by a further 0·5 b.m., finally leaving just 1·0 min of unoccupied time.

From the above it is clear that the addition of successively more machines will reduce the unoccupied time of the first machine by an amount equal to the total work for all the other machines, or (total work) × (number of machines excluding the first). This is the background to the concept of load factor, which is the ratio of the total work on one machine to its cycle-time, and describes the proportion of the cycle-time employed in carrying out inside and outside work.

For the above problem the load factor is:

$$\frac{\text{total work}}{\text{cycle-time}} = \frac{0\cdot 5}{2\cdot 5} = 0\cdot 2$$

For three identical machines the load factor is $0\cdot 2 \times 3 = 0\cdot 6$. In the usual way unoccupied time would be found by subtracting the total work of all machines from the cycle-time for one machine. Total work is $0\cdot 5 \times 3$ or $1\cdot 5$ b.m., which when deducted from the machine-running time of $2\cdot 5$ gives a period of U.T. of $1\cdot 0$ min. Alternatively this may be found from the load factor as follows:

U.T. is equal to $1 -$ load factor, which in this case is $1 - 0\cdot 6 = 0\cdot 4$. Because this is the proportion of the cycle-time which is unoccupied, in this case it is equal to $0\cdot 4$ of $2\cdot 5$, or $1\cdot 0$ min as before.

The proof of the above formula is as follows:

$$\text{Proportion unoccupied} = \frac{\text{unoccupied time}}{\text{cycle-time}} = \frac{\text{cycle-time} - \text{total work}}{\text{cycle time}}$$

$$= \frac{\text{cycle-time}}{\text{cycle-time}} - \frac{\text{total work}}{\text{cycle-time}}$$

$$= 1 - \frac{\text{total work}}{\text{cycle-time}}$$

$$= 1 - \text{load factor}$$

In the situation just analysed the operator is still faced with $1\cdot 0$ min of U.T. Load factor may be used to determine the number of machines which he is able to attend without causing interference.

From the formula derived above it is clear that to eliminate U.T.

$$(1 - \text{load factor}) \text{ must equal zero}$$

If L denotes the load factor for all machines

then $\qquad 1 - L = 0$ for no U.T.

so $\qquad 1 = L$

but L is made up of the load factor per machine (p) multiplied by the number of machines attended (n), i.e. pn

thus $\qquad 1 = pn$

and $\qquad n = \dfrac{1}{p}$

In the problem under discussion, the load factor (p) was $0\cdot 2$, therefore the number of machines which one operator is able to supervise is $1/0\cdot 2$ or 5. This result may be verified from Figure 12.2.

12.4 THE NEED FOR INTERFERENCE ALLOWANCE

Before discussing interference in detail it is necessary to derive a general formula for this quantity.

From Figure 12.4 it can be seen that $ow + c + I$ is equal to 3 times W, where W is the outside plus inside work, c is the machine-running time, I the amount of interference, and ow is the outside work.

Thus $\qquad ow + c + I = 3(iw + ow)$

if $\qquad\qquad\quad iw$ is inside work
$\qquad\qquad\quad ow$ is outside work

from this $\qquad I = 3iw + 3ow - ow - c$

therefore $\qquad I = 2ow + 3iw - c \qquad\qquad$ (A)

A general formula may be derived for any number of machines (n)

$$I = (n-1)ow + niw - c$$

In the example of Figure 12.4

$$I = (2 \times 0.3) + (3 \times 0.2) - 0.9$$
$$= 0.3 \text{ min}$$

The cycle time for this job is 1·2 min, from which it can be deduced that output is $60/1.2 = 50$ per hour per machine. But from Figure 12.4 the actual cycle occupied $0.3 + 0.9 + 0.3$, or 1·5 min, because of the effect of interference, which effectively lengthens the cycle-time.

If the operator works at 100 rating his maximum output will be $60/1.5$, or 40 per hour per machine. In order to attain 50 per hour he must work faster on the outside work; the machine-running time being invariable.

If $W = iw + ow$, for the cycle-time to be 1·2 min (i.e. for 50-per-hour production), then $3 \times W$ must equal 1·2,

therefore $\qquad\qquad W = \dfrac{1.2}{3} = 0.4 \text{ min}$

Therefore to achieve 50 per hour the operator must complete the work in 0·4 min by working at a rating of $0.5/0.4 \times 100$, or 125 rating. Notice that there is still some interference, because from formula (A)

$$I = 2ow + 3iw - c$$
$$= (2 \times 0.2) + (3 \times 0.16) - 0.9$$
$$= 0.06 \text{ min}$$

12.41 MAXIMUM OUTPUT

The maximum output which may be achieved is when the interference is zero. If the operator works faster, then this merely produces unoccupied time as he runs out of work. This occurs at a rating deduced as follows:

ALLOWANCES

From formula (A) $I = 2ow + 3iw - c$
for no interference $I = 0$
$$0 = 2ow + 3iw - c$$
$$0.9 = 2ow + 3iw$$

now $$\frac{iw}{ow} = \frac{0.2}{0.3}$$

therefore $$iw = \frac{2ow}{3}$$

$$0.9 = 2ow + 3\left(\frac{2ow}{3}\right)$$

$$4ow = 0.9$$

$$ow = \frac{0.9}{4}$$

$$= 0.225$$

therefore $$iw = \frac{2ow}{3} = 0.15 \text{ min}$$

The rating which must be achieved for this to occur is

$$\frac{0.5}{0.225 + 0.15} \times 100 = 133$$

and the output is $\frac{60}{0.375 + 0.7} = 56$ per hour approx.

12.43 DIFFERENT MACHINES

When the cycles of the machines are different the amount of interference may be calculated either from a man/machine chart such as that shown in Figure 12.4, or by purely algebraic means, or in the case of random attention-periods, by special formulae.

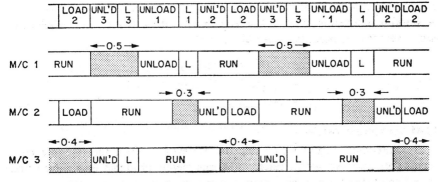

Figure 12.3 Calculation of interference for three dissimilar machines

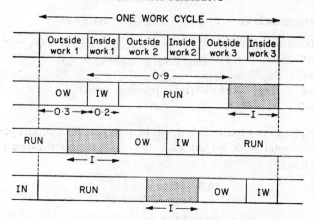

Figure 12.4 Calculation of interference for three similar machines

Under the above heading the two cases considered are:

(i) Wholly automatic machines for which the running times are constant, and the attention is both predictable and constant;
(ii) Machines whose running time is broken up by intermittent unpredictable attention occasioned by machine breakdown, as well as by regular loading and unloading.

Interference for the first situation may be solved by graphical or algebraic means, and this will be considered first. The second, more complex case has caused much concern, and several empirical or statistical formulae have been derived to analyse it.

The man/machine chart depicted in Figure 12.3 clearly shows that the total interference is $(0\cdot 5 + 0\cdot 3 + 0\cdot 4)$ min, or $1\cdot 2$ min. The interference effectively increases the cycle-times to $1\cdot 7$ min, producing a potential output of 35·3 per hour from each machine, or about 106 per hour total. If interference were eliminated and the work were performed at 100 B.S. rating, the potential output from the three machines is 50, 43, and 46 respectively, giving a total of 139 per hour. Thus the losses for which interference allowance must compensate are 15, 8, and 11 respectively.

The interference of $1\cdot 2$ min may be calculated from a formula derived as follows from Figure 12.3.

$$\text{Interference on machine No. 1} = I_1 = C - (w_1 + c_1) \quad \text{(B)}$$

where C is the overall cycle time

w_1 is the work on machine No. 1

c_1 is the machine cycle, or running time for machine No. 1

Similarly interference on machine No. 2 $= I_2 = C - (w_2 + c_2)$

and for machine No. 3 $= I_3 = C - (w_3 + c_3)$ (C)

and so on.

Total interference is $I_1 + I_2 + I_3 \ldots I_n$ for n machines

$$= [C - (w_1 + c_1)] + [C - (w_2 + c_2)] + \ldots [C - (w_n + c_n)]$$
$$= nC - (w_1 + c_1) - (w_2 + c_2) - \ldots (w_n + c_n)$$
$$= nC - (w_1 + w_2 + w_3 \ldots w_n) - (c_1 + c_2 + c_3 \ldots c_n)$$
$$= nC - \text{(total work)} - \text{(total running times)}$$
$$= nC - w_t - c_t \qquad (D)$$

But, by reference to Figure 12.3, it will be seen that the overall cycle-time C is equal to the total work for the man, thus $C = w_t$. Substituting C for w_t in equation (D):

$$I_t = nC - w_t - c_t$$
$$= nw_t - w_t - c_t$$
$$= (n - 1)w_t - c_t \qquad (E)$$

Equation (E) may be used in general situations for n machines.

Example
1. Using the example of Figure 12.3
$w_1 = 0.6$; $w_2 = 0.6$; $w_3 = 0.5$ min
$c_1 = 0.6$; $c_2 = 0.8$; $c_3 = 0.8$ min $n = 3$ machines
Therefore $I_t = (n - 1)w_t - c_t$
$= (2 \times 1.7) - (0.6 + 0.8 + 0.8)$
$= 3.4 - 2.2$
$= 1.2$ min, as before

2. Find the total interference time for the following situation:

Machine No.	work	running time
1	1·2 min	1·6 min
2	0·8 min	1·2 min
3	1·3 min	0·9 min
4	1·6 min	1·0 min

$I_t = (n - 1)w_t - c_t$
$= 3(1.2 + 0.8 + 1.3 + 1.6) - (1.6 + 1.2 + 0.9 + 1.0)$
$= 3 \times 4.9 - 4.7$
$= 14.7 - 4.7$
$= 10$ min

The effective cycle-time is equal to w_t, or 4·9 min.

12.431 Conditions for interference. Clearly for there to be interference, from formula (D) I_t must be greater than zero

i.e. $\qquad (n - 1)w_t - c_t > 0$
or $\qquad (n - 1)w_t > c_t$

The situation is complicated if unoccupied time in addition to interference is present. In actual cases the absence of unoccupied time may be verified by reference to a man/machine chart of the situation.

12.44 RANDOM ATTENTION

Much of the work with machines is complicated by breakdowns which occur at random intervals. In such cases the operator will not deal with the machines in a logical sequence but in a random order. At any one time he may be required to stop work on one machine to attend to another: he may even deliberately stop a running machine so that one which is easy to repair may be re-started.

An operator placed in such a situation will be in a position of having unoccupied time in periods when all machines are running, while being overworked during periods when breakdowns synchronize, hence causing interference.

It is possible, of course, to plot the situation as a man/machine chart, but this represents merely the order of attention pertaining *at that particular time* when the study was taken; it may not be representative of everyday working. However, it will be shown later that the representation provides quite a good fit when compared with a standard-formula result. The usual man/machine chart as shown in Figure 12.3 is inappropriate for random breakdown, as it will often indicate merely unoccupied time, whereas in fact the true situation is one of both U.T. and interference. Figure 12.5 shows the case for purely automatic

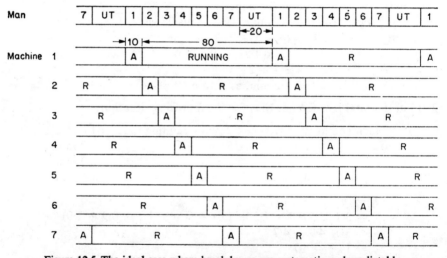

Figure 12.5 The ideal case where breakdowns are systematic and predictable

machines and indicates a U.T. of 20 min. Figure 12.6 illustrates how the same machines with the same running times and attention times produce interference when random attention is introduced. Notice that the attention times and running times are broken up in the case of random attention, but the sum totals for both cases are still the same.

The characteristics of random attention may therefore be summarized as follows:

1. The occurrence of interference is governed by the laws of chance.
2. The amount of interference will depend to a certain extent upon the speed

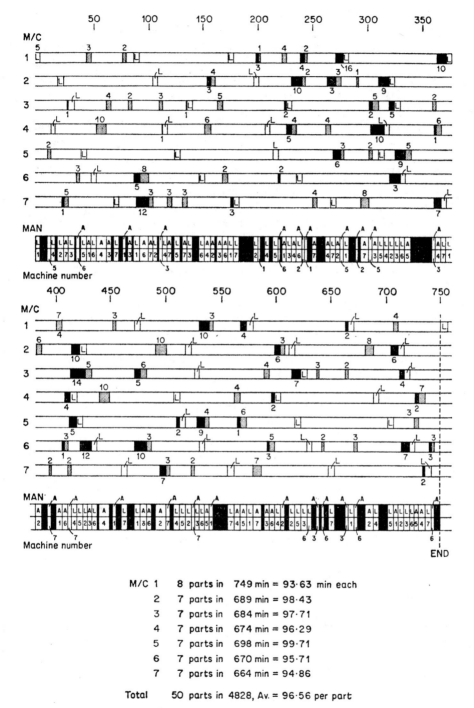

	M/C 1	8 parts in	749 min = 93·63 min each
	2	7 parts in	689 min = 98·43
	3	7 parts in	684 min = 97·71
	4	7 parts in	674 min = 96·29
	5	7 parts in	698 min = 99·71
	6	7 parts in	670 min = 95·71
	7	7 parts in	664 min = 94·86
Total		50 parts in	4828, Av. = 96·56 per part

Figure 12.6 A more realistic case of the machines of Figure 12.5, with random breakdowns

L

of movement of the operator, although the faster worker may increase his U.T.
3. Intelligent anticipation of impending trouble and good planning of work priority by the repairer or operator will reduce interference in some cases. However, the order of attention does not reduce interference as much as may be supposed, as a reduction on one machine may increase interference on another machine by the same amount.
4. The amount of interference depends on the ratio of attention time to running time: in general, the greater the attention time the more the interference.

12.45 METHODS OF SOLUTION

Three methods of solution will be discussed: one graphical and two by means of formulae.

Figure 12.6 shows the results of a production study on seven coil-winding machines. The machines are liable to break down owing to the wire snapping and other factors. There is also a regular loading and unloading operation of 5 b.m. Machine-running time is 80 min to produce one unit of production. The time to service the breakdowns vary according to the type of fault. However, the production study revealed an average time of 5 min per breakdown.

The department operates on a continuous shift-work scheme. From Figure 12.5 the following data are apparent:

Machine 1 Eight cycles completed between 0 and 749 min = 749 min
2 Seven cycles completed between 21 and 710 min = 689 min
3 Seven cycles completed between 31 and 715 min = 684 min
4 Seven cycles completed between 14 and 688 min = 674 min
5 Seven cycles completed between 12 and 710 min = 698 min
6 Seven cycles completed between 50 and 720 min = 670 min
7 Seven cycles completed between 71 and 735 min = 664 min

Total = 4,828 min

Thus 50 cycles took a total of 4,828 min to complete or 96·6 min per cycle. During 96·6 min (average) the seven machines produced seven coils, thus the effective cycle-time for the group of seven machines was 96·6 min per 7 coils.

12.46 ASHCROFT'S NUMBERS

It has been argued that the graphical methods of determining cycle-time and interference shown above suffer from certain limitations. Consequently several solutions by formulae have been proposed, prominent among which are those of H. Ashcroft and W. Wright. Ashcroft's approach will be considered first.

From formula (B) on page 302

$$I = C - (w + c)$$

The production from one machine results from that machine running for c hours. Therefore the running time for N machines is Nc machine-hours. Thus in one clock-hour there are Nc/C machine-hours effectively. H. Ashcroft,* working at the B.I.C.C. plant at Prescot, Lancs, gave this ratio the symbol A; thus $A = Nc/C$. The factor A is defined as the average number of machine-hours per hour. Using this ratio he developed a set of tables known as Ashcroft's Number of A, for various numbers of machines supervised by one operator. The work was in connection with the operation of *semi-automatics*.

From this formula
$$C = \frac{Nc}{A} \quad \text{(F)}$$

Substituting in formula (B) $\quad I_t = \dfrac{Nc}{A} - (w_t + c_t)$

Example
The data previously used in the coil-winding example will be used again (Figure 12.6).

> Load and unload time = 5 b.m.
> Running time = 80 min Number of machines (n) = 7
> Breakdown time (average) = 5 min

First a value p is calculated so that an appropriate entry may be made in the tables. The factor p is defined as the ratio of attention-time to running-time.

$$p = \frac{5+5}{80} = 0.125$$

From Ashcroft's Tables, $A = 5.76$

Now
$$I = \frac{Nc}{A} - (w + c)$$
$$= \frac{7 \times 80}{5.76} - (10 + 80)$$
$$= 97.2 - 90$$
$$= 7.2 \text{ min}$$

From Formula (E) the cycle-time is: $\dfrac{80 \times 7}{5.76} = 97.2$ min

This value for cycle-time obtained from Ashcroft's Numbers may be favourably compared with the value obtained from the graphical method just described.

For a full account of Ashcroft's analysis reference should be made to Ashcroft's paper quoted above.† A very useful account of the subject is contained in a monograph by T. F. O'Connor.‡

---

* H. Ashcroft, 'The Production of Several Machines under the care of One Operator', *Journal of the Royal Statistical Society*, Series B, Vol. XII, No. 1, 1950.
† Ashcroft, op. cit.
‡ T. F. O'Connor, *Production and Probability* (Manchester: Mechanical World Monographs No. 65, 1952).

12.47 WRIGHT'S FORMULA

A method for solving the problem of machine-interference time is suggested by W. R. Wright, who based his formula on the work of T. C. Fry. Wright's formula often gives values which are higher than those determined by other methods, often as much as 50 per cent above. A better fit to the true situation is obtained where the machine/operator ratio is greater than 6:1.

For purposes of comparison the previous example again will be used.

Wright's formula is: $I = 50\{\sqrt{[(1 + X - N)^2 + 2N]} - (1 + X - N)\}$

Where I is the interference as a percentage of the attention time
   N is the number of machines supervised
   X is the ratio of running time to attention time

In this case $N = 7$, and $X = 80 \div 10 = 8$

$$I = 50(\sqrt{18} - 2)$$
$$= 50 \times 2.24$$
$$= 112 \text{ per cent of the attention time}$$

As attention time is 10 min (including average of 5 min breakdown)

$$I = 112\% \text{ of } 10$$
$$= 11.2 \text{ min.}$$

Comparison with the other results shows this to be over 50 per cent higher.

12.5 CONTINGENCY ALLOWANCE

Much of the work performed may be analysed into elements which may be measured using the most appropriate technique. Basic times for elements which occur regularly and frequently should be measured with great care, as large relative errors in cycle basic-times will produce a misleading assessment of the day's performance. On the other hand to spend a great deal of time on measuring a job which occurs infrequently would be uneconomic as well as unrealistic. It is usual to spread the time for an occasional element (i.e. one which does not occur in all cycles) evenly over all the cycles. For example, it may be necessary to replenish a bin with components sufficient to make 500 parts. The time to fill the bin may be 0·25 b.m., which means that each cycle (during which one part is made) will be allocated 1/500th of the time to fill a bin, or 0·0005 b.m. Not only is this time negligible when working to two places of decimals, but also the implication is that it is possible to measure jobs to the nearest ten-thousandth of a minute.

Faced with the above situation the usual practice is to lump together all such insignificant elements or jobs into a *contingency allowance*. This allowance is defined by B.S.I. Term number 43028 as:

⌈A measured or estimated allowance of time which may be necessary for inclusion in the standard time to cover specified and legitimate work activities and/or unavoidable interruptions (not recorded waiting time) in work sequence. Contingency allowance is applied only when it is impractical to treat such items as occasional elements.

Contingency allowance is subdivided to cater for (*a*) legitimate and expected items of work and (*b*) legitimate and expected delays. The example quoted above would fall into the first subdivision; other examples being occasional grinding, sharpening, or adjusting of tools, rectifying the odd rejects, occasional adjusting of dials on equipment, and brushing down the work bench. The second subdivision includes machine delays, interruption by the supervisor or servicing operator, and time lost awaiting the results of a quality-control check. It is essential to realize that only very infrequent elements should be included in contingency allowance, which should not become a repository for all the odd occasional and variable elements which the practitioner cannot be bothered to time. Furthermore, it is necessary to state what is included in the contingency allowance to avoid subsequent allegations of omissions of elements.

It is often said that contingency allowance should always be confined to a very small percentage of the work; figures of 3 to 5 per cent being quoted. It is argued that if the jobs included in contingency allowance exceed 5 per cent some or all must be significant enough to measure and a standard time derived. In general this is true, but occasionally the contingency allowance may be comprised of many small items of work and delay which individually are not worth measuring, but which collectively add up to more than 5 per cent.

The extent of the contingency allowance may be determined from labour statistics, or obtained through production studies or activity sampling (*see* Chapters 13 and 7).

The contingency allowance should be subject to periodic reappraisal to detect and allow for variations in the work pattern. Contingency allowance is usually added to the total work content (standard minutes or standard hours) as a percentage of this time.

13 Production Studies

13.1 SCOPE OF THE PRODUCTION STUDY

Assessments of utilization of equipment or employees may be made using activity sampling as described in Chapter 7. Such surveys produce acceptable results; the accuracy may be predetermined and selected by the practitioner before the study begins. Under certain circumstances it is necessary to obtain a detailed account of a complete period of work such as a shift or day, from which data may be extracted for checking and audit purposes. A continuous study used for this purpose is known as a *production study*.

The British Standards Institution defined the production study in their Term number 41029 as follows:

A continuous study of relatively lengthy duration, often extending over a period of one or more shifts, taken with the object of checking an existing or proposed standard time, or its constituent parts or obtaining other information affecting the rate of output.

The alternative term sometimes used is *overall study*.

Data may be more appropriate for checking purposes if the operator being studied is rated so that basic times may be used in place of observed times. The rating is often an average over that particular period of study. When the period extends beyond, say, ten minutes, it is advisable to subdivide, or to rate every minute or so to obtain the average.

There may be many reasons for obtaining data through the use of a production study, and much incidental information emerges from such a study which highlights problems hitherto unrecognized. Potential sources of trouble may be anticipated and obviated through information fed back to management by the sharp-eyed observer.

Production studies may be used for:

1. Obtaining a detailed account of a representative period;
2. Validation of issued standard and allowed times;
3. Checking of contingencies for adequate coverage;
4. Assessing the amount of waiting time and other ineffective time;
5. Checking relaxation allowance and other allowances (such as interference and unoccupied time);
6. Checking levels of output and investigating unusual or unexpected changes in performance;
7. Validation of synthetic-data formulae and values in the working situation;

8. Routine checking and auditing of existing standard and allowed times as a regular service, up-dating where necessary;
9. Generally ensuring that the standard time adequately covers all work contained in the job and that nothing has been omitted from the work specification;
10. Ensuring that the frequencies of occurrence are correct.

Incidental information fed back to management may include observations on operator discipline (lateness, absence etc.), morale, ideas for methods improvements, line balancing, and other factors which may improve productivity or assist in motivation.

13.2 PRODUCTION-STUDY PROCEDURE

After the usual courtesies having been observed the practitioner agrees to be present at the workplace some minutes before the start of the observation period.

Times may be recorded on a specially designed form, such as that shown in Figure 13.1, or the normal cumulative type Observation Record Sheet (Figure 9.12) may be adapted for the purpose. The fly-back method of timing is suitable for recording the activities, but continuous or cumulative timing is often preferred. As an additional check the time from the departmental wall-clock may be recorded at half-hourly intervals.

Production-study times record changes of activity, or noteworthy occurrences, which is where this technique differs from time study, which analyses jobs into elements which are timed separately. Average ratings may be recorded also, but because some of the periods between occurrences may be protracted, the average is obtained from ratings made at half-minute or minute intervals, as for rated-activity sampling.

Reference to Figure 13.1 will show that several items of information are recorded, including: (*a*) the actual time, (*b*) the occurrence or activity, (*c*) rating where appropriate, and (*d*) the output for that period. From these data other information may be extracted subsequently, such as basic times for activities, and selected times from the cumulative times. It is advisable to keep a continuous check of production quantities, and if possible agree these with the operator. Such data should not be obtained from the operator's work-record at the end of the day as this may not be accurate.

Standard times may be checked using the data showing ratings, times, units of production completed, and allowances, but often it is more instructive to incorporate a check study (such as P.M.T.S. or a time study) inside the production study. This check study may be performed during a long run of uninterrupted work when the observer has time to spare.

The production study may be used for checking machine and process speeds from which potential output may be calculated.

PRODUCTION STUDY

Operator's name: Mrs E Robson Date of study: Jan 23
Department: Varnishing
Equipment: No. 3 (first dip) machine – Lloyd 35
No. 5 (varnishing) machine – Bridges
Details of operation: Dipping brush handles in gelatine finish
Dipping brush handles in varnish
Product: Domestic 50mm and 65mm paint brush handles

Clock time	Obs. time	Rating	Basic time	Out-put	Observation description
8·30·00				racks	Study commences. Operator absent.
32·10	2·10				Operator arrives and prepares workplace.
35·10	3·00				Starts dipping operation (gelatine)(Batch 3872).
42·30	7·20	90	6·48	10	Adds more gelatine to bath.
45·70	3·40	85			Resumes dipping operation
51·50	5·80	90	5·22	7	Operator moves boxes of rejects to stack
54·20	2·70				Resumes dipping operation
59·00	4·80	75	3·60	5	Supervisor instructs operator on a point
9·02·45	3·45				Resumes dipping operation
08·15	5·70	90	5·13	7	Tunnel jammed with work. Operator loads dipped work to trolley instead of tunnel.
22·92	14·77	80	11·82	17	Trolley full. Operator goes off to find another
27·66	4·74				Arrives back with trolley. Goes off for another.
29·28	1·62				Arrives back. Resumes dipping. (End of batch 3872)
30·78	1·50	90	1·35	2	Operator checks quality of 10 handles
32·20	1·42	90			Resumes dipping. (New batch 3869)
39·44	7·24	80	5·79	9	Operator leaves workplace (personal needs).
45·02	5·58				Returns and resumes dipping
50·04	5·02	90	4·52	6	Tunnel free. Operator unloads work from trolley to tunnel
54·36	4·32	80	3·46		All trolleys cleared. Resumes dipping
10·01·12	6·76	85	5·74	8	One bath out of action due to timer fault. Operator calls supervisor
02·15	1·03				Operator dips again using only one bath
12·84	10·69	80	8·56	6	Other bath in action. Resumes using both baths
20·35	7·51	90	6·76	11	Operator runs out of work. Moves over to varnishing
22·00	1·65				Starts dipping on varnishing equipment.
25·30	3·30	80	2·64	3	Stops work ready for coffee break at 10·30
44·08	18·78				Returns from break and starts dipping again
52·55	8·47	80	6·78	9	Supervisor queries operator's time sheet
54·21	1·66				Resumes varnish dipping

Figure 13.1 An observation form

Clock time	Obs. time	Rating	Basic time	Output racks	Observation description
58·00	3·79	90	3·42	4	Stops work to talk to friend from other dept.
11·01·45	3·45				Starts dipping again
13·08	11·63	85	9·89	13	Supervisor moves operator back to gelatine dipping
14·22	1·14				Commences dipping
20·40	6·18	95	5·87	8	Leaves workplace for personal needs
26·04	5·64				Operator returns and resumes dipping (end of 3869)
31·05	5·01	90	4·50	6	Operator checks the quality of 10 handles
32·41	1·36				Resumes dipping (new batch 3874)
35·10	2·69	80	2·16	3	Bath fault again. Operator calls supervisor
35·22	0·12				Reverts to one bath again
42·30	7·08	90	6·37	4	Other bath in action. Operator resumes using both.
59·16	16·86	100	16·86	26	Tunnel choked again. Operator works to trolleys
12·11·06	11·90	90	10·71	16	Operator breaks off for lunch
15·00	3·94				Study ends for morning
13·00·00					Study resumes. Operator already at workplace tidying the area
04·75	4·75				Operator starts unloading trolleys to tunnel
07·28	2·53	75	1·90		Operator resumes dipping
12·14	4·86	90	4·37	6	Operator tidies boxes around work area
14·20	2·06				Resumes dipping
17·10	2·90	100	2·90	4	Operator checks quality 10 handles
18·30	1·20				Resumes dipping (end of batch 3874)
23·11	4·87	90	4·38	6	Operator breaks off work to speak to neighbour
25·71	2·54				Resumes dipping (new batch 3876)
38·10	12·39	90	11·15	17	Operator calls supervisor to look at the quality of dipping. Stops work
40·84	2·74				Resumes dipping
44·22	3·38	80	2·72	4	Supervisor returns to instruct operator to inspect for bad work
44·63	0·41				Operator breaks off to sort bad handles into bin
58·10	13·47	65			Operator finishes inspection and removes rejects to reject area
14·01·00	2·90				Resumes dipping
13·70	12·70	90	11·43	18	Operator leaves workplace
19·50	5·80				Resumes dipping
24·40	4·90	90	4·41	6	Operator obtains and adds more gelatine to dip
27·10	2·70				Resumes dipping

for a production study

Clock time	Obs. time	Rating	Basic time	Output racks	Observation description
31.00	3.90	90	3.51	5	Operator breaks off to give letter to neighbour
33.14	2.14				Resumes dipping
37.10	3.96	90	3.56	5	Trouble with mechanism. Operator calls supervisor. Operator stops work.
43.80	6.70				Mechanic arrives and works on machine.
52.62	8.82				Operator resumes work on dipping
57.12	4.50	80	3.60	5	Operator breaks off for tea
15.15.09	17.97				Operator returns and resumes dipping (end of batch 3876)
24.10	9.01	80	7.21	11	Operator checks quality of 10 handles
25.25	1.15				Adjusts setting and dips a test rack
27.55	2.30				Checks test rack
27.90	0.35				Resumes dipping (new batch 3875)
43.12	15.22	90	13.70	22	Chats to neighbour (stops work to do so)
44.20	1.08				Resumes work
58.86	14.66	90	13.19	21	Operator breaks off for personal needs
16.04.18	5.32				Resumes dipping
16.26.66	22.48	100	22.48	35	Operator sorts through rejects and removes 'knobs' from 8 handles, placing these with the good ones for dipping
29.20	2.54	65			Resumes dipping
32.60	3.40	90	3.06	4	Operator starts clearing up for night
37.24	4.64				Operator stands by the machine ready to leave
39.32	2.08				Finishing bell rings. Operator leaves
40.00	0.68				Study ends. (Difference of 0.68 min. between study clock and official finishing bell)

Figure 13.1—*continued*

13.3 ANALYSING THE STUDY

The actual form of the analysis depends on the objective of the production study. An analysis for the example which follows is given, augmented by Figures 13.1 and 13.2.

Many analyses are produced in the form of a report to management, and contain the main headings:

1. The situation revealed by the study.
2. Conclusions.
3. Recommendations.

DAILY WORK RECORD

Name: E.Robson
Department: Varnishing
Date: 23 Jan.
Supervisor: E.J.Tomms

Batch number	Operation description	Volume or time	Code	Std time	s.m.
3872	Dipping in gelatine	48 rks.			
-	Moving boxes of rejects	10 min			
-	Quality check				
3869	Dipping in gelatine	44 rks			
-	Collecting trolleys	10 min			
-	Quality check				
-	Unloading trolleys	5 min			
3882	Dipping in varnish	28 rks			
3869	Dipping in gelatine	14 rks			
3874	"	68 rks			
-	Unloading trolleys	3 min			
3876	Dipping in gelatine	71 rks			
-	Inspecting rejects	15 min			
-	Quality check				
3875	Dipping in gelatine	81 rks			
-	Sorting and removing knobs from handles	8 hand			

	WAITING TIME				
2.35pm	Machine breakdown	20 min			

	OTHER COMMENTS				
3869	Using one bath only	10 min			
3874	Using one bath only	10 min			

Figure 13.2 A daily work record submitted by the operator in the case-study

These may be subdivided as follows:
  (i) The purpose of the study.
  (ii) Production figures, and comments on these.
  (iii) Basic times, with comparisons against issued basic times.

(iv) Relaxation allowance. Comparison of allowed R.A. against the amount of time actually taken by the operative.
(v) Contingency allowance: analysis of contingencies to validate the issued contingency allowance.
(vi) Analysis of breakdowns and other machine or process failures.
(vii) Analysis of ineffective time and waiting time.
(viii) Operator indiscipline, i.e. lateness, absence, early finishing, time-wasting, and so on.
(ix) Incorrect recording of time and volumes produced on the time-sheet.
(x) Any other points of recommendation to be brought to the attention of management (i.e. methods improvements, new tools or machinery, etc.).

13.31 MATHEMATICAL ANALYSIS

The validity of an existing standard or basic time can be assessed mathematically using tests of significance based on the statistical concepts outlined in Chapter 6.

The purpose of the production study may be to check the existing standards which have been in use for a considerable period, or to confirm a basic time which has just been set by work measurement and which is suspect in some way.

In the former case, if the particular basic time had been considered satisfactory for a significant time then the sample size can be taken to be very large; thus the comparison is between the sample obtained from the production study and the existing standard.

When a basic time has to be validated in this way it is necessary to measure some of the cycles individually to obtain a sample from which the standard deviation of the cycle basic times may be calculated. Suppose the existing basic time for a job were 2·1 b.m., and the production study yielded a mean basic time of 1·90 b.m. based on a sample of 80 individual cycles. The significance test measures the separation on the two basic times, not in minutes but in standard errors (s.e.).

From Chapter 6, the standard error = (standard deviation) $\div \sqrt{}$ (sample size). The standard deviation is calculated from the 80 individual cycles, as described in Section 6.13 of Chapter 6. Suppose this were 0·3 b.m., then the standard error is

$$0.3 \div \sqrt{80} \text{ or } 0.034 \text{ b.m.}$$

The significance test is performed by determining how many standard errors separate 2·1 from 1·9 b.m., i.e. 0·2 $\div$ 0·034, which is approximately 6 standard errors.

Because the criterion for significant difference is 2 s.e. the separation of 6 s.e. is very highly significant. In other words there appears to be a significant discrepancy between the original standard and that produced by the production study. This warrants further investigation, and it could mean that the new standard of 1·9 b.m. should replace the existing one.

If the production study were designed to check a basic time which had just been set it would be necessary to calculate the standard deviation of the basic times for the cycles studied by the work-measurement application. For example, let the results of the time study be: mean basic time = 1·98 b.m., standard deviation = 0·28 b.m., and number of cycles studied = 36. Suppose the results of the production study were as in the previous example. This time, because there are two standard deviations, one for each sample, a 'pooled standard error' must be used. (In the first example the sample of the existing standard was considered to be very large and thus the standard error by the above formula negligible.) The pooled standard error or *standard error of difference* is found by adding the variances (i.e. the squares of the standard deviations).

$$\text{The s.e. of difference} = \sqrt{\left(\frac{\sigma_1^2}{n_1} + \frac{\sigma_2^2}{n_2}\right)}$$
$$= \sqrt{\left(\frac{0\cdot 0784}{36} + \frac{0\cdot 09}{80}\right)}$$
$$= \sqrt{(0\cdot 0022 + 0\cdot 0011)} = 0\cdot 058 \text{ b.m.}$$

The significance test is (1·98 − 1·90) ÷ 0·058 = 1·38 s.e. As this is less than 2 s.e. there is no significant difference and the time-study result of 1·98 b.m. is confirmed.

It must be remembered that these analyses are *statistical* indications and they do not show the *reasons* for differences.

13.4 AN EXAMPLE OF A PRODUCTION STUDY

The following is an account of a production study in report form. The actual recordings are shown in Figure 13.1 with a Daily Work Record (Figure 13.2).

Report on a Production Study Carried out in the Coating Room (Paint-Brush Section)

Review of the situation
The existing standard times were agreed and introduced on 20 September 1970 by the Industrial Engineering Department (Work Study Section), using time study. The two jobs involved (with their Work Study References) were

Op. No. 286 Varnishing brush handles   0·87 s.m. per rack of handles
Op. No. 294 Gelatine-dip brush handles   0·72 s.m. per rack of handles

Although the standard times were agreed at this time, production had been consistently below the anticipated levels for these operations. Complaints had been received from operators engaged on these tasks that they were unable to earn a reasonable amount of bonus payment. The supervisor for the section was sympathetic, and through his Department Manager had requested a production study to validate the standard times. Management felt that insufficient time had been allowed for incidental work and interruptions.

Consequently a production study was conducted on 23 January to cover the whole day from 08·30 to 16·40. The study is analysed in the following report.

The purpose of the study

To check the validity of the standard times in the current situation, and check the allowances for relaxation, contingencies, and process times; further, to investigate possible causes for restrictions in working.

Workplace arrangement

A flow diagram and multiple-activity chart for the work are given in Appendix I of the production study report. (*Note:* They are not reproduced in this book.)

SECTION 'A': FINDINGS OF THE PRODUCTION STUDY

1. Production

The tasks for which the operator is engaged are dipping handles in gelatine, and varnishing handles. She accumulated a total of 284·5 s.m. while engaged on this work (*see* Section 'B', Part I). Her production was 320 racks of dipping and 29 racks of varnishing. This, together with ancillary work, produced a departmental performance of 72 per cent.

2. Basic times

The production-study times for the operation indicate that both basic times are 'tight'. If the accepted limits are taken as ± 5 per cent, both basic times are outside this limit. The production-study basic time for dipping in gelatine was 12 per cent higher than that issued, while the basic time for varnishing was 10 per cent higher than the existing basic time.

3. Relaxation allowance

The relaxation allowance was checked using the standard Tables of Relaxation Allowance, and the existing standard was compared with this value, and with the actual relaxation taken by the operator. An interview with the operator subsequently revealed that she was not unduly taxed during the study, and she felt that she had taken adequate relaxation.

The analysis of her relaxation is as follows:

Operator away from her workplace	22·34 min
Operator takes coffee and tea breaks	36·75
Operator stops work to talk to neighbour	9·21
Total	68·30 min

The above figures do not include late starts and early finishing, which are dealt with separately in this report.

The calculation of percentage relaxation time was based on total working, and not just working on operations 286 and 294.

Total attendance	Ineffective	Working	Relaxation	Relaxation %
445 min	26 min	419 min	68 min	16

4. Contingencies

The contingencies for the work amounted to 13 per cent, of which 6 per cent is considered to be usual for this work. Periods of work involving the use of trolleys is not considered to be part of the usual operation and is not included in the contingency allowance recommended. Checking quality should be removed from contingencies and included in the basic time as an occasional element.

Contingencies observed in the study are listed below.

	Observed	Allowable
1. Prepare to work, and clean up at the end of the day	7·64	10·00
2. Add gelatine to the baths twice per day	6·10	6·00
3. Check quality four times a day	7·78	
4. Consult with supervisor (about the work)	7·74	6·00
5. Move boxes of work	2·70	
6. Get trolleys	6·36	
7. Unload from trolley to tunnel	6·85	
8. Move between operations (dipping to varnishing etc.)	2·79	3·00
9. Tidy up after lunch	4·75	
10. Tidy up boxes	2·06	
Total	54·77	25·00

Working time = 419 min.   Contingencies = 55 min = 13 per cent
                                         Allowable   = 25 min = 6 per cent

5. Ineffective time—breakdowns
During the study there were three breakdowns caused by equipment failure. On two occasions it was possible for the operator to continue working, using one bath only. On one occasion the whole equipment was inoperative, during which time 6·7 min were lost while the operative awaited the arrival of the mechanic, and a further 8·82 min due to the mechanic working on the fault. One or both baths were out of action for 7·0 per cent of the day.

6. Ineffective time—interruptions
Of the several interruptions, most were unavoidable, being part of the job. The remaining interruption was by the supervisor questioning the operative about her work-sheet; an insignificant time of 1·76 min.

7. Ineffective time—indiscipline
A total of 8·12 min was lost through bad timekeeping, i.e. late start in the morning (2·10 min), early finish at lunch time (3·94 min) and early finish at the end of the day (2·08 min).

8. Ineffective time—additional work
The tunnel was observed to be jammed on two occasions because of the inadequate capacity. A total of 13·21 min was lost through the operative searching and fetching trolleys, and subsequently unloading these trolleys to the tunnel. The difference in basic time between loading work to the tunnel and loading work to trolleys introduced a further slight loss of time. This

difference of 0·015 b.m. per rack (on 50 racks) amounted to a negligible 0·75 min.

9. *Incorrect recording*

Several discrepancies were observed between the data recorded by the observer and those recorded by the worker on her Daily Work Record. These are summarized in Section 'B'—Analysis of Results, Part VIII. Some errors were in the operative's favour, but one was to her disadvantage, and was obviously a mistake in omitting to record a quality check.

SECTION 'B' ANALYSES

PART I: OPERATOR RATING AND PERFORMANCE

The arithmetic means of the ratings of the operator were as follows:

$$\text{Op. No. 286 Varnishing brush-handles} = 84$$
$$\text{Op. No. 294 Dipping brush-handles} = 89$$

The departmental performance was calculated as follows:
   The total production on dipping is 320 racks
   The total production on varnishing is 29 racks

Standard times (amended as in Section 'B' Part II) are 0·820 and 0·968 respectively.

Thus standard times for measured work are:

$$(0·820 \times 320) + (0·968 \times 29) = 290·5 \text{ s.m.}$$
$$\text{Ancillary work (assessed at actual time)} = 31·3$$
$$\text{Total work} = 321·8$$

$$\text{Departmental performance (as per B.S.I. definition)} = \frac{321·8}{445·0} \times 100$$
$$= 72\%$$

Possible output at 100 performance:

    Op. No. 294 Dipping brush-handles   543 racks per day
    Op. No. 286 Varnishing brush-handles 460 racks per day

PART II: ANALYSES OF STANDARD TIMES

Operation: Dipping brush-handles in gelatine	Prod. study	Existing	Var. %
Basic time per rack	0·665 b.m.	0·596	12%
Add Quality check (1 in 136) at 2·0 b.m. per check	0·015	—	
Relaxation allowance (at 14% of the basic time)	0·094	0·083	
Contingency allowance (see Part IV below)	0·046	0·041	
Standard time with quality check included	0·820	—	—
Standard time without quality check	0·803	0·720	12%

Operation: Varnishing	Prod. study	Existing	Var. %
Basic time per rack	0·784	0·713	10
Add quality check (1 in 120) at 2·0 b.m. per check	0·017	—	
Relaxation allowance (at 14% of the basic time)	0·112	0·100	
Contingency allowance (see Part IV below)	0·055	0·060	
Standard time with quality check included	0·968	—	—
Standard time without quality check (for comparison)	0·948	0·873	10

Operation: Dipping and loading to trolley		
Basic time per rack	0·680 b.m.	

PART III: RELAXATION ALLOWANCE

Total basic minutes of work assessed at 88 rating (see Pt. VII)	290·0
Total allowance in minutes (at 14% of the total basic time)	40·6
Actual R.A. taken by operator	68·0
Variance in minutes	27·4
Variance in percentage	67·5%

PART IV: CONTINGENCY ALLOWANCE

Total contingencies observed in the study	54·77 min
Contingencies allowable (observed in study)	24·27
Contingencies at the issued 6%	25·00
Variance	0·73

PART V: ANCILLARY WORK

Moving boxes of rejects to stack	2·70 min
Obtaining trolleys	6·36
Unloading work from trolleys to tunnel	6·85
Tidying up boxes	2·06
Sorting handles for rejects	8·76
Removing rejects to reject area	2·90
Sorting rejects	1·65
	Total 31·28 min

Dipping using one bath (not included in the total)	14·93 min

PART VI: INEFFECTIVE TIME   *Prod. study Existing Var. %*

1. *Breakdowns etc.*
Awaiting mechanic   6·70
Mechanic working on apparatus   8·82   15·52   3·5% of day
(one bath under repair)   14·93

2. *Interruptions* (other than work)
Supervisor consults about time sheet   1·66

3. *Discipline*
Late start   2·10
Early finish at lunch   3·94
Idle at end of day   2·08   8·12

4. *Study error*
Difference in cumulative watch-reading and works clock   0·68
                                                        ─────
                                                        25·98

PART VII: TIME ENGAGED ON DIPPING AND VARNISHING

		Weighting	*Rating*	*Weighted rating*
Time engaged on dipping	= 207 min	207·0	89	18,423
Time engaged on varnishing	= 22·7	22·7	84	1,907
Total times		229·7		20,330

The weighted average rating is $\dfrac{20{,}330}{229\cdot7} = 88$ rating

Total work, calculated at 88 rating:

Dipping	*Varnishing*	*Dipping (one bath)*	*Ancillary*	*Work contingencies*	*Total work*	*Basic mins at 88 rating*
207·0	22·7	14·93	31·28	54·07	329·98	290

PART VIII: RECORDING

	Dipping	*Varnishing*
Quantities produced during the study	320	29
Quantities recorded on work-sheet	326	28
Variance in operator's favour	+6	−1

	Operator's	*Observer's*	*Variance*
Moving boxes of rejects	10·00	2·70	+7·30
Collecting trolleys	10·00	6·36	+3·64
Unloading trolleys (first occasion)	5·00	4·32	+0·68
Unloading trolleys (second occasion)	3·00	2·53	+0·47
Inspecting rejects	15·00	13·88	+1·12
Quality check	not recorded	1·95	−1·95
Machine breakdown	20·00	15·52	+4·48
Net variance			+15·74

Section 'C': Conclusions and Recommendations

1. Production
Although the operative is engaged for dipping and varnishing, the many interruptions experienced, and engagement on other ancillary work, caused her to produce a low departmental performance of 72 per cent. It was noticeable that the operator's average rating rose to 100 on the occasion when she was able to work for a sustained period of 22 min. It is recommended that the work be better organized to reduce the number of interruptions and allow the operator longer runs at the work. Even at 90 rating she should be capable of producing 490 racks of dipping, and 423 racks of varnishing, or any combination between these values, compared with the actual output of 370 racks produced during the present study.

2. Standard times
It is considered that the standard times are 'tight' and it is recommended that the jobs be re-measured to confirm the production-study times of:

Op. No. 294 Dipping in gelatine   0·820 s.m. (with quality check included)
Op. No. 284 Varnishing   0·968 s.m. (ditto)

3. Relaxation Allowance
Although the worker actually took 16 per cent relaxation (18 per cent if bad timekeeping be included), the 14 per cent R.A. has been rechecked and is considered adequate for this work.

4. Contingencies
It is recommended that the quality check be removed from contingency allowance and included in the basic time as an occasional element. In this event the contingency allowance of 6 per cent is adequate.

5. Breakdowns
One or both baths were out of action for 7·0 per cent of the day, and in view of this it is recommended that the equipment be investigated and if necessary overhauled. The possibility of planned maintenance should be investigated.

6. Interruptions
This was dealt with in Section 'A'.

7. Indiscipline
A check made on the clock-cards for operatives in this section revealed that lateness was not uncommon. It is recommended that time-keeping at starts and finishes of periods of work be subject to a review by management.

8. Additional work
While it is appreciated that performance of some additional work is unavoidable, some of this could be delegated to labourers, and the rest could be eliminated by reorganization. In any event additional work should be kept to a minimum.

9. Incorrect recording

The apparently lax discipline in the section is reflected in the recording, an analysis of which is made in Section 'B' Part VIII. It is recommended that this problem be discussed with the operatives to improve the accuracy of recording work produced and other data.

10. General conclusions

The conclusions of the observer were as follows:

The operators were unable to earn sufficient bonus because:

(i) The standard times are considered to be tight due to 'drift';
(ii) The operators are not allowed to gain their rhythm because of the lack of opportunity for sustained effort.

Management's allegations of insufficient time for contingencies and interruptions are unfounded; the presence of interruptions is a management organization responsibility and should be drastically reduced or eliminated.

11. Acknowledgments

The observer would like to record his appreciation of the co-operation of the supervisor and operative throughout the study.

14 Analytical and Comparative Estimating

14.1 MERITS AND DEMERITS

Of the techniques of work measurement those in the category of predetermined motion–times offer most in terms of precision and consistency. However, the advantages of these systems in certain circumstances may be offset by other disadvantages, the most important of which is the time required to set standards by these means, and hence the cost involved. Even measurement by *timing* may be inappropriate in some cases, such as 'one-off' jobs, and where a standard time is required before the work commences.

In circumstances where the use of timing or predetermined motion–times is obviated through reasons of economy or lack of sufficient time, and for jobs which require a time-standard derived from a 'visualization' of the work before it is actually performed, the technique of *estimating* may offer the only practical means of measurement.

Estimating is not accepted universally as a valid method of work measurement; indeed the purist may reject the concept entirely on the grounds of inconsistency and subjectiveness. Nevertheless the technique is widely used, and incorporated in many of the so-called proprietary systems.

The definition preferred by the British Standards Institution for estimating (BS 3138:1979) is:

> Assessment of the time required to carry out work, based on knowledge and experience of similar types of work (Term number 41004). Estimating may be done without a detailed breakdown of the work into elements and their corresponding times at a defined level of performance.

While it may be used to obtain a rough assessment of the time required to perform a job, in its crude form estimating does not provide the necessary consistency or precision, and because it is not necessarily assessed at any particular pace of working the time may not be appropriate, being too slack or too tight. In this form estimating does not offer a reliable means of establishing time standards for most management-control purposes.

Despite these inherent drawbacks, estimating still retains the attractive qualities of simplicity and speed of application, and attempts have been made to introduce methods designed to inject an element of greater precision and consistency into the unsophisticated basic method of estimating. Three main subdivisions have evolved from long practice and experience of using the technique, and these are generally known as analytical estimating, comparative estimating, and category estimating. The three terms are separately defined in the respective sections of this chapter devoted to them. The first technique, analytical estimating, will now be described.

14.2 ANALYTICAL ESTIMATING

Analytical estimating may be described as a combination of estimating and synthesis. The synthetic data employed may be in recorded form or exist as mental images in the mind of the estimator.

The technique is defined by the British Standards Institution as:

> A work measurement technique: being a development of estimating, whereby the time required to carry out elements of a job at a defined level of performance is estimated partly from knowledge and practical experience of the elements concerned and partly from synthetic data (Term number 31004).

The procedure has its main uses in engineering maintenance and allied trades, the methods being determined and the times assessed for particular jobs by estimators who are skilled tradesmen trained in work study.

Briefly the estimator analyses each job, breaking down the work into elements. Data are used to establish the basic time of as many elements as possible. However, often there will be some elements for which no data exist and the estimator will draw upon his training and experience to estimate the basic time of these.

The technique is based upon the following assumptions:

1. That a skilled tradesman, properly trained in work-study techniques, can make a reasonable estimate of basic times for elements of work with which he is familiar;
2. The breaking down of the job into elements ensures that while the accuracy of any one element-time may not be very good, the errors will be random, so that it is probable that the overall accuracy for each job will be within acceptable limits (say $\pm$ 10 per cent), due to compensating errors; and
3. That errors of up to $\pm$ 10 per cent in particular jobs should themselves be random, thus reducing the error over a week's work to negligible proportions compared with 'job to job' errors.

14.21 USES OF ANALYTICAL ESTIMATING

The more important uses of analytical estimating are:

1. The provision of reasonably reliable job-target times which may be used to improve the planning and control of the work;
2. Provision of data which may be used to evaluate the cost of labour employed on specific jobs;
3. To assist method study in the improvement and standardization of methods;
4. To provide standards which may be used as a basis for incentive schemes.

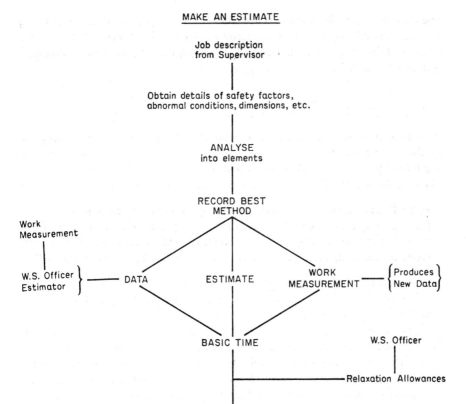

Figure 14.1 Procedure for analytical estimating

14.22 MAKING AN ESTIMATE

The basic steps are outlined in Figure 14.1. They can be described in concise terms as:

1. Decide on the best method of carrying out the job.
2. Obtain details of safety factors, abnormal conditions, lengths, areas, quality, size, distance, and other parameters.
3. Analyse the job into convenient and appropriate elements.
4. Apply available synthetic data to as many elements as possible and adjust as necessary for abnormal conditions.
5. Carry out work measurement on repetitive elements where economically justified.
6. Estimate remaining elements to obtain the comprehensive basic time for the whole job.
7. Add appropriate allowances, and calculate the allowed time.

The details of these steps are summarized below:

1. Decide on the best method of working

A broad description of the job is given to the estimator by supervision. Before an estimate can be made it is essential to decide upon the proper method by which the work should be carried out. The estimator benefits by being a skilled tradesman who can capably perform the work himself.

It is possible, therefore, to visualize each step to be taken. He will know which steps have a variable work content; for example, nuts can be well greased, slightly rusted, or completely corroded, so that they may be anything from easy to impossible to remove by spanner (the latter case demands other means of removal). Such steps he will check by inspecting the job, where possible, before it is started, so that the method of working can be decided.

The estimator now uses his knowledge of work study to decide on the best method to be used. Very often in engineering-maintenance work formal method study is not economic unless the job is repeated at frequent intervals.

Alternatively a multiple-activity chart may be prepared quite quickly for a team job, resulting in significant savings, so in all cases the estimator must use his own judgment in deciding the extent to which method study (and indeed other disciplines) may be employed.

2. Analysis into elements

Prior knowledge of the work to be undertaken will enable the estimator to analyse the job into appropriate elements of work. Normally these elements will be of longer duration than those selected for use with other techniques of work measurement. The actual lengths will vary according to the form of the work being studied, but an element of 1 min is typical in a job lasting an hour or more, with elements of 5 min being the maximum which can be estimated with any degree of confidence.

The estimator is now able to write brief elements descriptions based on the agreed method of working, and for this purpose a standard recording sheet such as that illustrated in Figure 14.2 may be used. The descriptions should be sufficiently comprehensive to instruct the operator in the method, and to distinguish the different items of work and the sequence in which they are to be performed. It is inevitable that in the interests of economy some details of logical sequence must be sacrificed. For example, if a ceiling is to be painted and this involves moving steps four times, the estimate will not show the laboured description:

1. Position steps at 'A'.
2. Paint 10 sq.m of ceiling.
3. Position steps at 'B'.
4. Paint 10 sq.m of ceiling.
5. Position steps at 'C'.
6. Paint 10 sq.m of ceiling.
7. Position steps at 'D'.
8. Paint 10 sq.m of ceiling.
9. Remove steps.

ANALYTICAL AND COMPARATIVE ESTIMATING

A more concise description in standard data-form, as it would appear on the estimate sheet, is:

1. Position steps four times.
2. Paint 40 sq.m of ceiling.

ESTIMATE SHEET

Schedule   Sheet number
Job   Date
Location   Estimator

No.	DESCRIPTION	Basic mins
	MOVE TO SITE, PREPARATION, CLEAR SITE	

Basic time =   mins.   STANDARD TIME =   s.m.
R.A.   =   mins.
                                 ALLOWED TIME =   s.m.

THE TIMES QUOTED ABOVE INCLUDE TIMES FOR GETTING AND CHANGING TOOLS, WIPING HANDS, PREPARATION, UNLESS OTHERWISE STATED.

Figure 14.2 An estimating sheet

3. Applying the data

The next task of the estimator is to allocate the appropriate time-standards to the elements. When making the element breakdown the proficient estimator would have chosen elements which coincided with the standard elements appearing in the synthetic-data tables where this were possible. This ensures that the need for estimating is minimized. Provided the initial data-collection for element breakdown was carried out effectively, sufficient additional information to satisfy the demands of the variable elements (for example, area to be painted, length of pipeworks, distances walked, and so on) would have been noted. This requires experience of the work, and a degree of foresight.

It is unlikely that labour, materials, and conditions under which the work is executed will exactly correspond with the descriptions of the standard elements. Significant variances may be assessed and due allowance made to accommodate such differences.

4. Applying work measurement

Occasionally a job may contain an element of high repetition. A small error in the production-unit time will be magnified by the successive repetition of the element, causing the cumulative error in the complete job-time to be considerable. Such cases could warrant the use of synthetic data generated from a predetermined motion–time system or from time study, whichever is more appropriate. This action is justified if the element or job is likely to recur at future times, in which case it may be incorporated into the synthetic-data bank.

5. Estimating the remaining elements

In engineering maintenance certain jobs will occur for which synthetic data are not available and which do not justify the use of timing or predetermined motion–time systems. Examples of such elements are 'Wait for storeman', and 'Remove grease from surface area'.

While it may be possible to collect sufficient data and subsequently to measure the basic time for average conditions, because of factors such as low frequency of occurrence or excessive variability requiring very high sample sizes, orthodox measurement may be rendered uneconomic. The estimator will be required to resort to estimating the basic time relying on his experience as a tradesman and his knowledge of work study to produce a reliable estimate.

6. Calculating the allowed time

The allowed time is computed from the basic job time from the summation of all elemental times (adjusted by their frequencies of occurrence) and addition of appropriate relaxation allowances. To this, the standard time, may be added relevant bonus increment or policy allowance. The whole process follows the principles described in previous chapters.

7. Provision of instructions to the operator

A duplicate copy of the estimate sheet may be provided for information to the operator. The purpose of the duplicate sheet is to instruct the operator in the correct working method, but should not include individual-element times.

On completion of the job the sheet should be returned to supervision and hence to the bonus clerk (where incentive schemes are operated) or any other interested party.

Individual-element times are excluded, to obviate loss of confidence in the estimator through the occurrence of errors in individual elemental times. Because of the randomness of errors the total basic time will be much more accurate.

14.23 DATA FOR SYNTHESIS

Synthetic data are provided to facilitate the assessment of the work estimate, and to improve the consistency and speed with which the estimate is prepared. Thus it is essential that the data be issued in such a way that particular values can be quickly located. The importance of clear descriptions of elements, working methods, and details of tools has been stressed previously. Values may be given as averages for a range of conditions, or as 'best' conditions, but these must always be stated. In many cases it is advantageous to give the average and extreme values at each end of the range.

For example: Loosen $\frac{1}{2}$ in. Whitworth nuts.

$$\text{Rusty} = 0.5 \text{ min} \quad \text{Average} = 0.2 \text{ min} \quad \text{Good} = 0.1 \text{ min}$$

Synthetic data should be obtained in the most suitable way, using the most appropriate method of work measurement.

If time study be chosen, particular care must be exercised in selecting the break-points so that elements are complete in themselves and can be used in the synthesis of larger work-values. Because of the non-repetitive nature of the work it will often be necessary to study several different jobs to obtain one common element rather than to study one job many times. Provided the objective is clear, then useful and accurate data will result.

When predetermined motion–time systems are used for measurement the precise motion pattern will be clearly defined, and any variations from this will become apparent. It is essential, however, that the work-study practitioner should be experienced in the trade of the job and should have had formal training in the particular system.

Other techniques may be used to augment these methods of measurement, such as activity sampling and Basic Work Data. One limiting factor of such second-level systems as Basic Work Data is the loss of accuracy on short elements compared with that provided by the first-level systems of predetermined motion–times.

To use the data effectively the estimator is obliged to understand the implication of the definitions given to the elements of work so that he is able to make quite accurate modifications to the figures to cater for different conditions.

14.24 SPECIAL CASES OF ESTIMATING

1. Estimating for scheduled maintenance

Where it is possible to introduce preventive maintenance jobs will tend to become repetitive and pre-planned. Repeated use of standard data will demand more reliable standards. Whenever possible work-measurement techniques will be used for these jobs, and relaxation allowance will be assessed separately for each element. Particular attention should be paid to:

(i) The use of the most suitable tools and equipment;
(ii) The best practicable method of working;
(iii) The most efficient manning for any job requiring more than one man.

The above requirements should be specified clearly in a standard work-specification. Amendments to these specifications will almost certainly necessitate re-study.

2. *Estimating for breakdowns*

Inevitably there will be urgent jobs which will be started before the estimate can be produced.

(i) Where repairs are effected at the workshop and the extent of the damage may not be apparent or it may not be possible to diagnose the fault, the supervisor will not be able to assess the amount of work involved. It will be necessary to deal with the job in stages. In such circumstances the estimator will prepare an estimate in advance for the known work. At the end of the job a supplementary estimate will be prepared for the previously unknown content.

(ii) Alternatively, the engineer may be sent out to breakdowns. Supervision will not know the amount of work involved until after the completion of the job, on the return of the engineer. Again, an estimated-time standard in retrospect (post-estimating) must be prepared. An essential requirement of post-estimating is that the supervisor pass to the work-estimator all instructions which he has given to the tradesman and ensure that the work estimator is in full possession of the relevant facts. The supervisor should pay particular attention to significant details when he enters a description of the job in such written records and log books as he is required to keep.

It has been generally accepted that post-estimating should be minimized, since with large numbers of post-estimates the operators appear to lose the incentive to work and become easily dissatisfied.

Certain systems employ 'diagnostic standards' which are standard times for diagnosing faults of unknown origin. The standard time is essentially an average time developed from past records and experience in the work. Usually the fault-finder is required to report back to his supervisor after a certain fixed time if he has failed to locate the source of the trouble. On reporting he may be given a further standard time, or additional assistance in solving the problem. Individual forms of this diagnostic standard vary, Two important versions of this concept are found in MTM-2 Maintenance Data (Minor Job Ticket), and Universal Maintenance Standards (Troubleshooter's standard).

3. *Estimating for routine work outside the department*

Most building and construction work (for example shop-fitting) will require the estimator to visit the site before work is started so that he may acquaint himself with the nature of the work and prevailing conditions at the site. This is particularly necessary with maintenance work at branches. The estimator will make a tour of his area with the supervisor at regular intervals, and all work to be executed will be noted and assessed.

4. *Extra work*

On occasions the actual conditions and work will vary from those specified in the estimate. The temptation to allow extra working time should be resisted unless it can be demonstrated that the additional work has not been covered in the estimate, and confirmation of this given and agreed by management.

5. *Travel time*

It is not possible to estimate or measure the times for all journeys to, from, and about the site. Certain of the more frequent journeys, such as between head office and garage, could be measured and synthetic data established. It should be possible to arrange a system of geographical zones between which standard times for travelling may be established.

Using the accepted standard rate of walking, or some other agreed pace, it is possible, with the aid of large-scale plans of the area, to assess the travelling times between specified points.

14.25 ACCURACY OF ANALYTICAL ESTIMATING, AND METHODS OF CONTROL

The accuracy of a work estimate set by this means (compared with an equivalent one established through the use of, say, MTM) depends on several factors such as the provision of sufficiently reliable information about the job. The ability of the estimator to visualize the best method of working to produce a complete list of elements is a further important factor. Clearly the estimate is as reliable as the ability of the estimator to make accurate estimates of basic times for each element.

Estimating accuracy may be controlled through the use of statistical checks installed by the work-study officer, the checks being designed to include the estimator. Methods of control are described below.

A random sample of work estimates which have already been prepared can be taken and a *time study* then taken of the job. The results of the comparison are then entered on a control chart suitable for the purpose.

If the circumstances so demand, a *production study* may be made, with periodic ratings of individual workers. The production study is described in Chapter 13.

Predetermined motion–time systems are useful for the purpose of checking estimates, and while time-consuming to perform, the studies provide the degree of consistency sometimes demanded. Normally the systems would be reserved to check elements rather than complete jobs. One possible advantage of this method is that elements so tested may become standard data for future use.

Companies employing one of the second-level systems such as Basic Work Data have at their disposal a quick and convenient way of determining the basic time for elements of five to ten minutes' duration. With elements of shorter length the accuracy tends to reduce, while the use of the method with exceedingly long elements involves the analyst in an excessive amount of work. However, this may be offset by the advantage of producing further synthetic data.

Standard estimates are of value where jobs are reasonably repetitive. A standard, compiled from time study or a predetermined motion–time analysis, is established for a complete job. The estimator is then given the job description and asked to produce an estimate. The estimate can be compared with the standard estimate element by element.

Another, rather dubious method, requires estimates, chosen at random, to be scrutinized by a panel comprising management, supervisors, and work-study personnel. Experience has shown this to be a most unsatisfactory method, leading to ill feeling and accusations of bias leading to rate-cutting.

The only value in this check is in the discussions of methods of working, which encourages participation.

14.3 ESTIMATOR TRAINING

The content of any training syllabus for estimators will be dictated by the ability and previous experience of the trainees. The programme specified below is a comprehensive syllabus intended for potential estimators with virtually no experience.

The course incorporates classroom training and practical training under actual working conditions. The initial practical training of about two weeks is carried out under close supervision on highly repetitive work to familiarize the estimator with the manual dexterity which can be achieved on such jobs. Subsequently the trainee graduates to studying work in the department for which he was engaged.

The final training-period includes instruction in labour balancing for the most efficient method of working.

It is essential that the trainee be closely supervised during his training period, and indeed during subsequent initial installations. He may assist the work-study officer to collect the data and to estimate times, but these should be checked by the experienced work-study staff.

Syllabus

Week 1. Method study, recording procedure, critical examination, methods improvement, consultation, and installation.

Week 2. Job specification, operation and element breakdown, element description, study procedure, and predetermined motion–times.

Week 3. Work measurement, rating, time study, synthetic data, allowances, performance, indices, incentives.

Week 4. Activity sampling, rated-activity sampling, analytical estimating, operator training.

Week 5.⎫ Practical work on the shop floor involving all work-study procedure
Week 6.⎭ on highly repetitive operations.

Week 7.⎫ Practical work on the shop-floor involving all work-study procedures on maintenance work. The emphasis will be on visualizing (from a brief job-description) all the elements of work necessary in doing the
Week 8.⎭ job. This will be achieved initially by writing detailed descriptions of short jobs such as replacing the wiring on a plug and inspection lamp.

Week 9. Preparation of estimates and collection of data under supervision.
on

14.4 COMPARATIVE ESTIMATING

Another method of improving the consistency of estimated times has been developed under the title of *comparative estimating*. The technique relies on the precisely measured *bench-mark jobs* which provide the criteria with which tasks to be estimated may be compared.

Probably the first system to formally employ the technique was Universal Maintenance Standards, but subsequently other workers have incorporated comparative estimating into their systems of measurement.

Comparative estimating is defined in B.S.I. terms as:

a work measurement technique in which the time for a job is evaluated by comparing the work in it with the work in a series of similar tasks—benchmarks—the work contents of which have been measured. The arranging of tasks into broad bands of time is referred to as 'slotting' (Term number 41006).

The above definition also provides a succinct but comprehensive description of the method.

As with most other forms of measurement of this type two main phases may be distinguished: the static one of setting and maintaining the comparative standards (bench-marks) and the dynamic phase of applying the standards. The first phase of setting up the system demands the services of work-measurement specialists capable of establishing adequate standard times with the appropriate degree of precision. It will be appreciated that the system cannot be put into operation until a comprehensive library of typical bench-mark jobs with their respective standard times has been assembled. This is a laborious task involving hundreds of jobs and months of work. It is for this reason that often companies resort to the purchase from firms of consultants of their data-banks, the results of many years of research and thousands of work-measurement studies.

As it is measured, each bench-mark is allocated, or 'slotted' to its appropriate *time band* or category (see Figure 14.5). Thus, using the categories of Figure 14.4 as examples, a bench-mark job of four hours' duration will be allocated to category 'F', having a range of 2·5 to 4·6 hr with a mid-point of 3·35 hr. All other bench-mark jobs are slotted similarly, and tabulated with short job-descriptions.

Once the standards have been derived and slotted the continuing task of application may be introduced. This responsibility is delegated to the line supervisor in charge of the section, or a senior craftsman. When he has occasion to estimate a time for a new task, using his experience and knowledge of the work he searches the bench-mark descriptions for those jobs of similar work content to the one to be estimated. On locating these, the category in which they occur is assigned to the new job, which is given the category mid-point as its standard time.

14.41 ADVANTAGES OF COMPARATIVE ESTIMATING

The advantages of the system favour its use in the fields of maintenance, and similar work of non-recurring nature. The main merits are:

1. Minimum of training

Only a limited number of jobs need be measured. The application requires only an ability to compare the work content of jobs; a detailed knowledge of other work-measurement techniques is not necessary, although experience in the type of work studied is essential.

2. Economy of staff

Often the estimator is the maintenance superintendent or supervisor, but even where a special estimator is employed the tradesmen/estimator ratio is much higher with this technique than with other forms of measurement, including analytical estimating.

3. Simplicity of application

Although the bench-marks must be established with care by experienced trained staff using recognized work-measurement techniques, application of the system requires no equipment such as stop-watches, and obviates the need for complicated tables of time standards which require detailed analyses and intimate knowledge of application rules.

14.42 COMPARATIVE ESTIMATING PROCEDURE

The preliminary steps, vital to the success of a project, are selection of the areas for study, and consultation with all interested parties, both of which have been described elsewhere in this book.

Comparative estimating may be adapted for use in most indirect areas (*see* Chapter 15), but it is most prevalent in maintenance departments of medium to large organizations. In such areas it is usual first to make a survey of the organization of the maintenance function, to ensure that its form will be compatible with the new system, and that there exist adequate recording and reporting documents which may serve as a basis for the control or application phase. Work measurement should not exist in isolation, divorced from other related specializations. Consequently it may be decided to install a new system (or to adjust an existing system) of planned and preventive maintenance concurrently with the measurement system, facilitating the use of common docu-

ANALYTICAL AND COMPARATIVE ESTIMATING 337

ments and procedures. Similarly job-descriptions necessary for defining benchmarks will be useful in detailing the correct working methods for routine tasks.

14.43 SELECTING THE CATEGORIES

The successive cycles of the type of work which may be described as 'repetitive' do not repeat in exactly the same way on each occasion they are performed. The deviations may be caused by minute subtle changes of method almost invisible to the eye, or may be large variations due to the existence of highly variable elements in the cycles. Changes in the pace of working from cycle to cycle may also help to vary the actual cycle-times.

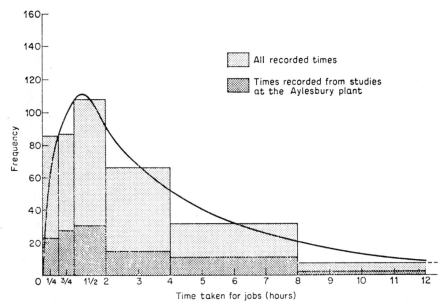

Figure 14.3 A typical frequency-distribution for job times with log-normal curve superimposed

If a large number of cycles of a job be observed and the actual cycle-times attained plotted as a frequency-distribution curve, provided there is no trend or bias present the distribution will be in the form of a 'log normal' curve. Figure 14.3 illustrates the form of this curve superimposed on some clinical data obtained during a recent study. The 'closeness of fit' of the data to the log normal curve will be seen to be very good. This association is important in deriving the mid-points of comparative estimating categories.

The log normal curve is characterized by the marked right-hand skewness. If the logarithms of the values are plotted on the x axis in place of the actual values (Figure 14.4) the well-known normal distribution is formed (*see* Chapter 6).

The dimensions of the categories suitable for comparative estimating depend on the time-ranges of the jobs to be covered by the scheme. Some systems employ

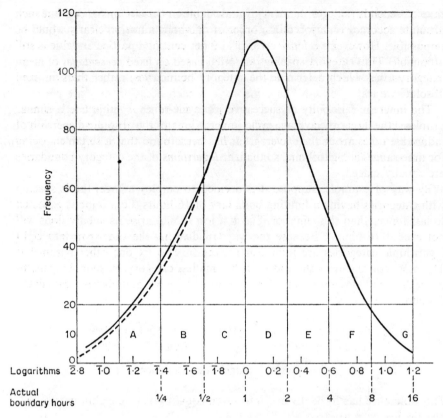

Figure 14.4 The log-normal curve of Figure 14.3 plotted to a logarithmic scale

a dozen categories while others prefer as few as four. The optimum number is generally accepted to lie between five and eight. The fewer the categories the coarser the system, while the larger the number the more difficult becomes the estimating. Clearly it is easier to decide that a particular job should take between 1·00 and 1·45 hours than to allocate it to a narrower band chosen from 1·00 to 1·15, or 1·15 to 1·30 hours, or 1·30 to 1·45 hours.

In estimating maintenance-engineering work, job-times may vary between a few minutes to a few weeks in duration. For reasons of adequate control it is desirable to restrict the categories to a maximum of between 12 and 20 hours, to reduce the demands on the estimating ability of the assessor. Longer jobs may be analysed into smaller stages, which may then be estimated separately.

Once the limiting job-time has been chosen and the number of categories decided, the spacing of the time-ranges may be determined. Category spacings will not be equal, but will increase non-linearly for the following reason, illustrated by way of example. An estimator may be expected to assess the time for the completion of a job of about a minute duration correctly to within a few seconds of the actual time. An estimate of one minute for a job which should take 55 seconds (i.e. +9 per cent) is reasonable in the circumstances, but such absolute accuracy on a job of the order of one-hour duration clearly would be impossible. However a *relative* error of +9 per cent of a particular value is still attainable. Thus category spacings are often based on fixed *percentages* of mean category-times which determine the category boundaries, rather than on fixed absolute values.

The inherent variability of successive job cycle-times is subject to a similar problem: the larger the average cycle-time the greater the absolute dispersion of individual times around the average. It has been found that a suitable spacing for time-ranges is logarithmic, so that the *logarithms* of the category boundaries are equally spaced.

By way of example, suppose it is decided to employ a logarithmic spacing with categories having a limiting boundary of 16 hours (i.e. no job is expected to last longer than this time, or if such a job should arise its subdivisions will not exceed 16 hours). Because the ways of dividing the time-span into eight logarithmic categories are infinite it is necessary to fix one other parameter. The most convenient is the width of the smallest category. A suitable value to cover the small jobs is 0·2 hours. The spacing of category boundaries is calculated as follows:

$$\begin{aligned} \text{Highest boundary} &= 16\cdot 0 \text{ hr} \quad \text{Log } 16\cdot 0 = 1\cdot 204 \\ \text{Lowest boundary} &= 0\cdot 2 \text{ hr} \quad \text{Log } 0\cdot 2 = \bar{1}\cdot 301 \\ \hline \text{Algebraic difference} &= 1\cdot 903 \end{aligned}$$

This difference has to be divided into seven equal parts (the eighth part, 0–0·2, has already been fixed).

$$1\cdot 903 \div 7 = 0\cdot 272$$

Starting with the first value of $\bar{1}\cdot 301$, the other values are found by adding 0·272 successively to each new value thus:

Log	$\bar{1}\cdot 301$	$\bar{1}\cdot 573$	$\bar{1}\cdot 845$	0·117	0·389	0·661	0·933	1·205
Antilog	0·200	0·370	0·699	1·309	2·449	4·581	8·570	16·03
Boundaries (rounded)	0·2	0·4	0·7	1·3	2·5	4·6	8·6	16·0
Categories		0–0·2	0·2–0·4	0·4–0·7	0·7–1·3	1·3–2·5	2·5–4·6	4·6–8·6	8·6–16

It has been shown that jobs are distributed according to a log normal distribution. The boundaries' mid-points to be used will be the antilogs of the log mid-points

Log	$\bar{1}\cdot 301$	$\bar{1}\cdot 573$	$\bar{1}\cdot 845$	0·117	0·389	0·661	0·933	1·205
Mid-points	$\bar{1}\cdot 150$	$\bar{1}\cdot 437$	$\bar{1}\cdot 709$	$\bar{1}\cdot 981$	0·253	0·525	0·797	1·069
Antilog	0·1415	0·2735	0·5117	0·9572	1·791	3·350	6·266	11·72
Rounded (2 sig. figures)	0·14	0·27	0·51	0·96	1·8	3·4	6·3	12

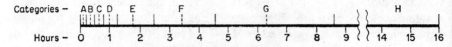

Figure 14.5 Illustration of the categories used in the text

Figure 14.5 illustrates the categories and their mid-points. It is the mid-points which are used as the basic or standard job-times for the purposes of calculating earned time, utilization ratios, and performances.

In some systems the category averages are medians, modes, or often even arithmetic means.

A further method of constructing the mid-value exploits the properties of the normal distribution and its associated probability theory. A working knowledge of the statistical method of Chapter 6 is necessary to appreciate the mathematical theory.

After deciding the number of categories to be employed the mean time for all jobs (obtained from a large representative sample) is used to establish the middle category. This method was used in the application from which Figures 14.3 and 14.4 were constructed. The procedure for establishing the categories is identical with that just described, but working from the middle category instead of from the smallest (0–0·2) subdivision.

The calculations will be illustrated by an example:

(i) Before the mid-points of the categories can be found it is necessary to calculate the standard deviation for the sample job-times so that the areas of the normal distribution enclosed by the category boundaries and hence the probabilities may be calculated.

$$\text{s.d.} = 0.35$$

(ii) The distance of the second boundary to the right of the mean value is measured in standard deviations (i.e. no. of s.d.s $= \dfrac{\text{log boundary} - \bar{x}}{\text{s.d.}}$)

$$\frac{0.661 - 0.253}{0.35} = 1.16 \text{ std deviations.}$$

(iii) The area enclosed by the mean and this number of standard deviations (Figure 14.6a) is extracted from tables of areas under the normal curve to be found in many textbooks on statistical method.

Area for 1·16 standard deviations $= r = 0.377$

(iv) Similarly the area enclosed by the mean and log boundary 1 (p) is also extracted from tables.

$$\frac{0.389 - 0.253}{0.35} = 0.38 \text{ s.d.}$$

Area for 0·38 s.d. $= 0.148 = p$

ANALYTICAL AND COMPARATIVE ESTIMATING 341

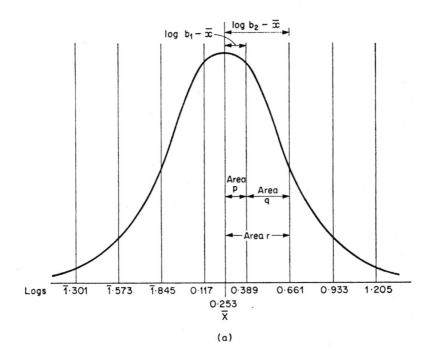

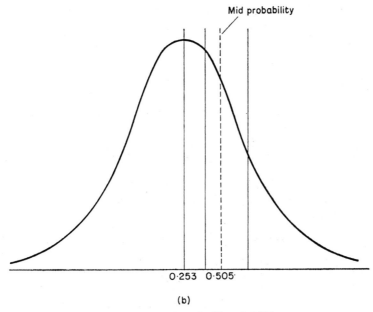

Figure 14.6 Calculation of mid-probabilities

(v) The area q is obtained by subtracting the two areas, $r - p$.
$$\text{Area } q = r - p = 0{\cdot}377 - 0{\cdot}148$$
$$= 0{\cdot}229$$

(vi) The mid-point is that which halves this area (or probability).
$$\frac{0{\cdot}229}{2} = 0{\cdot}115$$

(vii) The whole process is now reversed. The half-value is added to the area p and the number of standard deviations for this total area extracted from tables.
$$0{\cdot}148 + 0{\cdot}115 = 0{\cdot}263$$
$$\text{No. of s.d. for } 0{\cdot}263 = 0{\cdot}72$$

(viii) The number of standard deviations is translated into the log value of the mid-point (using a transposition of the formula in (ii) above, i.e.
$$\log_{\text{mid-pt}} = \text{s.d. (no. of s.d.s)} + \bar{x}$$
$$\log_{\text{m-p}} = (0{\cdot}35 \times 0{\cdot}72) + 0{\cdot}253$$
$$= 0{\cdot}252 + 0{\cdot}253$$
$$= 0{\cdot}505$$

(ix) The antilog gives the actual mid-point value.
$$\text{Mid-point} = 3{\cdot}19$$

(x) Other mid-points are found by repeating these steps for other categories.

The suitability of categories should be tested by taking a sample of, say, 200 to 300 jobs, and constructing a frequency distribution of their operation times. Jobs lasting over 16 hours should be subdivided into smaller tasks. The distribution should approximate to a log-normal distribution (alternatively statisticians may prefer to perform a 'chi-square' test on the data—actual versus theoretically predicted). Usually a compromise will be reached, the final boundaries being so defined that they satisfy the foregoing criteria, and also facilitate estimating by avoiding the whole numbers. For example, an assessor will estimate to the nearest hour for a job of about seven hours. If the category boundary is at 7 hours he will be undecided between the category to the left or the right of the boundary of 7. Careful choice of categories such as those in the foregoing example (e.g. 4.6. to 8.6) avoid this difficulty.

14.44 ESTABLISHING BENCH-MARKS

The establishing of bench-marks and their basic times is a laborious task. Separate work-measurement studies must be made on each bench-mark, and these may number hundreds.

It is necessary to identify the main areas of work, the bench-marks for which may be derived individually. Typical areas are building maintenance, decorating, construction, mechanical repair, electrical repair, and routine maintenance and lubrication.

Bench-marks must be assembled and exhibited in a form which facilitates rapid and easy reference. For this reason much careful thought should be given to the subdivision of the main headings, and the successive subdividing into smaller and smaller groups of like jobs, until finally the individual jobs within the subdivisions may be allocated to their appropriate categories (Figure 14.7).

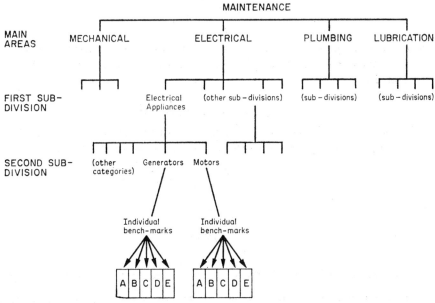

Figure 14.7 A responsibility chart for a maintenance department

Individual bench-marks must be described clearly but concisely and arranged under their respective categories on appropriately designed charts. These charts are often referred to as 'spread sheets'. An example of one spread sheet is given in Figure 14.8.

Jobs which commonly occur and which are well known to the staff in the department make the best potential bench-marks. These jobs must be carefully defined in terms of purpose and method, and then measured with equal care. The technique of measurement to be selected depends on the usual criteria of availability of techniques and their suitability in the particular circumstances. Synthetic data, predetermined motion–times, and time-study standards will figure predominantly in the final compilation of bench-marks. It is essential that bench-marks be accurately measured to ensure that they are correctly categorized.

14.45 APPLICATION OF COMPARATIVE ESTIMATING

The vehicle for initiating action takes the form of a works order or equivalent document (excepting those cases of routine maintenance). On receipt of the order the first task of the estimator is to ascertain the trade and subdivision into which the requested work will fall. Having identified the task-group he is able to extract the appropriate spread sheet.

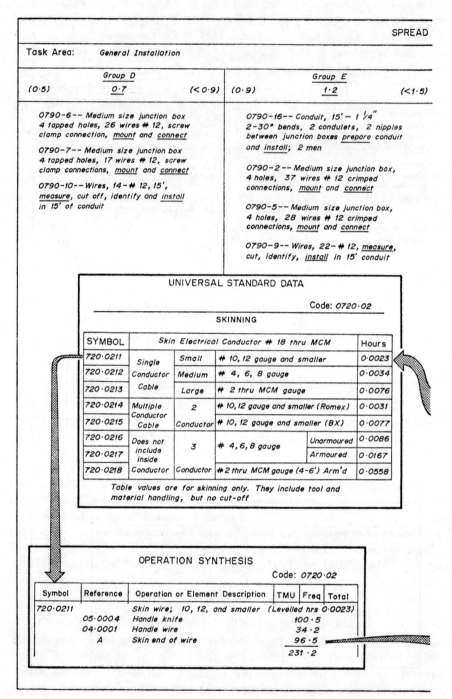

Figure 14.8 A spread sheet

ANALYTICAL AND COMPARATIVE ESTIMATING

SHEET Code: *0795*
Craft: *Electrical*

(1·5)	Group F 2·0	(<2·5)	(2·5)	Group G 3·0	(<3·5)
	0790-15-- *Conduit, 35'-2", 2-30° bend, 2 condulets, 2 nipples between junction boxes, prepare conduit and install; 2 men* 0790-17-- *Conduit, 15'-1½", 2-30° bends, 2 condulets, 2 nipples between junction boxes, prepare conduit and install; 2 men* 0790-8-- *Wires, 37-#12, measure cut, identify, install in 35' conduit*			0790-3-- *Medium size junction box 4 holes, 85 wires #12 crimped connections, mount and connect* 0790-11-- *Wires, 54-#12, measure cut, identify, install in 80' then 50' conduit* 0790-19-- *Medium size junction box, splice, #12 wire, make 54*	

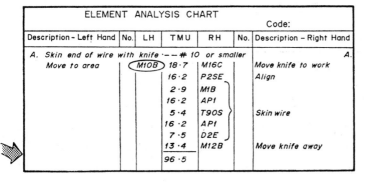

BENCH MARK ANALYSIS SHEET
Code:

Description: *Medium size junction box, 4 holes, 85 wires #12 crimped connections, mount and connect*			Date: *26/4/--*		BM *0790-3*
			Craft: *Elect; Gen. Instdll*		
		No. of men: *1*	Anal: *W.M.*		Sht: *1 of 1*
Line	Men	Operation Description	Reference Symbol	Unit time	Freq	Total time
1		Mount medium size box	750·0207			0·3243
2		Select proper wire	13·0002	0·0035	85	0·2975
3		Move marker on wire	720·0660	0·0094	85	0·7990
4		Cut off 85-#12 wires	720·0101	0·0021	85	0·1785
5		Skin 85 #12 wires	720·0211	0·0023	85	0·1955
6		Connect 85 #12 wires	720·0323	0·0110	85	0·9350
Notes:			Bench Mark Time			2·7298
			Standard Work Group			G

ELEMENT ANALYSIS CHART
Code:

Description – Left Hand	No.	LH	TMU	RH	No.	Description – Right Hand
A. *Skin end of wire with knife -- #10 or smaller* *Move to area*		M10B	18·7	M16C		A. *Move knife to work*
			16·2	P2SE		*Align*
			2·9	M1B		
			16·2	AP1		
			5·4	T90S		*Skin wire*
			16·2	AP1		
			7·5	D2E		
			13·4	M12B		*Move knife away*
			96·5			

(by permission of H. B. Maynard and Co. Ltd)

A search of the descriptions of bench-marks on the spread sheet will reveal those which bear the greatest similarity to that which is to be estimated. If none of the bench-marks is similar he may be obliged to compare work-content for those jobs which contain a similar amount of work. On locating the relevant group of jobs it is a simple matter to identify the category basic-time, and hence by the addition of allowances, the standard time. To verify his choice the estimator may consider the bench-marks in the categories immediately above and below the one selected. Those in the former should, in his estimation, have a higher work-content and those in the latter a smaller work-content. If these conditions are not satisfied he should reconsider his choice.

One main purpose of comparative estimating is to equip the departmental management with available control-data. The control procedure and its accompanying documents are described in Chapter 15, the procedure being common to most work measurement.

In the interests of reliability and to combat drift the provision of an adequate audit procedure is essential. Periodic work-measurement checks are made on selected bench-mark jobs to verify that they are still applicable. Alternatively operator performance may be assessed through the medium of the production study, the results being compared with the performance calculated from the category standards. Discrepancies are to be expected because of the tolerances to which the bench-marks are subject and the inherent variability of the actual jobs. However, differences should not exceed the expected values (say 5 to 10 per cent).

Further audit checks may be made by plotting histograms of random job-times and following the procedure outlined for constructing the categories.

Clearly, individual bench-marks may be checked using separate work-measurement studies.

14.5 CATEGORY ESTIMATING

An extension of comparative estimating, category estimating, is alleged by some practitioners to be more reliable and cheaper to install because it does not depend on the establishment of a large bank of bench-mark job times. Furthermore, the consistency of estimating is continuously monitored. Several versions have been described, appearing under different names, including category estimating.* The early stages of the techinque follow closely those of comparative estimating, including derivation of the job categories.

The first step in setting up the system is to obtain a representative sample of jobs of all sizes. A clear picture of the distribution of job times may be obtained through the use of rated systematic-activity sampling, or by employing the production study. Actual times must be extended and hence standard times calculated after the addition of appropriate relaxation allowances. These sample jobs are then allocated to their respective job categories.

* P. W. Kirby, 'Category Estimating', *Work Study and Management Services*, April 1970.

ANALYTICAL AND COMPARATIVE ESTIMATING

		1·23				
		1·23				
		1·18				
	0·76	1·44				
	0·67	1·23				
	0·59	1·03				
	0·94	1·16				
	0·93	1·07				
0·40	0·82	1·72				
0·28	0·63	1·37				
0·25	0·68	1·20				
0·27	0·77	1·12				
0·41	0·90	1·12				
0·46	0·68	1·20				
0·27	0·56	1·48				
0·31	0·76	1·37				
0·46	0·64	1·07	3·54			
0·37	0·98	1·53	2·86			
0·26	0·66	1·07	3·18			
0·18	0·98	1·18	2·50			
0·41	0·91	1·56	3·59			
0·49	0·67	1·35	3·72	4·56		
0·36	0·97	1·09	2·21	7·63		
0·40	0·53	1·17	3·67	4·29		
0·38	0·57	1·16	2·17	4·11		
0·47	0·96	1·02	3·67	5·25		
0·30	0·98	1·20	3·36	5·69		
0·22	0·78	1·05	2·89	4·19		
0·38	0·75	1·30	3·64	4·30	10·90	
0·37	0·64	1·32	2·69	4·96	9·82	
Job times (hours) up to ½	½–1	1–2	2–4	4–8	8–16	over 16
Frequency 22	27	30	14	9	2	0

Figure 14.9 Typical results obtained during a study in a toolroom (by permission of D. G. Harris, and United Carr Ltd)

Provided that a sufficiently large sample of jobs be studied, usually the log-normal character of the distribution may be demonstrated; a typical set of results is illustrated in Figure 14.9. On the emergence of this log-normal shape a grand average for these job times may be deduced. The existence of this log-normal distribution may be checked mathematically using the chi-square test for significance, comparing actual frequencies in each category against the expected frequencies for these categories.

The records may be supplemented by data supplied by the operatives themselves. Initially they will be required to record actual times taken for their jobs, using a job ticket designed for this purpose (Figure 14.10). It is necessary for some form of levelling to be employed in order that actual times may be extended to basic times. Once again this may be achieved through the use of rated-activity sampling carried out concurrently with the collection of the sample times.

JOB TICKET		
TOOLROOM JOB REQUEST TICKET	PART NUMBER	CATEGORY
NAME		CLOCK NUMBER
DATE STARTED		DATE FINISHED

TIME ON	TIME OFF	DESCRIPTION OF JOB
		FOREMAN'S SIG.

Figure 14.10 Job ticket for an installation in a toolroom
(by permission of D. G. Harris, and United Carr Ltd)

The average rating for the control period is applied to each mid-category time to extend it to basic and hence to standard times. The respective category standard times are used later in the calculation of department or section performances.

It is in the application of the technique that the greatest difference between this form of estimating and comparative estimating is apparent. Armed with the information about category boundaries the supervisor/estimator is better equipped to allocate each job to a specified category as he instructs the tradesmen and issues jobs to them. Whereas previously he would attempt to assign an *absolute* job time, even to within a few minutes, in a vain attempt to achieve accuracy, now he is merely required to differentiate between relatively widely spaced categories. Once he may have been unsure whether the replacement of a certain length of copper water-pipe would take $1\frac{1}{4}$, $1\frac{1}{2}$, or $1\frac{3}{4}$ hours. With the new system he need only decide that, for example, Category B—$\frac{1}{2}$ to 1 hr—is obviously too short for the amount of work involved, Category D—2 to 4 hr —is clearly too liberal, therefore Category C—1 to 2 hr—must be the most appropriate.

Two other phases are important in the operation of category estimating; assessment of performance, and monitoring of the foreman's estimating ability.

Regular control-periods are instituted for continuous monitoring purposes. During this period the supervisor/estimator is required to record the numbers of jobs allocated to each category. The department or section performance is calculated by multiplying the number of jobs in each category by the respective standard times previously established for each category. The performance is readily calculated from the usual basic formula, or from variations of this:

$$\frac{\text{total standard hours earned}}{\text{total actual hours worked}} \times 100$$

It may be necessary to install periodic checks on the overall rating for the section, and on other conditions, to obviate tendencies to drift.

When the system was originally established, data were obtained when the log-normal characteristic of the work was apparent. If the estimates of the supervisor are consistent and appropriate then the log-normal form of the distribution is preserved, any bias inherent in the estimating having the effect of

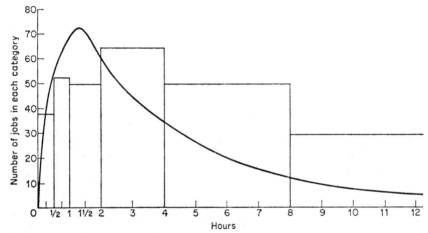

Figure 14.11 A set of job times with the theoretical log-normal curve superimposed, showing the discrepancies
(NOTE: frequencies are indicated by the height measured on the y axis, not by the areas of the blocks)

distorting the shape of the distribution. This phenomenon may be exploited as a check on the accuracy of the foreman's estimating aptitude, by constructing a histogram of the category times upon which superimposed the theoretical log-normal curve. Statisticians again may prefer to use the chi-square test. Deviations, which may be seen clearly (Figure 14.11), may be due to biased estimating, or, less likely, to a radical change in the nature of the department's work which significantly alters the average job time and standard deviation of job times. The latter reason, of course, would be obvious from other considerations, such as changes in product or in the company's policy.

Typical forms used in category estimating by one large company are shown in Figures 14.12, 14.13, and 14.14. The first form has been designed to assist the calculation of the various statistical measures to which reference was made

CATEGORY ESTIMATING

Dept./Group.................... Data for Period:....................
Sheet 1. DISTRIBUTION AND SAMPLE TESTS X^2 Test for log-normal distribution

JOB CLASS (HOURS)	(b) LOG OF BOUNDARY	(x_0) LOG OF CLASS MID-MARK	(f) OBSERVED FREQUENCY	t	ft	ft^2	$f(t+1)^2$	(b) LOG OF BOUNDARY	DEVIATION FROM MEAN $x = \dfrac{b-\bar{x}}{S_x}$	PROB OUTSIDE BOUNDARY	PROB WITHIN CLASS	EXPECTED FREQUENCY E	OBSERVED FREQUENCY O	$\dfrac{(O-E)^2}{E}$
0 – 0.3	–.52288	–.67340						–.52288		Table 3				
0.3 – 0.6	–.22185	–.37237						–.22185						
0.6 – 1.2	+.07918	–.07134						+.07918						
1.2 – 2.4	+.38021	+.229695						+.38021						
2.4 – 4.8	+.68124	+.530725						+.68124						
4.8 – 9.6	+.98227	+.83176						+.98227						
9.6 – 19.2	+1.28330	1.13279						+1.28330						
> 19.2	CLASS INTERVAL C = .30103	1.43382	Σ								Σ 1.000			

CHARLIER'S CHECK

Σf =

$2\Sigma ft$ =

Σft^2 =

$\Sigma f(t+1)^2$ =

Antilog S_x =
Antilog $\bar{x}$ =

$\bar{x}$ MEAN $= x_0 + C \dfrac{\Sigma ft}{\Sigma f} =$

S_x STD. DEVIATION $= C \sqrt{\dfrac{\Sigma ft^2}{\Sigma f} - \left(\dfrac{\Sigma ft}{\Sigma f}\right)^2} =$

V VARIANCE $= S_x^2 =$

DO SAMPLES COME FROM SAME POPULATION? (TICK BOX)

$\dfrac{V}{r}$ last sample $= \quad \Sigma \dfrac{V}{r} = \quad 2\sqrt{\Sigma \dfrac{V}{r}} = \quad \dfrac{V}{r}$ this sample $=$ YES

$\bar{x}$ last sample $= \quad |\bar{x} - \overline{\overline{x}}| =$ NO

Is $|\bar{x} - \overline{\overline{x}}|$ less than $2\sqrt{\Sigma \dfrac{V}{r}}$?   Safe sample? YES / NO

MINIMUM SAMPLE $= \left(\dfrac{\text{Antilog } S_x}{0.05 \times \text{Antilog } \bar{x}}\right)^2 =$

Number of classes – 3

= [ ] DEGREES OF FREEDOM

grouped classes count as one class

X^2 at 95% ....
      at 5% ....
(from table 8) — $\Sigma \dfrac{(O-E)^2}{E}$

FOR GOOD FIT $\Sigma \dfrac{(O-E)^2}{E}$ MUST LIE WITHIN THESE LIMITS. IS THIS SAMPLE A GOOD FIT?

YES   NO

Figure 14.12 A category estimating sheet for distribution and sample tests

CATEGORY ESTIMATING

Sheet 2 STANDARD TIME CALCULATION

Dept.....................
Calculated by.....................
Date:.....................

Rate whilst working .... A ⎫ Overall rate .... C = A × B
% time working .... B ⎭

$\bar{x}$ =
S_x =

Relaxation allowance .... % (D)

CAT	BOUNDARY	LOG OF BOUNDARY (b)	DEVIATION FROM MEAN $x_1 = \dfrac{b_1 - \bar{x}}{S_x}$	PROB. OUTSIDE BOUNDARY FROM TABLE 3 (E)	PROB. WITHIN CATEGORY (F)	MID–PROB OF CATEGORY G = ½F	PROB AT MID-BOUNDARIES	DEVIATION FROM MEAN Tab 3 rev. x_2	LOG OF MID BOUNDARY $b_2 = x_2 S_x + \bar{x}$	MID VALUE OF CATEGORY H = antilog b_2	BASIC HOURS J = H × C	STANDARD HOURS = J(1 + D/100)
A	0.3											
B	0.6											
C	1.2											
D	2.4											
E	4.8											
F	9.6											
G	19.2											
Tail											—	—

Σ 1·00000

Figure 14.13 A category estimating sheet for standard time calculation

M/C BUILDING. MONTHLY BONUS CALCULATION SHEET

Category estimating Group scheme Period ending

WEEK ENDING	TOTAL STD. HOURS EARNED FROM WEEKLY SHEETS	TOTAL CLOCKED HOURS FROM ACCOUNTS	TOTAL HOURS U.M.W AND W.T.	TOTAL HOURS LEFT OUT SCHEME
1				
2				
3				
4				
5				
Totals	A.	B.	C.	D.

	Item		
E	Standard hours earned	$= A$	
F	Hours on measured work	$= B - (D + C)$	
G	Measured work performance	$= A \div F$	
H	Geared performance	$= 50 + \frac{1}{2} G$	
J	Geared standard hours	$= \frac{H}{100} \times F$	
K	U.M.W. and W.T. credits	$= C \times 0.75$	
L	Total standard hours	$= J + K$	
M	Total hours in scheme	$= F + C$	
N	Pay performance	$= L \div M$	
P	Bonus percentage (from table)	$=$	

Remarks

Figure 14.14 A category estimating sheet for monthly bonus calculations

above. The result appears in the bottom right-hand corner as a simple 'Yes' or 'No' to the question about the validity of the sample. Figure 14.13 is used to calculate the standard time, and the sheet depicted in Figure 14.14 is to aid the assessment of bonus earnings.

15 Measurement and Control of Indirect Work

15.1 BACKGROUND

In previous chapters the basic techniques which constitute work measurement have been described. With few exceptions the techniques were developed to measure repetitive, and in most cases, short-cycle jobs. Only recently has greater interest begun to focus on the so-called 'indirect areas', which previously had been almost totally ignored by practitioners and managements.

This earlier apathy and lack of interest was engendered by the two main factors of (*a*) technical difficulty—the inadequacy of existing methods to measure work of highly variable content, and the difficulties in assessing throughput of work; (*b*) psychological reactions—fear of managements of greater resistance from 'white-collar' workers for whom work measurement is not traditional, a fear which may be real or imagined.

The term 'indirect labour' appears to be an invention of the cost accountant, who needs to differentiate between labour engaged directly on producing the company's products and those who support production for the purposes of costing and apportioning overheads. To the work-measurement practitioner the distinction matters little; what is important is the *variability* of the elements of the job. Generally the more variable the work the more complex is the measurement of that work. Thus the term 'indirect work' is not a suitable one to use in this context, but as it has gained wide acceptance its use will be perpetuated here.

The field of indirect work is extremely widespread, and growing all the time. The areas include maintenance, janitorial, storekeeping, and inspection jobs, but by far the largest group is comprised of the clerical and office, or 'white-collar' work. In the United States white-collar workers already outnumber 'blue-collars' (Figure 15.1), and it is expected that this will happen in the United Kingdom by the 1980s. In the United Kingdom the rate of increase of clerical workers is about twelve times that of other workers, according to a recent estimate.* In 1970 the U.K. figure stood at $3\frac{1}{2}$ million, when it was about 35 million in the United States.

A concise generic term is needed for 'those techniques developed for the measurement and control of indirect work'. The term *varifactor synthesis* has been widely accepted in many countries to serve this need.†

---

* *Growth of Office Employment*, Ministry of Labour Manpower Studies No. 7 (London: 1968).
† D. A. Whitmore, *Measurement and Control of Indirect Work* (London: Heinemann, 1970), pp. 56–69.

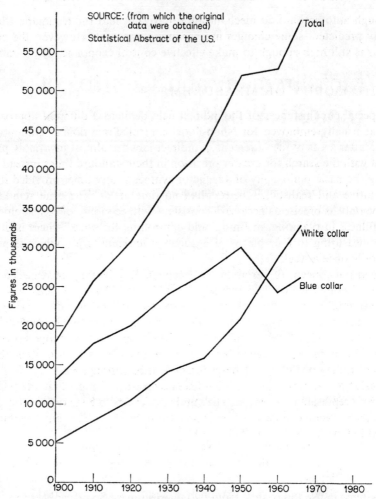

Figure 15.1 Growth of white-collar workers in the United States
(source of original data: *Statistical Abstracts of the United States*)

15.2 THE NEED FOR MEASUREMENT

It has been found that before the introduction of objective controls, utilization of staff engaged on indirect work is low compared with that of other workers on whom comparable measurements have been made. Estimates range from 35 per cent for maintenance workers to 74 per cent in offices. The recent emergence of methods of controlling indirect work was stimulated by two factors:

1. The need to contain the disproportionate increase of indirect to direct workers; and
2. The development of new techniques of measurement and control designed specifically to meet this situation.

Although automation and mechanization have not had the traumatic effect originally predicted, some changes are becoming apparent. However, the cost of labour is still high enough to make effective control economically desirable.

15.3 PHILOSOPHY OF MEASUREMENT

The special circumstances of the indirect field demand a different approach from that usually employed for 'shop-floor' control. Practitioners graduating from the latter areas to measurement of indirect jobs are almost invariably preoccupied with the search for greater precision in their standard times instead of accepting the basic philosophy of varifactor synthesis: any time, provided it is representative and realistic, is better than no time at all. The time standard should be only as precise as necessary consistent with economy of measurement and usefulness in planning, manning, and controlling the work. There is little point in measuring to four places of decimals a ten-minute job which is performed only once a week.

Provided the concept of variability is accepted, all jobs may be regarded as repetitive, although the replications may differ in detail. Thus the element 'type a memorandum' may be repeated by a typist many times in a day, but each memorandum is unique in itself. The concept is not as foreign to the practitioner as may at first appear; the so-called 'shop-floor' jobs contain many variable elements. It is necessary to regard work as repetitive in this way in order that *units of count* may be derived as bases for measuring throughput.

The basic times for these units of count are necessarily averages of many cycles taken over a representative period. The number of cycles to be taken to ensure a representative average time may be found from the formulae and methods described in Chapter 7.

15.4 SYSTEMS

The systems in use today stem from two main sources. Some evolve from the basic work measurement already being practised by companies who develop their own forms and control documents which support their system. The second group is composed of the proprietary techniques developed by management-consultant firms, who provide 'ready-made' systems (and some standard times) which may be tailored to suit the particular circumstances, and which offer experience, know-how, and validated, field-tested standards. The high initial cost of employing the proprietary techniques may be defrayed in the long term by the time and expense saved in setting up a system without outside help.

The standard times offered by the proprietary techniques are valuable and have the backing of experience and expertise, but all systems are based on one or more of the three basic methods of work measurement previously described: timing, estimating, and predetermined motion–times. Moreover the documents employed are remarkably similar in principle, even if not in detail. Typical documents and forms are described later in the chapter.

The proprietary systems appear to fall into three clearly definable groups, for either (*a*) general application, (*b*) the clerical and office field, or (*c*) maintenance and repair work. A list of some tasks which have been covered is given in Table 15.1.

TABLE 15.1

EXAMPLES OF SOME TASKS WHICH HAVE BEEN MEASURED USING THE TECHNIQUES DESCRIBED IN THIS CHAPTER

Accounting	Foundry planning	Purchasing
Accounts payable	General files	Quality control
Accounts receivable	General ledger accounting	Receiving
Addressograph		Repair and maintenance
Advertising	Housekeeping	Research and development
Auditing	Inspection	
Bank teller	Insurance	Retail and field sales
Branch banking	Janitorial work	Sales and service
Budgeting	Jig and tool making	Sales distribution
Calculating	Job-shop production	Sales invoicing
Cashier operation	Key punching and verifying	Sales office
Central typing pool		Secretarial services
Checking by variables and attributes	Library work	Security and police
	Machine accounting	Service bureau
Cleaning duties	Mailing (mail room)	Service claims
Clerical (general)	Maintenance engineering	Shipping
Complaints department	Manual bookkeeping	Sign making
Comptometer pool	Marketing	Spares provisioning
Computer operations	Materials control	Stationery provision
Contracts department	Materials handling	Statistical charting
Cost accounting	Mechanical stores	Stock control
Cost estimating	Office services	Storekeeping
Credit control	Order filling	Supervising
Customer service	Order processing	Tabulating
Data processing	Packaging	Tax accounting
Dealer correspondence	Parts stores	Technical estimates
Designing	Payroll and salaries	Technical publications
Draughting and tracing	Pensions and welfare	Teletype room
Duplicating and reprography	Policy writing and assessing	Time keeping
		Tool cribs
Engineering	Product development	Tool room
Engineering services	Production control	Traffic control
Estimating records	Production scheduling	Transport
Expediting	Project costing	Trouble-shooting
Filing and archives	Proof-reading	Typewriting
Financial accounting	Property accounting	Warehousing
Foreign exchange	Pro-rating revenue	Work study
Forms control	Purchase invoicing	

15.5 MEASUREMENT AND CONTROL

There are two distinct phases in varifactor synthesis, as indeed there are in 'shop-floor' work measurement. These are:

1. The setting of the time standards for the jobs, and updating of these;
2. The continuing phase of control and the supply of data upon which control is effected.

The theory of control is described elsewhere in this book, but in basic terms it has four main parts of monitoring the output, interpretation of the monitored

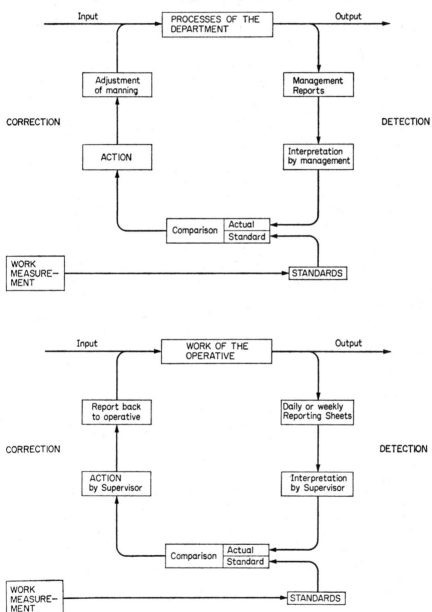

Figure 15.2 Control loops for control at (*a*) corporate level, and (*b*) operative level

data, comparison with standards, and corrective action to restore the required direction.

Control is a continuous action, and thus may be depicted as a loop, the loop being the vehicle for feedback of data. Figure 15.2 shows how the information is fed back, interpreted, and compared with the standard times to determine the variances and hence the utilization ratios. Also shown is the completion of the loop, the corrective action. It will be seen that control loops exist on two levels, and that they interact.

The operative-level control loop monitors the throughput of the individual, group, or section through the medium of the reporting sheet. Correction is effected at the source of variance. The media for control are discussed in the next section. The corporate-level control loop uses the accumulated data from sections and departments, corrective action being taken on the level of manning required. The data are used for other purposes of planning and costing.

The supervisor is the 'middleman' in the management hierarchy, providing liaison between operative level and higher management. Thus it is not surprising that he is the linch-pin at the interaction of the two control loops, his function being to translate the data fed back from the operatives into concise and relevant management reports upon which appropriate action may be taken.

The importance of the supervisor in the programme is due to three factors:

1. He is the person responsible for the work-flow and productivity of his section, and as such is immediately concerned with his subordinates' performances.
2. He should know exactly how his section is performing, and by collecting and processing data the objective is achieved.
3. The supervisor must be seen to support and favour the programme, and by processing results and acting upon the results he is shown to be in control.

Because of his importance the success or failure of the system is in his hands. It is too easy for the unscrupulous supervisor, identified too closely with his subordinates, to 'adjust' the utilization ratios one way or the other to suit his own purposes. The supervisor, therefore, is in a position of greater trust, a situation which of course is consistent with the contemporary theories of motivation (*see* Chapter 4).

The hub of the reporting system is the collection of data from the staff. The two principal ways of monitoring throughput are self-recording and work allocation. Having obtained the necessary standards account must be taken of the volumes of work produced.

Self-recording is based on the Daily or Weekly Work Record upon which the operative or group of workers report the volume of output produced under each standard.

Work allocation. This method is employed where work is normally issued in batches or as a complete job. The amount of work issued in terms of standard minutes (or hours) may be calculated from the schedule of standard times.

15.51 SELF-RECORDING

This is a widely used method of reporting, employed in most of the proprietary systems.

The basic document for the purpose is shown in Figure 15.3, and carries the following information:

1. A short description of each job performed;
2. The units of production (or count) to be recorded;

FLUORESCENT FITTINGS COMPANY VETS DAILY REPORTING FORM					
Department: *Stores* Section: *Inspection*		Name: *W. Rowe* Job: *Inspect goods received*		Date: *June 10*	
WORK PERFORMED		Count unit	Quantity	Std. min	Earned min
1.	Visual inspection of large components (housing etc)	item	20, 16, 30 = 66	0·3	19·8
2.	ditto small components. (brackets etc)	item	55, 25, 30 = 110	0·1	11·0
3.	Mechanical inspection – 1 or 2 measurements	item	15	0·2	3·0
4.	3 or 4 measurements	item	15, 35 = 50	0·4	20·0
5.	5 or more	item	18	0·6	10·8
6.	1 or 2 thread tests	item	—	0·6	—
7.	Electrical test on lamps	item	30	0·6	18·0
8.	Electrical test on ballast units for fluorescent tubes	item	30	0·8	24·0
9.	Test sheet metal delivery (gauge, etc)	sample	5	8·0	40·0
10.	Collect sample, sort C.R. Note and return sample	sample	ℍℍℍ ℍ ll = 22	5·0	110·0
11.	Unpack and repack sample to be inspected	sample	ℍ l +10 = 16	1·6	25·6
12.	Unwrap and rewrap protected items	item	10	0·7	7·0
13.	Write Reject Note: sign G.R., enter in book, staple to material, book into Kardex file	sample	4	2·0	8·0
14.	Supervise execution of 100% inspection	occ.	—	5·0	—
15.	Trips to drawing office, etc	trip	ℍ l = 6	5·0	30·0
16.	Queries	occ.	ℍ = 5	15·0	75·0
Hours worked: 7·9 (39½ per week)			Earned minutes:		402·2
			Total actual minutes:		474
			PERCENT UTILIZATION:		85

Figure 15.3 A daily reporting form for self-recording

3. The standard time (or basic time) per unit of production for each of the jobs.

The foregoing are pre-printed on the form. Provision is also made for the following to be entered:

4. Volume of work under each heading which is produced. This is entered by the operator.
5. Earned time (in earned minutes or hours). This calculation is made by the supervisor, analyst, or record clerk, and entered on the form by him.

The system has been criticized as vulnerable in two ways: the recorder may make genuine mistakes, or he may be tempted to falsify the entries. Self-recordings must be made as soon after the event as possible in order to ensure accuracy. In cases where systems have been allowed to degenerate, workers are sometimes expected to think back as much as a week and are faced with the impossible task of remembering all they have done.

Self-recording satisfies the theory of the behaviourists who argue for giving people more responsibility and demonstrating that they are in a position of trust. This is more effective where more control of their own work is allowed in the form of job enrichment.

15.52 WORK ALLOCATION

Work which is usually allocated from a central position may be suitable for control through the method of work allocation, sometimes known as 'batching', or alternatively as *short-interval scheduling*.

The five steps in work allocation are:

1. Work is issued from the central control-point by the supervisor, or a specially appointed control-clerk. The amount of work issued is usually that which should occupy one or two hours of the worker's time.
2. Simultaneously a 'batch control-ticket' is issued on which information about the work, its volume, time of issue, and expected time of return are written. The issuing clerk is provided with a schedule of standard times for the purpose.
3. The work issued is entered on a Gantt-type chart or planning board (see Figure 15.4) which indicates availability of staff and facilitates allocation of work.
4. On the return of the completed work the supervisor or issuing clerk notes the time, and from this and the expected return time he may calculate any variance, which he may decide to investigate if excessive.
5. Batches of work are prepared in order of priority.

Figure 15.4 shows the general form of the batch ticket.

Work allocation is useful in such areas as typing pools, and in maintenance departments where an informal system may already be in operation. One advantage of the system is that immediate action may be taken in cases of high variance between expected and actual times, a situation which is not possible in

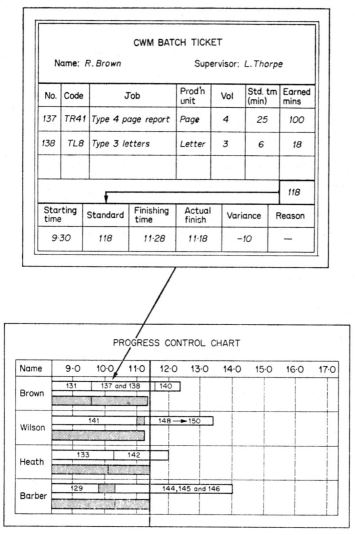

Figure 15.4 A batch ticket, and Gantt chart

the retrospective method of self-recording. Furthermore the supervisor is aware of how his staff is performing, and of any problems which may be causing high variances.

15.53 UNITS OF PRODUCTION

Before a standard can be issued it is essential that the unit of production be clearly defined to avoid confusion and ambiguity. It is necessary for the person who is recording the throughput to be clear about exactly what he is to record, i.e. what he is supposed to count. For this reason the unit of production is sometimes known as the 'unit of count'. A document called 'counting instructions' (*see* Section 15.72) is often provided for the guidance of staff.

The choice of suitable units of production is an important consideration. The analyst must be certain that no room for doubt is left and that the unit is not ambiguous (*see* Section 15.72). Moreover, the unit of production should be selected so that a minimum of additional work is required of the operative; existing counting points should be utilized where possible. Sometimes serial numbers on documents may be used to establish the number processed, or mechanical counters may be employed. The unit should cover as much of the work as possible consistent with the desired precision, to avoid excessive counting and repetitive recording. Standards for a typing pool, for example, would not employ the most obvious unit of 'min per key-stroke', partly because of the prodigious task of counting and calculating involved in evaluating utilization, and partly because the speed of typewriting is not the sole factor determining overall effectiveness. Copy-typing from hand-written notes, and audio-typing, are far more exacting and time-consuming than straightforward retyping of printed documents. Examples of units of production are given in Figure 15.11.

To summarize, it may be said that to be effective, units of production should be selected according to certain principles.

1. The unit should be free from ambiguity. Thus 'type report' is too vague; a standard time for 'type report from typewritten original, double spaced on A4 paper' indicates to the typist exactly what she should record.
2. Existing counting-points should be used where possible to obviate the introduction of additional work of assessing throughput.

15.54 UTILIZATION RATIOS

The culmination of the control phase is the determination of individual or group utilization-ratios. Basically these are identical in all systems, being based on the operator, departmental, or overall performance ratios defined in Chapter 3.

The general form of the utilization ratio is:

$$\frac{\text{earned time}}{\text{attendance time}} \times 100,$$

earned time being the product of the number of jobs performed and their respective time standards, and attendance time varying between actual time spent on the work, to total clock hours of attendance at the company, depending on the particular ratio employed. For purposes of calculating utilization the work is usually separated into *measured work* for which standards are available, and *unmeasured work* for which the actual time taken to perform the work is used.

The Group Capacity Assessment system refers to ratios of efficiency and of realization, while the Clerical Work Improvement Programme calculates effectiveness. Performance is an alternative term also used, particularly in Master Clerical Data.

15.6 BASIC TECHNIQUES OF MEASUREMENT

The basic techniques of work measurement in their original forms are often unsuitable for measuring indirect work and must first go through some form of modification. Synthesis into suitable standard data or amendment of the basic principles are the usual adjustments.

Time study is widely used in initial applications by those entering the field with little prior experience, as it is the most obvious device in the transition from repetitive to highly variable indirect work. Time study is extremely limited in more complex situations, where the rating factor is completely ineffective. These applications include jobs which require 'thinking time', decision-making, conversation and discussion, and work with a degree of creativity. In such cases either the elements which include the foregoing restrictions must be assumed to be performed at standard rating, or a modified form of time study must be used. Averaging of actual times achieved by experienced workers provide sufficiently reliable target times in many cases, or alternatively some practitioners use a levelling system with, say, four gradings of performance. Many of the amendments depart so far from the accepted B.S.I. definition that the technique no longer qualifies as time study.

In the field of predetermined motion–times the first and even second-level systems (i.e. MTM-1 and MTM-2) are totally unsuited to much of the long-cycle and non-repetitive work, and are usually formed into more manageable standard data, as for example the synthesis of Motion Sequences and Data Blocks from MTM-2 motions. Again, these systems suffer from the limitations imposed by mental work, being capable of measuring only extremely simple binary decisions or confined to eye motions (which many do not regard as mental processes).

Some advances have been made by other systems of predetermined motion–times in the measurement of mental work, but this still remains an area which defies measurement in all but the simplest cases of binary decisions. The only effective way of covering this work is by overall timing, which embraces all elements of the work without unnecessarily attempting to separate mental from manual work. Some practitioners of predetermined motion–times carefully compute the time to dial telephone numbers to an accuracy of five places of decimal hours, and are then forced to resort to watch-timing the calls when they could have timed the whole process from start to finish by overall timing in the first place. This is but one of many such examples of vain attempts at precision.

15.61 INSTALLATION PROCEDURE

There are two main stages in a varifactor-synthesis application: measurement and control. The actual procedure adopted for a complete installation will depend on the circumstances pertaining in any particular case.

WORK MEASUREMENT

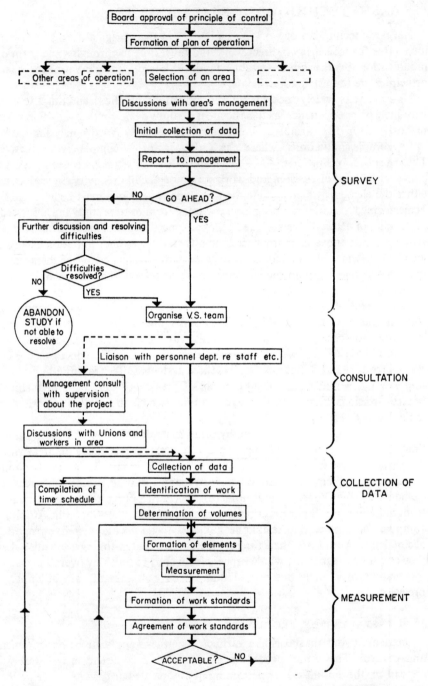

Figure 15.5 Phases in the

MEASUREMENT AND CONTROL OF INDIRECT WORK

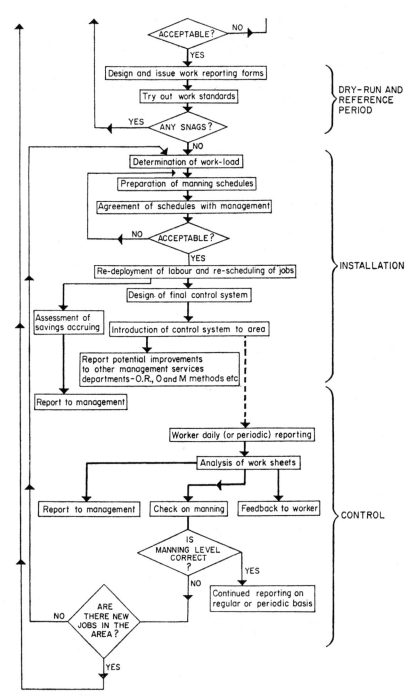

installation of varifactor synthesis

Method study has been described in six stages, and similarly the varifactor-synthesis installation may be separated into a distinct number of phases, in this case, seven. As with method study, the phases must not be considered as watertight compartments, but only as a convenient way of showing their logical sequence (*see* Figure 15.5).

The purpose of the *survey* is to investigate the situation and make an assessment of the benefits likely to accrue from the project against the costs of its installation. This involves the collection of certain data and information about the areas to be covered, which includes details of personnel, wages, overtime, backlogs of work, labour turnover, peak and slack working periods, bottlenecks in the work-flow, and a mental picture of morale, attitudes, and pace of working of the section. The survey culminates in the Report to Management which sets out the pros and cons, and recommendations on the action to be taken.

The second phase is one of *liaison* with line management and the personnel department, and *consultation* with workers, trade-union representatives, and others concerned with the application. In common with any other work-study project it is essential that adequate preparations are made before beginning the installation. At this stage some induction training and courses of appreciation will be undertaken to acquaint those concerned with the details of the system.

The data collected in the third phase provide the necessary information for the compilation of organization charts of authority and responsibility.

Additionally layout drawings, job descriptions, and other data not acquired during the survey stage are obtained. Close liaison with line management, and personnel and work-study departments, is clearly necessary. Records of targets agreed, perhaps by management-by-objectives, should be made available.

At this stage sufficient information will have been collected to enable the analysts to draw up a time-schedule for the project. Often a Gantt Chart is used to display in pictorial form the proposed phases of the project.

The methods of collecting data concerning volumes of throughput and details of all work performed by the sections are not new to the experienced work-study practitioner. Collection is facilitated by the provision of diary reporting-sheets upon which the staff record short descriptions of their jobs as they perform them, and the volumes handled over a representative period. Typical diary sheets are shown in Figures 15.8 and 15.9. In certain cases methods such as production studies or activity sampling may be appropriate.

The *measurement phase* is the one which more than any of the others distinguishes the particular system in use from the other systems. All of the systems in current use, whether proprietary or not, are inevitably based on timing, estimating, or predetermined motion–times. Some of the most common are described briefly in Section 15.8. The main objective of this phase is to determine time standards for the units of production which were defined and derived from the third-phase data-collection.

The purpose of the fifth phase is two-fold. It affords the analyst the opportunity of testing and validating his standards during the two- to four-weeks trial period (also known as the 'dry-run'). Furthermore it is used as a reference period, the new standards being applied to the present volumes processed in order to establish the current utilization ratios and levels of work. Anomalies and ambiguities present in the definitions of units of production, and other problems, may be resolved during this period.

Installation, the sixth phase, is concerned with the determination of manning

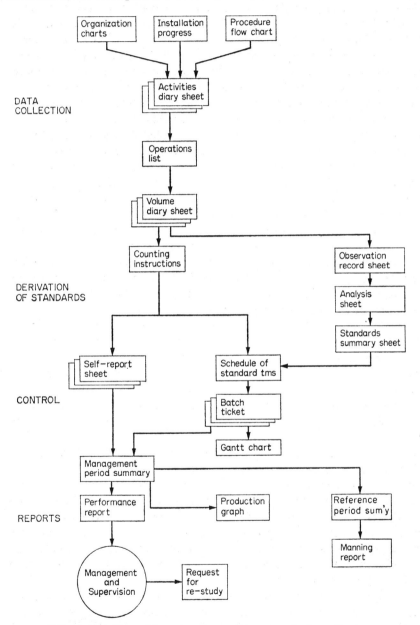

Figure 15.6 Flow of documents in a varifactor-synthesis application

WORK MEASUREMENT

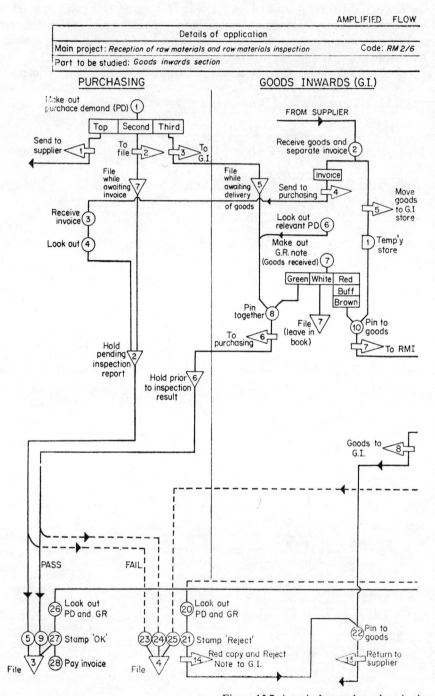

Figure 15.7 A typical procedure chart in the

MEASUREMENT AND CONTROL OF INDIRECT WORK 369

PROCESS CHART

Analyst(s): J. Hawkins and J. Ibbetson	Date commenced: 14 July
Dept. Manager: I. Lambert	Date completed: 22 June
Section Supervisor: K. Scott	

RAW MATERIALS INSPECTION (R.M.I.)   STORES   STORES OFFICE

form of an amplified flow-process chart

levels and hence manning schedules. Some redeployment of labour and rescheduling of jobs is inevitable, and these will need to be agreed with management and union representatives. Reports are made to management and to other management services such as methods study, organization and methods, operational research, and training.

The final phase is the dynamic one of continuous reporting by staff. Documents for control are described in the following section. Reporting may be continued without break, or may be installed on a spasmodic basis, introduced for limited periods at regular intervals as a monitoring device, or when it is thought that the situation has changed and the manning levels require adjustment. Control was described in Section 15.5.

15.7 DOCUMENTATION

The documents required to support a system of measurement and subsequent control must be designed specifically to meet the requirements of the individual applications. However, just as it is possible to standardize, or at least to lay down certain principles for form design in the case of general work-measurement, so it is possible to outline the general forms of the documents used in varifactor synthesis.

The flow of documents may be illustrated by Figure 15.6. This diagram does not show the documents for any particular system, but rather illustrates typical forms and paperwork required, and found in most systems of measurement at present in use. For convenience' sake documents may be separated into four groups of (i) data collection, (ii) derivation of standards, (iii) control documents, and (iv) reports.

15.71 DATA-COLLECTION

The collection of data for a work-measurement study is not new to the experienced practitioner, and there is no reason why the usual documents and methods used, augmented by those employed in O. & M., should not be appropriate in varifactor synthesis.

Suitable charts include flow-process charts, flow diagrams, procedure charts (Figure 15.7), and multiple-activity and travel charts. It is usually necessary to collect personal data about the employees concerned, and this may be facilitated by the use of pictorial charts in the form of organization 'family trees'.

Information about the work itself may be collected by direct observation, but much time is saved by delegation of this routine task to the employees concerned. Initial data-collection is accomplished through the medium of the self-recording diary sheet. This Activities Diary Sheet is distributed to the staff involved in the installation programme, who are required to record the work they perform over a set period in their own words. Figure 15.8 is typical of the return which can be expected. It is not unusual for the issue of the form to be accompanied by a letter from the department manager explaining the reason for

MEASUREMENT AND CONTROL OF INDIRECT WORK 371

the form. The letter is intended to supplement rather than replace the usual consultation and induction meetings.

The staff would not have been trained in work measurement, so it is inevitable that the form of the elements will not be suitable for the needs of the analyst. Elements will require regrouping into the standard forms of the measurement system. From this analysis two further documents will be raised.

ACTIVITIES REPORTING FORM		
Name: C. Kingston Job: Material Controller		Date: 10 Nov.
Short description of work performed	Frequency of work (daily, weekly, etc)	Estimated average volume
1 Write code numbers on summary sheet	daily	40
2 Make summary sheet up to date	daily	6 or 7 times
3 Ask about things on requisition not understood	daily	about 10
4 Help with the stocktaking	every two months	—
5 Write out new stock cards	daily	2 or 3
6 Answer telephone	daily	not known
7 File old stock cards	daily	5 or 6
8 File memos	daily	about 12
9 Go to production chaser's dept.	daily	about 6 times
10 Go to main stores	daily	about twice
11 Go to production meetings	weekly	once

Figure 15.8 A diary sheet or activities reporting form

The *Operations List* is a schedule of all work performed by the individual, group, or section being studied. The units of production are also stated. The second document, the *Volume Activity Sheet* (Figure 15.9) is the revised version of the Activities Diary Sheet, and is issued weekly to staff over a representative period. The purpose of this form is to acquaint the analyst with the frequency of occurrence of the jobs, and with the volumes of throughput; frequency and

VOLUME ACTIVITY SHEET

Name: C. Kingston
Job: Material controller
Date: 12 Nov.

DESCRIPTION OF JOB	Unit of Count	Tally	Number
1. Write code numbers on summary sheet	per number	||||| ||||| ||||| ||||| ||||| ||||| ||||| ||	37
2. Make summary sheet up to date	occasion	||||| ||||	9
3. Query on requisition	query	||||| ||||| |||||		16
4. Make out new stock record card	card	||||	4
5. Answer telephone	call	||||| ||||| ||||| ||||	19
6. Make telephone call	call	||||| |||	8
7. File old stock cards	card	||||| ||	7
8. File memos	memo	||||| ||||| ||	12
9. File other documents	document	||||	4
10. Make visit to other areas	visit	||||| ||||| |||||	15
11. Attend production meeting	meeting		0

Figure 15.9 A volume activity sheet

volume being proportional to the importance of the job in terms of precision of standard time. Infrequent and low-volume work does not need such careful measurement as does highly repetitive activities. Errors become insignificant when results are taken over a long period.

15.72 DERIVATION OF STANDARDS

Once the elements have been defined it is necessary (*a*) to describe very carefully, for the benefit of the staff recording their work, the exact definitions of the

COUNTING INSTRUCTIONS

Department	Section	Operation(s)
Production Planning	Materials Control	

Code	Abbreviated job description	Unit of production	Description of unit of production
42	Enter code number to summary sheet	line entry	Each code and its description entered on the sheet counts one unit
45	Up-date summary list	line entry	Each amendment which is made to the summary list counts one unit
59	Process component shortage list	shortage list line entry	Each item entered on the shortage list counts one unit
78	Compile shortage list	shortage list	Each page of the shortage list completed (maximum of 18 items per page) counts one unit
80	Amend shortage list	amendment	Each line amended on the shortage list counts one unit

Figure 15.10 Counting instructions

units of production, so that there is no confusion or ambiguity. For the item 'open mail', does this require the operator to record as 'one unit' each time she attends to the mail, or each time she opens one letter? Does 'open mail' include perusing it and deciding its route? Explicit descriptions of elements, sometimes known as *Counting Instructions*, are provided to answer these questions (Figure 15.10).

It is necessary also (b) to provide suitable recordings and analysis forms appropriate to the system of measurement employed. Usually these are standard time-study, P.M.T.S., or other work-measurement forms which have been already described in previous sections of this book.

15.73 CONTROL DOCUMENTS

Reference has already been made to the two reporting control-loops situated respectively at operative and management levels. At the lower level the purpose of the reporting documents is to record the throughput of work from which, using the respective standard times, the utilization ratios may be computed. The actual format will depend on the system for which it is intended, and the two main systems of self-recording and batching were described in Section 15.5. Figures 15.3 and 15.4 depict the two documents which form the nucleus for output data-collection.

The reporting form for self-recording carries its time standards already preprinted on the sheet, but the batch-ticket is usually blank and requires the time standards to be inserted by the issuing clerk. A *Schedule of Standard Times* must be prepared from which the relevant standards may be extracted at the time of work allocations. The schedule carries a short description of each job in the area, with its unit of production, and the appropriate time standard (Figure 15.11).

15.74 REPORTS

In his role of data-processor the supervisor has two main functions. The control-loops of Figure 15.2 show that information is collected by means of one of the worker reporting-forms, and after interpretation, action is taken. This feedback of results to operatives is usually verbal rather than written, although in rare cases results sheets may be posted in the department. The oral feedback, which may also include reprimand, is mainly concerned with determining the causes of variances from the time standards, and of low utilizations. This is the first function of the supervisor in this capacity.

Secondly, he is vested with the responsibility of analysing the data, and for transmitting the relevant parts in summary form to higher management for departmental or corporate action to be taken. The formal documents for this are described below.

Management summary-reports appear in two forms, issued either weekly or monthly. Some systems (for example, G.C.A. and M.C.D., listed in Section 15.8) adopt a *Weekly Time and Production Record*, which is an intermediate document

SCHEDULE OF STANDARD TIMES
PRODUCTION PLANNING DEPARTMENT

Number: 438/D   Job: Materials control

No.	Description of element	unit of count	standard
42	Enter code number to summary sheet	line entry	1.4
45	Up-date summary list	line entry	0.8
46	Process material requisition	requisition	2.8
49	Process release note	release note	4.2
51	Process demand note	demand note	3.3
56	Process rejection note (for mechanical materials)	reject note	3.6
57	Process component rejection note	reject note	4.8
58	Process rejection note (for electrical materials)	reject note	4.4
59	Process component shortage list	shortage list line entry	0.7
60	Check item against shortage list	item	0.4
61	Deal with query on release note	query	4.5
62	Deal with query on demand note	query	5.5
64	Deal with query on materials requisition	query	6.3
65	Deal with query on shortage list	query	2.1
66	Deal with query on delivery note	query	6.0
67	Deal with query on purchase order	query	10.0
70	Raise purchase order	order	6.5
72	Raise new stock card	card	5.2
74	Raise release note	note	3.5
78	Compile shortage list	shortage list	12.0
79	Amend stock records	amendment	2.0
80	Amend shortage list	amendment	1.4
82	Enter stock record amendment on to stock card on bin	entry	2.0
84	Make telephone call	call	6.5
85	Answer telephone query	query	6.0
86	Answer telephone (outside call)	call	4.0
90	Attend production meeting	actual time	----
91	Attend other meetings	actual time	----
95	General filing	document	0.3
96	Compile letter or memorandum	letter/memo	12.0
99	Make trip out of area	trip	10.0

Figure 15.11 Schedule of standard times

used to analyse and assemble the data, and from which the management report is compiled.

Performance reports are usually compiled by the supervisor from the information he has collected, and include data on:

1. The total standard minutes (or hours) earned by the workers in his area while working on measured work;
2. An analysis of productive and non-productive hours worked;
3. Hours spent on non-measured work, and on 'one-off' or special jobs for which time standards are not available;
4. Total attendance hours;
5. Hours spent on overtime and used by staff borrowed from other areas;
6. Backlog of work in terms of hours (estimated from standards);
7. Various ratios of utilization for individuals and department;
8. Numbers of staff in different categories (measured, unmeasured, supervisors, assistants, etc.);
9. 'Lost time': holidays, sickness, training, lateness, accidents, union business, meetings, etc.

This report to management exists under various titles in the proprietary systems such as Weekly Performance Report, Weekly Management Report, Performance control sheet, Performance Summary Report.

One of the stages in a programme of measurement and control is the reference period. The purposes of this important phase are, first, to allow the collection of current data from which the present utilizations may be calculated, for the purpose of providing bases for comparison and measurement of productivity increases resulting from the programme; and secondly for testing the validity of the new standards. The document which exhibits the reference data is the *Reference Period Summary*. It provides a week-by-week analysis of the standard hours earned by the department as a whole, and an analysis of backlog, manning levels, non-measured time worked, and utilization ratios. The summary shown in Figure 15.12 is typical of those used in the proprietary systems.

A comparison of the manning and utilizations before and subsequent to the programme is set out in *Manning Tables* (Figure 15.13). These tables also detail the job changes and reallocation of work which have taken place.

15.8 SOME IMPORTANT SYSTEMS

Many systems have been developed from the basic methods of measurement, some of which are for general application, and some for specific fields, the two most important being clerical and maintenance. The earliest systems date from the late 1950s.

While the reporting systems are similar, the methods of measurement differ considerably, some being developed directly from Methods–Time Measurement and other systems of predetermined motion–times. Some other techniques are

MEASUREMENT AND CONTROL OF INDIRECT WORK 377

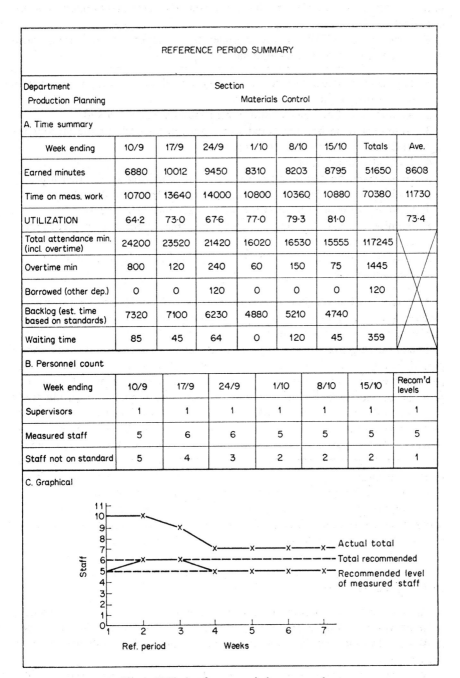

Figure 15.12 A reference period summary sheet

	PRESENT			PROPOSED NEW SCHEDULES		
				PROPOSED		
Name	Present job	Util.	Name	Proposed job	Details of alterations	Util.
C.Blackmore	Requisition clerk	64		Process order/ requ'n clerk 2 posts	Train two of the present clerks to do requisitions and process orders, this flexibility abolishes one post.	95
D.Smith	ditto	72				
E.Robinson	Proc. order clerk	55				
C.Kingston	Material controller (comps)	88		Mat. controller 2 posts	One material controller to be relocated to other work.	95
V.Cashman	ditto (elect.)	48				
E.Jacobs	ditto (mechanical)	55				
J.Cobelli	Material progress clerk	64		Post abolished	Senior materials controller to take responsibility for the technical aspects, junior clerk to take routine parts.	85
L.Snow	Junior clerk	44		Junior clerk		
L.Dill	Senior material controller	**		Snr.material controller	Takes on technical aspects of Mat. progress clerk's work.	
J.Hicks	Senior negotiator/estimator	**		Snr.Neg/est.	Post remains	
Present utilizations taken from reference period					Theoretically possible utilizations on revised schedules, based on standard times.	

** = not measured

Figure 15.13 Manning tables

MEASUREMENT AND CONTROL OF INDIRECT WORK 379

more flexible, employing the most convenient method of measurement appropriate to the circumstances obtaining at the time.

Some of the many systems in current use are listed below.

1. Basic Work Data (B.W.D.)—for maintenance work (P.M.T.S.)
2. Clerical Work Data (C.W.D.)—for clerical and office work (P.M.T.S.)
3. Clerical Work Improvement Programme (CWIP)—for clerical work (P.M.T.S.)
4. Group Capacity Assessment (G.C.A.)—general application, mainly clerical
5. Master Clerical Data (M.C.D.)—for clerical and office work (P.M.T.S.)
6. MTM-2 Maintenance Data—for maintenance work (P.M.T.S.)
7. MTM Data System for Clerical Work—for clerical work (P.M.T.S.)
8. Office Modapts—for clerical work (P.M.T.S.)
9. Universal Maintenance Standards (U.M.S.)—for maintenance work (P.M.T.S.)
10. VeFAC Programming—for general application
11. Wofac Mento-Factor System—for mental work (P.M.T.S.)

From this selection, systems representing the various types of measurement will be described.

15.81 GROUP CAPACITY ASSESSMENT (G.C.A.)

Group Capacity Assessment was developed in the United States of America in the middle 1950s from the integration of industrial engineering, and costing and budgetary control. The first installation was for Lockheed, followed by Boeing, who were interested in cost-reduction programmes on their 707 aircraft. The first United Kingdom application was in Rolls-Royce Ltd in 1962, introduced by Arthur Young & Company. The system is designed for general application, but finds its main use in the clerical field.

G.C.A. procedure follows the pattern previously described, but essentially it has been designed to assess the performance of small *groups* of workers through composite time-standards known as Factor Times. Much of the team-work which is characteristic of clerical and other indirect work relies for its cohesion on the interrelating of jobs by common flow-lines formed by the products or documents. Where these pass from member to member within the group it is unnecessary for all members to record the number of products or documents processed. The counting of units can be restricted to relatively few key positions in the flow-line, the volumes of throughput at these points being the output for the whole group. This procedure obviates the repeated counting of the same units by different workers. Moreover, staff reporting is confined to those members in the key positions rendering individual reporting superfluous. The concept is illustrated in Figure 15.14.

The variability of individual cycle-times has less effect on short-term evaluation of performance when included in composite Factor Times because of the effects of compensating errors.

380   WORK MEASUREMENT

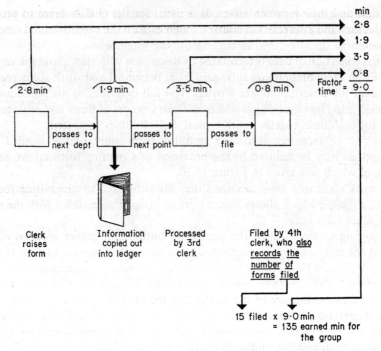

Figure 15.14 Pictorial representation of the calculation of a G.C.A. Factor time

The composite Factor Time is the sum of the standard (or basic) times of the jobs in the flow-line as shown in Figure 15.14.

The staff/analyst ratio depends on the methods of measurement used, but on average each analyst can expect to cover up to 100 staff in a year.

15.811 Outline of the system. In common with other systems, Group Capacity Assessment has the two main phases of standards-setting and subsequent control. The objective in the measurement phase is the synthesizing of the Factor Time, and G.C.A. is flexible enough to accommodate most techniques of measurement. A typical application could include methods from:

1. Multi-minute measurement, a combination of time study and work sampling, which allows the simultaneous study of more than one job;
2. Time study confined to the study of one job at a time;
3. Activity (or work) sampling, which includes high-frequency work sampling;
4. Standard data synthesized from other time standards.

Before actually making the study the analyst observes the usual proprieties, consulting the interested parties. Workers performing the jobs to which G.C.A. is appropriate are often of higher status than those ususally studied by work measurement in the manual-work areas. Preservation of this status, and demonstration of management's concern, may be achieved through thoughtful liaison with staff and their representatives. It is usual for the G.C.A. team to provide introductory and appreciation courses which explain the objectives and describe the system.

Initial collection of data is facilitated by discussion with staff about the various jobs. From these preliminary talks and from personnel and work-study records individual job descriptions are written. The job descriptions are subsequently compiled into Operations Lists which also carry ranges of times and average job-times for each task, together with typical daily volumes produced.

The risk of incorrect recording resulting from ambiguously worded job-descriptions may be reduced by the provision of Counting Instructions, an example of which was given in Figure 15.10.

As work standards are established they are submitted to supervision for approval, a liaison which allows supervisors to become acquainted with the standards at an early stage.

Reporting systems follow the procedure outlined in earlier sections of the present chapter. Information collected by supervisors is summarized on Weekly Time and Production Reports. Each section submits to management Weekly Performance Reports from which department managers are able to identify trends and assess the size of the work-force required.

An important feature of G.C.A. is the provision of facilities for budget meetings during which management and work-study representatives discuss manpower requirements and work-load.

The continued success of an installation depends upon the periodic review and adjusting of standards to accommodate changes in the work patterns. Supervisors are responsible for checking Weekly Performance Reports, ensuring that non-measured work is contained at a level which does not materially affect the reliability of the calculated performances. Similarly they should review manpower budgets at regular intervals, auditing the returns of volumes of throughput, standards, and factor times.

15.82 VeFAC PROGRAMMING

About 1938 the Work-Factor Company (now Wofac Company, a member of the Science Management Corporation) of Moorestown, New Jersey, introduced its Work-Factor System of Elemental Motion–Times. Realizing that this system was not appropriate to measurement of long-cycle work, Wofac initiated research in 1958 and produced, in 1959, a system which would embody the essential features for coping with the inherent variability of indirect work. Originally known as Variable Factor Programming (V.F.P.), VeFAC Programming made its debut at the Warwick Manufacturing Company, Chicago. Subsequently it has been installed in many countries throughout the world, including over 1,000 applications in the United States.

VeFAC Programming is a system of general application, being suitable for most indirect work.

Throughout the application of VeFAC Target Times the main objectives are the elimination of ineffective time, control of work flow, and the provision of a means of evenly distributing the work load. VeFAC Target Times provide the means by which each individual's contribution to group output may be fairly evaluated.

Throughout the applications due regard is paid to worker morale and job-satisfaction, and the establishing of a situation which stimulates motivation. Initial collection of data is facilitated by the completion of Daily Reporting Forms by the staff concerned. Short descriptions of tasks and volumes handled are entered. Other information is collected by the analyst from management, supervision, and wages, personnel and work-study departments.

The system is not based on any one exclusive method of work measurement, but on the one most suited to the type of work being studied, and the importance (volume of throughput) of the task. High-volume work requires more precise measurement.

VeFAC Co-ordinators are not restricted to any one technique of measurement, the system being flexible enough to accommodate many methods. Some of these are:

1. *Observation.* Actual times taken by operatives, or estimates from supervisors, may be used in the compilation of VeFAC Target Times. This method of deriving Target Times is valuable because it allows maximum participation of those concerned intimately with the project. This in turn stimulates interest in the programme, producing co-operation and realistic times. Target Times are mutually agreed before incorporation into the system.
2. *Work-Factor Time Study.* Where the work is highly repetitive it is possible to use a more precise form of measurement. In VeFAC Programming this takes the form of Ready or Abbreviated Work-Factor. The advent of the computerized system of Wocom has reopened the possibility of using such systems as Work-Factor on relatively long-cycle jobs.
3. *Standard data.* Standard data may be developed for use in VeFAC applications for controlled processes or machines. Wofac Company has compiled its own comprehensive library of proprietary data which covers a wide variety of operations which may be encountered in a VeFAC Assignment.

15.821 Employee reporting. The method of reporting is specifically designed to suit each application. However, two major systems may be identified. One of these, called Task Programming, ensures that the supervisor is in complete control of the work by allocating Target Times with the jobs as these are issued.

The supervisor is provided with a list of Target Times which cover the jobs in his section. Work is issued to the selected operative together with a Target Time extracted from the Schedule. This is entered on a Task Ticket which also carries a short description of the work, the volume to be processed, and a starting time. On returning the work and Task Ticket the operative may be made aware of the time taken for completion. This time, when compared with the appropriate Target Time for the given volume of work, gives the utilization. Variances of the actual times taken from the Target Times are investigated by the supervisor

MEASUREMENT AND CONTROL OF INDIRECT WORK 383

when these are excessive. The Task Ticket makes provision for the entry of reason for variances.

In circumstances obtaining in many areas the system of Task Programming may be impracticable, and recourse is had to the Signout Procedure. This method is useful in areas where it is not possible to allocate batches of work.

Signout Procedure requires the operative to enter on his Daily Signout Sheet the volume of work processed against each job specified on the Signout Sheet. Target Times are printed in the appropriate column. It is the responsibility of the supervisor to calculate the Earned Minutes by extending the volumes produced, multiplying them by their respective Target Times. From the total Earned Minutes the utilizations may be calculated in the usual way.

The supervisor is also responsible for collating the Signout Sheets and summarizing the information on Weekly Management Reports.

VeFAC Programming is a system of general application and of flexible format which must be tailored to fit the client's needs by trained and experienced co-ordinators. A full application includes a comprehensive training programme for co-ordinators, management, supervisors, and others concerned with the system. Human relations, consultation, and motivational concepts are incorporated to stimulate full co-operation and participation.

15.83 CLERICAL WORK IMPROVEMENT PROGRAMME (CWIP)

One subject of contention in the application of predetermined motion-time systems is the validity of adding elements consecutively to produce a complete job time. Some researchers have challenged the method on the grounds that such elements are not necessarily additive because of interactions between coincident elements.

In the United States during the early 1940s Paul B. Mulligan developed a system of standard data for use in the measurement of office work. His system utilizes methods of overall times obtained from the detailed analyses of cinefilms of the particular jobs being studied.

Mulligan's method enjoys wide popularity in the United States. In Australasia and Europe rights of marketing have been vested in a company of management consultants, W. D. Scott and Company. The first application in the United Kingdom was in 1962.

The data are assembled in the Scott–Mulligan Manual of Standard Time Data. These formidable volumes contain thousands of standard times covering all forms of clerical work, and specific makes of office equipment. New data are being continually added to augment the existing standards.

The approach is threefold—work simplification, human relations, and of course, work measurement.

CWIP installations follow the general principles outlined previously in this chapter. Time standards are derived from the elemental times extracted from the Scott–Mulligan Manuals. The use of the standard data does not preclude the use of other methods of measurement. A typical data-sheet from the manuals is illustrated in Figure 15.15. It will be seen that time standards are expressed in

STANDARD TIME DATA for the OFFICE

SECTION 16 PAGE A2c.1

COPYRIGHT 1951, 1957   PAUL B. MULLIGAN & COMPANY

I.B.M. ALPHABETICAL DUPLICATING PUNCH
TYPE 031

Element Time Values

Description of Elements	Man-Hours per Occurrence
Make-Ready	
Fill card hopper with blank cards	.00262
Insert skip bar	.00220
Transfer card from card rack to master card rack (includes inspecting card)	.00283
Manual	
Key punch on numeric keyboard, per keystroke:	
Good Source Records	.000065
Fair Source Records	.000085
Poor Source Records	.000120
Move hand between numeric and alphabetic keyboards	.00009
Key punch on alphabetic keyboard, per keystroke:	
Good Source Records	.000061
Fair Source Records	.000073
Poor Source Records	.000120
(cards prepared as set)	.00040
duplicating gate	
Automatic	
Feed and eject card	.00212
Skip, per column	.000004
Duplicate, per column	.000032

Figure 15.15 A typical sheet from the CWIP manual of standard times

decimal hours. Included in these comprehensive standard times is a relaxation allowance of $16\frac{2}{3}$ per cent.

Clerical work is separated into seventeen major categories. These categories cover manual actions, and operations performed using mechanical equipment.

General activities common to all work including such activities as walking, bending, and climbing stairs are covered by the category of *Personal Actions*. Specific activities are described under various categories; *Paper Handling* is concerned with the handling of cards, documents, charts, dockets, and various items of paperwork. *Longhand Writing*, and *Reading and Comparing*, are dealt with separately under other sections of the manuals. *Paper Fastening* time-standards cover such activities as collating, de-collating, stapling, taping, clipping, and other methods of attaching papers, in addition to the various ways of filing papers in binders. *Filing* is included under a separate section, and this includes filing papers in cabinets. The category *Communication* embraces the many methods of transmitting information, including making telephone calls, operating keyboards, and use of recording apparatus.

In the CWIP system there are four categories reserved for calculating, and these are *Adding, Subtracting, Multiplying,* and *Dividing*. Standard times are provided for manual, and for mechanically aided calculations. A comprehensive set of data for specific makes of machines is given under the category *Bookkeeping Machines*, standards being available for most of the well-known makes of mechanical and electronic equipment. This category does not include data-processing equipment for input of data, which are separately covered in the *Punched Card Machine* section.

Equipment and mechanical aids which cannot be classified under any of the foregoing categories are found in the comprehensive section called *Miscellaneous Equipment*. This category includes such equipment as cheque-writers, microfilming recorders and readers, making of offset aluminium plates, coin counters and sorters, and labelling machines.

A very large section provides standard times for the remaining group of machines. The *Duplicating Machines* category deals with the operating of the many items of equipment designed for duplicating, reproducing, and copying from originals or from prepared masters. This includes embossing, addressing (such as Addressograph), wax-stencil duplicating, photocopying, hand printing, and wet and dry copying by electrostatic (Xerox) and other means.

Typewriting standards include handling paper and carbons, typewriting, dictating, correcting errors, making wax stencils, and type-setting on Vari-Typer models.

The last category concerns those jobs classified as *Mailing*. Such tasks include dealing with incoming and outgoing mail, addressing, using postage franking and letter-opening machines (various models), and folding and inserting machines.

The standards are applied, and from the resulting target hours a measure of utilization for the clerical group is calculated. This measure is known as *Percentage Effectiveness*, found from the formula:

$$\frac{\text{standard hours of measured work}}{\text{net hours spent on measured work}} \times 100$$

15.831 Control. Procedure for control follows that generally outlined previously in this chapter. The two main documents for control are:

1. The CWIP Office Performance Control Sheet upon which the total hours (regular, overtime, borrowed, and loaned) are set out. Other information such as holidays, training, illness, special jobs, and net hours of measured work assists the supervisor in calculating the effectiveness.
2. The CWIP Performance and Progress Report is a summary of the monthly hours, standard hours, variances, and Effectiveness compiled by the

supervisor for communication to management. Details of staff (measured and non-measured) are included.

15.84 MTM-2 MAINTENANCE DATA

In June 1966 a consortium of companies was formed under the auspices of the MTM Association of the United Kingdom for the purpose of financing research, and applying the data resulting from such research. A method of work measurement so developed is known as an MTM Consortium Data System, an organized body of MTM-2 data and information relating to a specific craft, trade, or type of work. Such a system is approved and issued by the MTM Data Consortium.

The first system to be fully developed was MTM-2 Maintenance Data, based on Methods–Time Measurement and incorporating a technique of tape-recording verbal analysis known as Tape Data Analysis.

The concepts upon which the system is based include:

1. Use of MTM-2 and MTM-2 Motion Sequences as the basis for the new system;
2. Development of an effective and simple observation technique using hand tape-recorders, combined with the use of Data Blocks and simple verbal codes;
3. Use of statistical method, particularly probability theory and the analysis of variation;
4. Presentation of the methodology in an intellectually rigorous yet easily understandable way, by the use of algorithms and other forms of decision models.

The four main phases of an MTM-2 Maintenance Data study may be identified as:

1. Rationalization of the organization of the maintenance department before measurement is attempted;
2. Training of analysts in the various methods and techniques of the system;
3. Acquisition of data which may be used later in the synthesis of Data Blocks;
4. Application of the standards and techniques in the particular areas for which they were designed.

15.841 Data acquisition. Times for maintenance work may be allocated in any of three ways: through the issue of a minor job-time (a comprehensive standard which covers all work of less than one hour), by way of comparative estimating, or as a time synthesized from MTM-2 Data Blocks.

The characteristic of maintenance work which distinguishes it from production-type work is the motion pattern for the work-cycle, which rarely repeats, the observer being obliged to make his analysis on the basis of a single observation. Further, the work-cycles are very much longer than in production-type work and thus impose a greater strain on the analyst. The Tape Data Analysis (T.D.A.) technique has been specially designed to overcome these two problems. The demands required of the solution were judged to:

1. Provide a means of recording which allows the analyst to observe 100 per cent of the time;
2. Create a method of recording which enables the activity to be described as fast as it is performed by the operator, and in such a way that permits the accurate reconstruction of the basic motions actually used;
3. Provide a simple technique which is cabable of being readily understood by the average industrial worker, and which has training costs no higher than those involved with existing conventional systems;
4. Ensure that the speed and accuracy of analysis are specified and within acceptable limits.

To satisfy the needs of the first two demands Tape Data Analysis was evolved, and to facilitate the verbal description special codes are used, examples of which are F5 for a small finger turn, S30 for a spanner turn of up to 15 inches, SCR for scrape, RUB, and RUB-W for rubbing against resistance.

The above examples are known as Motion Sequences, and are intermediate

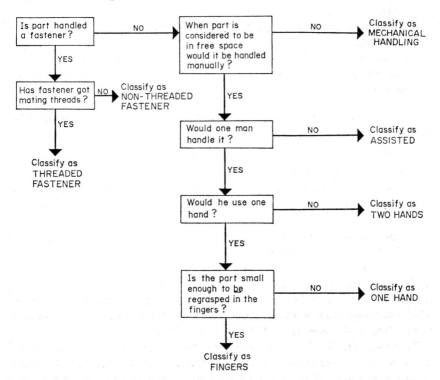

Figure 15.16 An algorithm for MTM-2 maintenance data (by permission of the MTM Association Ltd)

standards synthesized from groups of MTM-2 elements. Motion Sequences themselves may be synthesized into Data Blocks.

Data Blocks describe complete portions of work and are applicable to the work within the area for which they were developed. Consequently they may not be valid for use in other areas or companies because of local differences in methods, layouts, and equipment. Data Blocks may be of the order of hundreds of time-measurement units* to tens of thousands of t.m.u.s (about $\frac{1}{100}$ min to 10 mins) in duration. Examples from mechanical work are 'Fit part using two hands', 'Remove part using mechanical aids', or 'Fit a threaded fastener'.

The synthesis of a job time from Data Blocks is facilitated by the use of algorithms, an example of which is shown in Figure 15.16.† By tracing through the algorithm the analyst may select the correct Data Blocks for the work.

15.842 Application. The system just described provides the maintenance superintendent with a means of issuing a time standard for almost any type of work to be found in his department. His choice of method of measurement is assisted by a simple decision-model designed for the purpose:

1. IS THE FAULT KNOWN?
    No —issue minor job-ticket.
    Yes—make preliminary estimate.

2. DOES THE ESTIMATE EXCEED ONE HOUR?
    No —issue minor job-ticket.
    Yes—see 3 below.

3. IS JOB-TICKET REQUIRED NOW?
    No —do Data Block synthesis.
    Yes—do a comparative specification into a time-band which includes sample jobs known to be similar.

The minor job-ticket is issued for the purpose of providing a standard to cover the craftsman during fault-finding. It is a weighted average time over many jobs in the department of up to one hour's duration. Should the fault-finder be unable to diagnose the trouble within the hour he is required to report back for further instructions.

In cases where the cause of the trouble is known an estimate is made for the time to cure the fault unless the time is expected to exceed one hour. From the algorithm it will be seen that either a Data Block synthesis may be made, or an estimate may be made through comparison into a time-band of similar jobs.

Operator utilization may be computed in the usual way, using the information extracted from the job tickets.

---

* See Chapter 8, Methods–Time Measurement, for a definition of the time-measurement unit.

† Fred Evans, *Algorithms for the Processing of MTM-2 Tape Data Studies* (MTM Association Ltd, 1968).

16 The Uses of Work Measurement

16.1 INTRODUCTION

Chapters 5 and 6 described the techniques and mechanics of measurement. Several examples were given to illustrate the use of the techniques in various situations. From the descriptions it is clear that the characteristics ascribed to each restrict the usefulness to certain types of work and conditions. The purpose of this chapter is to illustrate the uses of work measurement in various areas, and in different industries.

It was shown in Chapter 3 how the work content of a job could be expressed in terms of work units (standard hours and standard minutes), and also how standard times could be used in the various manufacturing functions and controls. The vehicles for applying the standard times to manning, planning, incentive payments, and costing were the performance indices. For convenience sake the three most commonly used indices will be repeated here:

operator performance
$$= \frac{\text{total standard times for all measured and estimated work}}{\text{time on measured and estimated work (excluding diverted and waiting time)}} \times 100$$

department performance
$$= \frac{\text{total standard times for measured and estimated work}}{\text{time on measured and estimated work plus any waiting or diverted time for which the department is responsible}} \times 100$$

overall performance
$$= \frac{\text{total productive standard times for measured and estimated work plus productive uncontrolled work at assessed performance}}{\text{total attendance time excluding time on allocated work}} \times 100$$

Examples of the uses of standard times will be given in the following areas:

1. Costing and standard costing (Section 16.2);
2. Product design (Section 16.3);
3. Costing a service function (Section 16.4);
4. Planning, scheduling, and manning (Section 16.5);
5. Wage-payment schemes (Section 16.6);

6. Choosing capital equipment (Section 16.7);
7. Training (Section 16.8).

It must be appreciated that although the examples are applied to specific areas and industries, the techniques are almost generally applicable and may be tailored to suit individual circumstances.

16.2 USE OF WORK MEASUREMENT IN COSTING

As previously stated in Chapter 3, the total cost comprises the elements: labour cost, materials cost, and overhead cost. Analyses, for the purposes of cost control, may be assisted by the further division of these costs into the prime costs (direct labour, materials and expense), and overhead costs (indirect labour materials, and expense). Material costs may be obtained from the purchasing authorities within the company, and such costs include those materials which are converted into the product or service supplied by the company as well as the indirect materials (tools, documents, cleaning materials, and so on).

The overhead cost comprises the costs of running the organization (insurance, rent, advertising, clerical, and managerial costs), and of maintaining production (power, heating, telephone accounts, repairs and maintenance, etc.). Again, these costs may be obtained from the various accounts which have been paid.

Work measurement in this connection is concerned mainly with the costing of labour, and it is necessary to know both the hourly labour rate and the standard times for the operations. The method of finding the unit cost for work is described below.

The control of labour costs requires a feedback of information about variations between predicted and actual costs. The three main variances concerned are wages variance, wage-rate variance, and labour-efficiency variance. Suitable examples will illustrate the use of standard times set by work measurement in the calculation of variances.

The wages variance highlights the differences between the standard wages calculated as being appropriate for the given output and the wages actually paid. The wage-rate variance measures the difference between the wage rate specified for a job and the rate actually paid for the job. A measure of the effectiveness of the staff is given by the labour-efficiency variance, which is the difference between the time which should have been taken (if the operative had worked at standard performance) and the actual time taken for the same output.

It is useful in performing these calculations to determine a standard cost for a work unit (or unit of production) for the department. A departmental hourly wage rate may be established which includes overheads and incidental expenses personal to each employee. This value in a particular department may be of the order of, say, £7·10 per hour. Thus a standard minute is worth £7·10 ÷ 60 = 11·8p per s.m.

THE USES OF WORK MEASUREMENT

Once this value has been deduced the labour cost of any job may be found when the standard time is known. For example, an invoice passing through the several stages of processing may take a total of 9·5 standard minutes of work to complete the route. A value may be put on the operation of invoicing: 9·5 × 11·8p = 112·1p. Where very small amounts are occasionally involved, often companies will find invoicing to be uneconomical, to the extent of writing off the goods supplied.

Examples

In the above examples the standard department rate was £7·10, and this included overhead and machine rates personal to the operative. The actual standard wages rate upon which the wages are based may be £4·90 per hour, equivalent to 8·2p per minute.

Using this value with data given below the wages variance is found thus:

Standard time for job B 86	= 3·2 s.m.
total s.m.s for 350 parts	= 1,120 s.m.
standard wages (@ 8·2p/min)	= £91·81
If the time actually taken	= $22\frac{1}{2}$ hours,
and wages actually paid	= £110·70
then the wages variance is	£91·84 − £110·70 = −£18·86

In practice the wages variance may be calculated also for larger one-off jobs:

Standard time for job M 70	= $360\frac{1}{4}$ std hours
standard wages (@ 8·2p/min)	= £1772·43
If the time actually taken	= 383 hrs, and wages paid = £1964
the wages variance is	£1772·43 − £1964 = −£191·57

The *wage rate variance* for job M 70 is based upon the actual wage rate paid.

$$\text{standard rate} = £4·90 \text{ per hour}$$

$$\text{actual rate} = \frac{\text{wages paid}}{\text{hours worked}} = \frac{1964}{383} = 513p$$

$$\text{variance} = 490 - 513p = -23p$$

$$\text{wage-rate variance} = \text{actual hours worked} \times \text{actual wage rate}$$
$$= 383 \times (-23)$$
$$= -£88·09$$

The *labour-efficiency variance* for the same job:

at standard performance the time taken would have been $360\frac{1}{4}$ hours
variance in hours is $360\frac{1}{4} - 383 = -22\frac{3}{4}$ hrs
at 490p per hour the labour-efficiency variance is $-22\frac{3}{4} \times £4·90 = -£111·48$

A full account of standard costing, and of these and other variances, may be found in the textbooks on this subject, including *Standard Costing* by Dr J. Batty (Macdonald & Evans).

16.3 USE OF WORK MEASUREMENT IN DESIGN

Throughout the range of management services it is clear that the various techniques do not exist in isolation, and further evidence of this overlap is found in the application of time standards in the control of design costs. One method

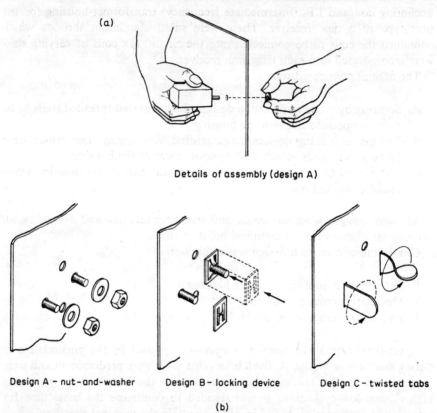

Figure 16.1 Three possible designs for anchoring coil housings

of evaluating the functions of product or procedure designs is known as *value engineering* or *value analysis*. These techniques are concerned with the assessment of the use function and esteem function (among others), with the purpose of achieving the required levels of this function at the lowest cost.

Much of the cost in the product or procedure is the cost of labour. This expense may be unnecessarily high if:

1. The methods employed in manufacture (or procedure) are not efficient, or

2. The design of the product or procedure is such that the time required for its production is longer than it need be, even though the method may be efficient in the circumstances.

There are other means of reducing these costs, through the use of less material, or alternative materials in the revised design, but it is in the two areas noted above that time standards may assist the design engineer.

The example is illustrated by Figure 16.1. The Value Engineering department of a large television-manufacturing company submitted three designs for the anchoring coil and I.F. (intermediate frequency) transformer-housing for the prototype of a new receiver. These were small aluminium shrouds which contained the coils to be connected into the circuit. Six coils of varying sizes were incorporated into each television receiver.

The designs considered were:

A. Securing by washer and nut, a design which required threaded studs to be fixed to opposite sides on the housing;
B. The use of locking devices manufactured from spring steel, which were pressed over studs attached to opposite faces of the housing;
C. The use of tabs which could be an integral part of the housing when initially blanked out.

All three designs were fabricated and tested for function and durability, all proving satisfactory from a technical point of view.

Clearly the cost of each design was a function of:

1. The cost of manufacture of the housing;
2. The relative costs of materials used;
3. The relative costs of assembly.

Apart from a small amount of compressed air used by the pneumatic tool during assembly in Design A, the labour costs in (3) were predominant and were proportional to time taken in each case. Because the nature of the work was high-volume–low-cycle time, it was decided to determine the basic time for assembly in each case by MTM in order to attain the required precision.

The analyses which resulted from the study are shown in Figures 16.2, 16.3, and 16.4. It will be seen that the respective assembly times for the three designs are:

$$
\begin{aligned}
&\text{Design A—basic time} = 0{\cdot}259 \text{ b.m. } (519{\cdot}4 \text{ t.m.u.}) \\
&\quad \text{and standard time} = 0{\cdot}295 \text{ s.m. } (14\% \text{ R.A.}) \\
&\text{Design B—basic time} = 0{\cdot}104 \text{ b.m. } (208 \text{ t.m.u.}) \\
&\quad \text{and standard time} = 0{\cdot}119 \text{ s.m. } (14\% \text{ R.A.}) \\
&\text{Design C—basic time} = 0{\cdot}075 \text{ b.m. } (150 \text{ t.m.u.}) \\
&\quad \text{and standard time} = 0{\cdot}085 \text{ s.m. } (14\% \text{ R.A.})
\end{aligned}
$$

MTM ANALYSIS

MTM-1 ✓
MTM-2
MTM-3

Job description:
 Assemble transformer to plate
 Method "A"

Sheet 1 of 2
Analyst PG
Date 2 March

El.	LH		tmu	RH	
	To washers	R14C	15.6	R14B	To transformer
	Grasp 1st washer	C4B	9.1	C1A	Grasp transformer
	Clear of container	M2B	—		Hold in box
	Palm washer	G2	5.6		"
	To 2nd washer	R2C	5.9		"
	Grasp washer	C4B	9.1		"
	Washers to area	M10B	16.9	M14C	Transformer to plate
			—		adjust grip to locate
			19.7	P28SE	Locate 1st. stud
			5.6	P18E	Locate 2nd. stud
	Washer to stud	M2C	5.2		Hold transformer
	Locate on stud	P28E	16.2		"
	Release washer	RL1	2.0		"
	Unpalm 1st. washer	G2	5.6		"
		M2B	—		"
	Washer to stud	M2C	5.2		"
	Locate on stud	P28E	16.2		"
	Release washer	RL1	2.0		"
	To nuts 2x	R14C	31.2		"
	Grasp nut 2x	C4B	18.2		"
	Nut to stud 2x	M14C	33.8		"
		G2	—		"
	Locate to stud 2x	P28E	32.4		"
	Back turn 2x	M1B	5.8		"
	Forward turn 2x	M2B	9.2		"
	Release 2x	RL1	4.0		"

Figure 16.2 An MTM

MTM ANALYSIS

MTM-1 ✓
MTM-2
MTM-3

Job description:

Method "A"

Sheet 2 of 2
Analyst
Date

El..			LH	tmu	RH	
	Turn nut onto	2x 3x	R2A	15.0		Hold transformer
	stud with	2x 3x	G1A	12.0		"
	fingers	2x 3x	M2B	27.6		"
		2x 3x	RL1	12.0		"
	To nut runner		R12A	9.6		"
	Grasp nut runner		G1A	2.0		"
	Nut runner to stud		M12C	15.2		"
			G2			"
	Locate to stud	2x	P28SD	50.6		"
	Operate start button 2x		MfA	4.0		"
	Operate nut runner 2x		P.T.	80.0		"
	Nut runner to 2nd. nut		M4C	8.0		"
	Nut runner clear		M4B	6.9		"
	Release nut runner		RL1	2.0	RL1	Release transformer
				519.4		

analysis for Design 'A'

MTM ANALYSIS

MTM-1 ✓
MTM-2
MTM-3

Job description:
Assemble transformer to plate
Method "B"

Sheet **1** of **1**
Analyst **P. G.**
Date **2 March**

El.		LH	tmu	RH	
	To spring retainer	R8C	11.5	R-E	Toward transformer
	Grasp retainer	G4B	9.1		
	Lift clear of container	M.B	6.4	R4B	To transformer
	and regrasp if required		2.0	G1A	Grasp transformer
			16.9	M14C	Transformer to plate
			19.7	P28SE	Locate 1st. stud
			5.6	P18E	Locate 2nd. stud
		M2C	5.2		Hold transformer
		P28D	21.8		"
		APB	16.2		"
		APA	10.6		"
	Contact release	RL2	—		"
	To spring retainer	R8C	11.5		"
	Grasp retainer	G4B	9.1		"
	Retainer to stud	M8C	11.8		"
	Locate retainer to stud	P28D	21.8		"
	Press home retainer	APB	16.2		"
		APA	10.6		"
	Contact release	RL2	2.0	RL1	Release transformer
			__208.0__		

Figure 16.3 An MTM analysis for Design 'B'

MTM ANALYSIS

MTM-1 ✓
MTM-2
MTM-3

Job description:
Assemble transformer to plate
Method "C"

Sheet 1 of 1
Analyst P.G.
Date 2 March

El.		LH	tmu	RH	
			14.4	R14B	To transformer
			2.0	G1A	Grasp transformer
	Pliers to work area	(M.B.)	16.9	(M14C	Transformer to plate
			—	~~G2~~	
			19.7	P28SE	Locate 1st. stud
			5.6	P18E	Locate 2nd. stud
	Pliers to tab	M4C	8.0		Hold transformer
	Open pliers	~~M1B~~	—		"
	Locate pliers to tab	P28SE	19.7		"
		M2A	2.0		"
	Regrasp pliers	G2	5.6		"
		SC3	2.2		"
	Twist tab	M3B3	6.0		"
	Open pliers	M1B	2.9		"
	Locate to 2nd tab	M2C	5.2		"
		P28SE	19.7		"
	Close pliers	M2A	3.6		"
	Regrasp pliers	G2	5.6		"
		SC3	2.2		"
	Twist tab	M3B3	6.0		"
	Open pliers	M1B	2.9	(RL1)	Release transformer
	Aside	(M-B)	—		
			150.2		

Figure 16.4 An MTM analysis for Design 'C'

398 WORK MEASUREMENT

Location	Area		Floors Times to:		Dust		Radiators		Empty bins etc.		Wash-		
	ft²	m²	Method	time (min)	Items	time (min)	Run m	time (min)	Items	time (min)	toil	ur.	basin
DAILY													
OFFICES													
Consultants	117	11·1	Vac (s)	4·4	11	7·7	2½	9·5	3	2·4			
Conference Room	240	22·8	Vac (s)	9·1	15	10·5	4	15·2	8	6·4			
Factory Manager	165	15·6	Vac (s)	6·2	8	5·6	2½	9·5	4	3·2			
Typists	224	21·2	Sw	3·8	15	10·5	4	15·2	6	4·8			
Indust. Engs.	154	14·6	Sw	2·6	11	7·7	2½	9·5	4	3·2			
Dept. Managers	154	14·6	Sw	2·6	12	8·4	2½	9·5	4	3·2			
Administration	154	14·6	Sw	2·6	14	9·8	2½	9·5	6	4·8			
Medical	154	14·6	Sw	2·6	0	0	2½	9·5	3	2·4			
Personnel Mgr.	100	9·5	Vac (s)	3·8	6	4·2	2½	9·5	3	2·4			
Personnel Secry.	100	9·5	Vac (s)	3·8	7	4·9	0	0	4	3·2			
CORRIDORS etc.													
Stairs (South)			Sw/Mop	25·0									
Stairs (North)			Sw/Mop	25·0									
Landing	116	11·0	Sw/Mop	10·9					1	0·8			
Passage (S.down)	278	26·4	Polish	21·0					3	2·4			
Passage (S up)	215	20·4	Polish	16·5	1	0·5			3	2·4			
Foyer (entrance)	252	24·0	Polish	19·2	6	3·0			2	1·6			
WASH-ROOMS													
Male (up)	80	7·6	Mop	6·2			2	7·6	2	1·6	3	2	3
Male (down)	152	14·4	Mop	11·7			4	15·2	2	1·6	5	6	6
Female (up)	110	10·4	Mop	8·5			2	7·6	2	1·6	4	0	4
Female (down)	240	22·8	Mop	18·5			8	30·4	8	6·4	16	0	12
Male cloaks	77	7·3	Sw	1·3			0		2	1·6			
Female cloaks	322	30·5	Sw	5·5			12	45·6	4	3·2			
OTHERS													
Overall stores	30	2·8	Sw	0·5									
Goods Inwards	180	17·0	(CLEANED BY STOREMEN)										
Stores (main)	255	24·2	(CLEANED BY STOREMEN)										
Special comps	1745	165·5	Vac (L)	59·5					12	9·6			
Spray	1745	165·5	Vac (L)	59·5					10	8·0			
Main Assembly	4950	469·5	Vac (L)	168					24	19·2			
Machine Shop	4950	469·5	Vac (L)	168					20 Swarf bins	16·0 26·0			
Tool Room	350	33·2	Sw	6·0									
Boiler Room	215	20·4	Sw	3·7									
Plant	215	20·4	Sw	3·7									
Switchgear	117	11·1	Sw	2·0									
Cracker gas plt	72	6·8	Sw	1·1									
Pumproom	240	22·8	Sw	4·1									
Air Conditioning	870	82·5	Sw	148									
Canteen	1385	132·0	(CLEANED BY CANTEEN STAFF)						Dispose of food Scraps bins	30·0			
KEY to abbreviations used at column headings	Square feet	Square metres					Linear run in metres				Toilet	Urinal	Hand wash-basin

Figure 16.5 Schedule of jobs carried out in an area

THE USES OF WORK MEASUREMENT

rooms				Total minutes work	WEEKLY			EVERY 6 WEEKS					
					Window frames		Polish floors and Misc.	Light fitting			Pipes and girders (min)	Scrub and seal (min)	Wash doors	
mir.	s.n.	roll towel	time (min)		No.	time (min)		Ty	No	M			No.	time (min)
				24·0	4	4·8		F	2	10			1	8
				41·2	2	2·4		F	6	30			1	8
				24·5	1	1·2	Polish floors	F	3	15			1	8
				34·3	2	2·4	17·0	F	4	20			1	8
				23·0	1	1·2	11·7	F	2	10			1	8
				23·0	1	1·2	11·7	F	2	10			1	8
				26·7	1	1·2	11·7	F	3	15			1	8
				14·5	1	1·2	11·7	F	2	10			1	8
				19·9	1	1·2		F	1	5			1	8
				11·9	0	0		F	1	5			2	16
				25·0	1	1·2		T	4	12			0	0
				25·0	1	1·2		T	4	12			0	0
				11·7	2	2·4		T	2	6			1	8
				23·4	0	0		T	8	24			1	8
				19·4	2	2·4	Clean brass	T	7	21			1	8
				23·8	1	1·2	11·0	T	4	12			2+1	32
3	0	2	21·4	36·8	2	2·4		T	4	12			1	8
6	0	4	43·0	72·4	3	3·6		T	4	12			1	8
12	2	4	44·0	61·7	1	1·2		T	4	12			1	8
16	4	6	98·4	153·7	4	4·8		T	14	42			2	16
				2·9	0	0		T	2	6		7·2	1	8
				54·3	4	4·8		T	15	45		29·7	1	8
				0·5	0	0	Overalls 71·0	T	1	3			1	8
					1	1·2		T	4	12			1+1	20
					2	2·4	Under benches	T	6	18			2	16
				139·1	6	7·2	70·0	F	40	200	160	220·0	4	32
				67·5	4	4·8	32·0	F	40	200	180	220·0	2+4	56
				187·2	11	13·2	160·0	F	90	450	360	625·0	4+2	64
				184·0	0	0	36·0	F	90	450	380	625·0	0	0
				32·0	4	4·8	Clean brass	F	6	30			1	8
				3·7	2	2·4	42·0	T	2	6			1+1	20
				3·7	2	2·4		T	2	6			1+1	20
				2·0	1	1·2		T	2	6			1	8
				1·1	0	0		T	2	6			1	8
				4·1	2	2·4		T	2	6			1	8
				14·8	0	0		T	6	18			3+1	44
				30·0	16	19·2		F	24	120		176·0	1	8
Mirror	Sanitary napkin equip.	Roller towel		1422·8		103·2	485·8	Type	Number off	1877 Time in minutes	1080	1902·9		520

(time standards were added later in the study)

The department labour–cost standard had been established previously as 3·2p per standard minute (see Section 16.2). Thus the relative labour costs could be assessed. The full comparison is analysed below.

Design	Cost savings (p) over Design A of attachments and assembly of housing	Labour costs of assembly	Labour cost saving over Design A (p)	Total saving over Design A (p)
A	0	0·94	0	0
B	0·14	0·38	0·56	0·70
C	0·42	0·27	0·67	1·09

It is clear that the greatest saving was on the cost of assembly labour, and although the assembly time could have been estimated, because of the high volume of work (three million coil housings) it was essential that the work content be measured more precisely. Thus the cost saving by adopting Method C as opposed to Method A is over £30,000.

16.4 USE OF WORK MEASUREMENT IN COSTING A SERVICE

In many areas of component manufacture, provision of services, and others, the decision must be made whether to provide the necessary goods and services from the resources of the company, or to buy them in from outside sources. In making such decisions, clearly management must be furnished with the relevant information about the costs of both strategies.

The example given here is concerned with the maintenance-cleaning (janitorial) work in a factory.*

At present the staff comprises three full-time workers and one part-time worker, who are employed in cleaning the factory according to the schedule prepared by the work-study team (Figure 16.5).

The jobs are measured by one of the methods outlined in Chapter 15, the standard times being listed in the Analysis Sheet (Figure 16.6). Using this list in conjunction with the schedule of work, the work-load is calculated as follows:

Daily work-load   = 1,422·8 min

Weekly work-load   =   589·0 min (or 117·8 min daily equivalent)

6-weekly work-load = 5,380·0 min (or 179·3 min daily equivalent)

The grand total for the daily work-load is 1,720 min, or 28·7 hrs.

Once this work-load has been established in terms of hours the job may be costed using the department-cost standard (see Section 16.2). The labour-cost

* Data are taken from Dennis A. Whitmore, *Measurement and Control of Indirect Work* (London: William Heinemann Ltd, 1970).

A.E.C. — FLUORESCENT FITTINGS COMPANY
ANALYSIS SHEET

Date: May 27   Analyst(s) E.K.L.

No.	Description	Description of unit of count	Abbreviated unit of count	Time standard (min)
1	Empty wall ashtray/rubbish bins	per tray/bin emptied	item	0·8
2	Empty ashtrays in offices	"	item	0·8
3	Issue overalls to employees	per employee dealt with	occasion	11·0
4	Pack soiled linen	each time this is done	occasion	20·0
5	Unpack nylon overalls	"	occasion	40·0
	Cleaning			
6	Dust chairs, tables, cabinets etc	count each item as <u>one</u>	item	0·7
7	Clean window frames (damp cloth)	each complete window is <u>one</u>	frame	1·2
8	Clean radiator (damp cloth)	per linear metre of run	metre	3·8
9	" overhead lamps	count number of lamps	lamp	3·0
10	" " fluorescent fittings	count number of fittings	fitting	5·0
11	" " girders	per linear metre of run	metre	4·6
12	" " pipes	"	metre	4·6
13	Remove and replace office chairs	count <u>one</u> each time	chair	0·6
14	" " " stacking chairs	a chair does a round trip	chair	0·5
15	Wash door (internal door)	wash one side of door	door	4·0
16	" " (external door)	"	door	6·0
17	Damp wipe under work-bench	per metre of run	metre	2·0
18	Scrub floor-mech: open area	per square metre	m²	0·28
19	" " -mech: between benches	"	m²	0·63
20	" " by hand	"	m²	1·80
21	Polish floor-mech: open area	"	m²	0·58
22	" " -mech: offices	"	m²	0·80
23	Sweep floor - broom	"	m²	0·18
24	Clean floor - damp mop	"	m²	0·81
25	Mop up after scrubbing	"	m²	0·27
26	Use squeegee	"	m²	0·36
27	Vacuum clean- domestic model	"	m²	0·40
28	" " -industrial model	"	m²	0·36
29	Seal floor using applicator – cleared area (time incl. prep.)	"	m²	0·70
	Toilets (wash-rooms)			
30	Clean hand-basin and top up soap dispenser	per basin cleaned	item	1·50
31	" toilet-basin and renew tissues	per stall cleaned	item	2·10
32	" urinal	per urinal cleaned	item	1·60
33	Renew auto. roller towels	per machine attended	item	2·20
34	Clean mirror	per mirror cleaned	item	1·00
35	Attend san. napkin equip. (disp's'r)	per machine attended	item	3·40
36	" (receptacles)	"	item	1·00

Figure 16.6 An analysis sheet for cleaning operations

standard for cleaners (including appropriate overheads) was assessed as £1·60. Thus the cost of cleaning was:

$$£(28·7 \times 1·60) = £45·92 \text{ per day}$$
$$= £229·6 \text{ per week}$$

Tenders were considered from several cleaning contractors, for a contract to be negotiated for maintenance of the present standard of cleanliness.

Estimates varied between £260 and £140 per week (the lower figure was found later to have been estimated on a lower standard). Eventually one estimate of £210 was selected. Armed with this information management was able to assess the situation, not only on financial grounds, but taking into account other factors.

The essential part taken by work measurement is to provide the information: management must make the final decision, as always.

16.5 SCHEDULING AND PLANNING

The management indices previously described, by definition, are capable only of measuring performance retrospectively. Unfortunately there is no absolutely dependable method of predicting future conditions or requirements. Forecasting relies on past records, and on statistical trends which are extrapolated to provide information about the future which will be reasonably reliable provided nothing occurs which will upset the predicted trend. As the marketing manager knows

SCHEDULE OF COMPONENTS REQUIRED							
DEPARTMENT: Machine Shop			Section: S.P. Lathes			Date prepared: 4/4/–	
COMPONENT	CODE	BATCH SIZE	STANDARD TIME (S.H./100)	STANDARD HOURS	SET-UP TIME S.H.	NUMBER OF DAYS' STOCK REMAINING	
Gland Body	42 A	500	2·16	11·80	2·00	2	
	42 C	500	2·18	11·90	2·00	4	
	113	200	3·05	6·10	1·60	6	
	28 M	2000	1·10	22·00	0·80	8	
Flange	F 20/60	350	2·00	7·00	1·15	9	
	F 20/80	350	2·10	7·35	1·15	8	
	F 25/80	400	2·25	9·00	1·15	12	
	F 25/102	100	2·42	2·42	1·15	7	
Spec. Bolt	SB 262	1000	0·60	6·00	0·60	6	
	SB 276	1500	0·70	10·50	0·60	5	
	SB 5102	200	0·40	0·80	0·60	6	
Nipple	NS 31	5000	0·42	21·00	0·40	8	
	NS 32	5000	0·40	20·00	0·40	9	
	NS 33	1000	0·45	4·50	0·40	11	
S/S Union	KU 10	4000	0·62	24·80	0·90	9	
	KU 12	4000	0·56	22·40	0·90	12	
	KU 24	12 500	0·60	75·00	0·90	10	
	KU 24A	4 500	0·60	27·00	0·90	9	
	KU 36	6 000	0·65	39·00	1·10	12	
				328·57	18·70		

Figure 16.7 A schedule of components required for the S.P. Lathe Section

only too well, the pattern of the past does not always repeat so conveniently. However, this is the best that is available, and one must ensure that as many eventualities as possible must be anticipated, and appropriate corrections built into the prediction model.

Although the only information the planner has at his disposal refers to historical data, he is able to plan and schedule, and establish the labour force required for his department, by calculating the present work-load, and hence predict his future requirements after taking into account expected variations in volumes and future sales. All schedules, of course, are dynamic and subject to continual updating.

The scope of the present book does not extend to a comprehensive explanation and description of the techniques of planning and scheduling, but the role of work measurement in this field will be illustrated by an example taken from a batch-production manufacturing unit.

A group of five special-purpose capstan lathes is engaged on producing small and medium batches of components. Current requirements are listed on a schedule together with the standard times for each job, for the guidance of the planning engineer (Figure 16.7). This schedule also states the number of days' stock of each item remaining, thus giving the planning engineer an indication of the priorities.

One of the most useful planning boards for the visual presentation of machine-loading is the Gantt Chart introduced at the turn of the century by Henry L. Gantt, a follower of Frederick W. Taylor (*see* Chapter 2). This chart, depicted in Figure 16.8, displays the planned machine-loadings in the form of bars, the lengths of the bars being measured on a *time* scale. Additionally the *actual* production achieved is plotted in bar form. In the type of work illustrated by this example, and indeed in most work connected with stock control, it is time rather than volume which is the most important unit to use in planning. It means little to say that 1,000 special bolts type SB262 and 500 gland bodies type 42A are required. What is more important to the planner is that the required bolts should take six hours to make while the gland bodies require 11·8 hours of machining time. It will be seen that stock is also quoted in terms of days' stock remaining because it is essential to know how long material will last rather than how much of it there is left.

The planning board carries a cursor which may be moved along to indicate the present time (in the example: 17·00 on Friday, 11 April). For those machines which are up to schedule their black progress-bars touch the cursor, for example machines 1 and 4 in the Figure. Those bars which fall short (machines 2 and 5) are those machines which are behind in their work. Machine 3 is four hours ahead of schedule, which means that Job F 20/60 may be brought forward to start on Monday morning.

The importance of time as the basis for the Gantt Chart emphasizes the need for reliable standard time-values set by work measurement. Standard times may be established by the usual methods, and when extended by the volumes the total standard hours of work may be found.

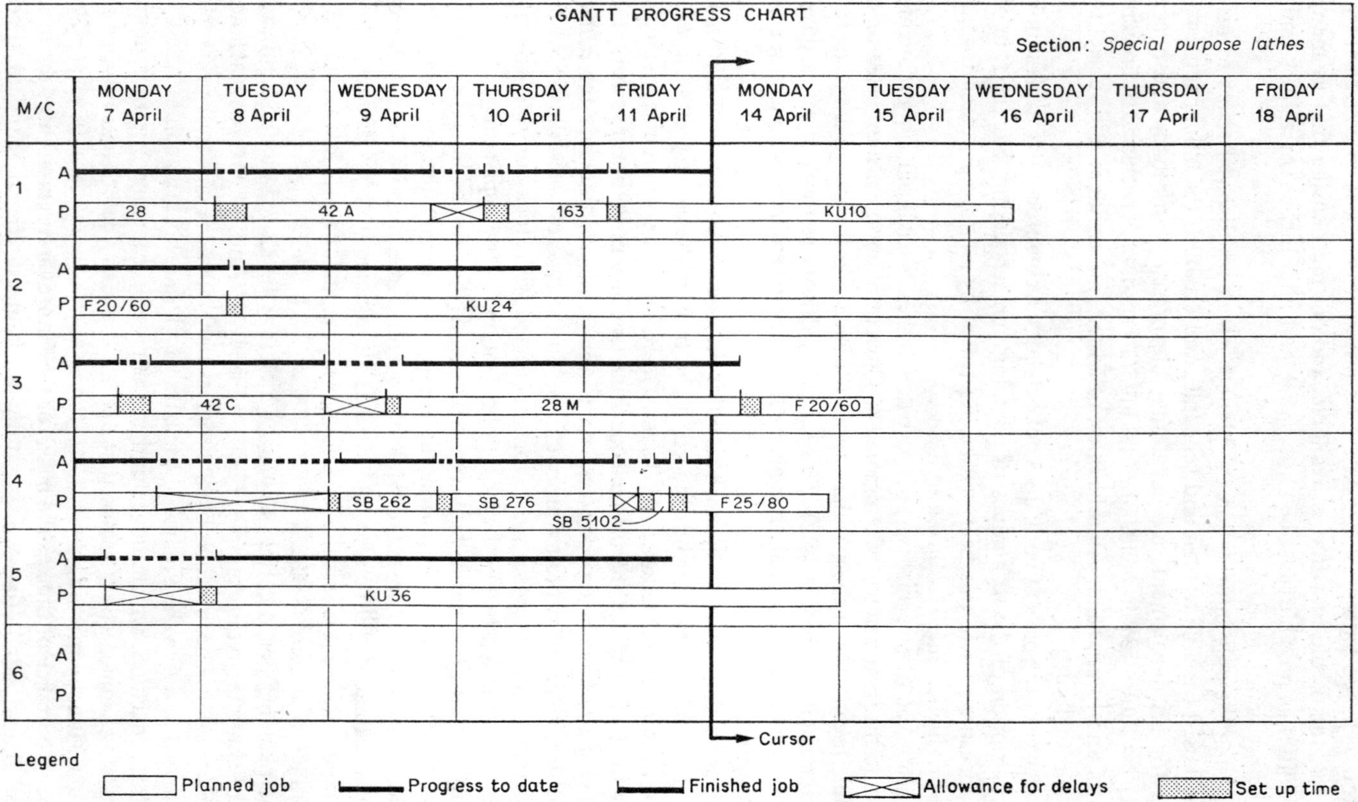

Figure 16.8 A Gantt Progress Chart for the S.P. Lathe Section

In this example the schedule shows that there are 347·27 standard hours to be allocated to the five special-purpose lathes, including setting-up times. This total does not include allowance for the various stoppages, but appropriate compensation may be built into the calculations. From past records the effectiveness of the group was found to be 72 per cent, thus a more realistic time of 34,727/72 or 482·3 hours could be used. At the time of issue of the schedule the s.p. lathe section was loaded with about 96 hours of work on average for each machine, or just over two weeks' production.

In the foregoing example each job was entirely independent of any of the others, but often they are interrelated and must be placed in their correct sequence. The sequencing of jobs is part of the discipline of Operational Research. One group of techniques widely used since its appearance in 1958–59* is known as *critical*

LIST OF ACTIVITIES FOR SPECIAL WAVE-FORM GENERATOR

Node Nos.	Duration	Activity	Start E	Start L	Finish E	Finish L
1–2	4	Prepare schedule	0	0	4	4
2–3	2	Allocate jobs	4	4	6	6
2–17	8	Design packing	4	20	12	28
2–20	20	Prepare instruction book	4	16	24	36
3–4	1	Order metal components	6	6	7	7
3–7	1	Order bought-out components	6	8	7	9
3–8	1	Order chassis	6	14	7	15
3–9	1	Order case and parts	6	11	7	12
4–5	10	Wind transformers	7	7	17	17
4–6	7	Form metal parts	7	10	14	17
5–6	0	Dummy	17	17	17	17
6–13	6	Plate	17	17	23	23
7–10	6	Obtain transistors	7	15	13	21
7–11	12	Obtain small components	7	9	19	21
8–12	5	Bend and drill chassis	7	15	12	20
9–15	12	Construct case	7	15	19	27
9–18	20	Obtain knobs and dial	7	12	27	32
10–11	0	Dummy	13	21	13	21
11–13	2	Sub-assembly panels	19	21	21	23
12–13	3	Plate	12	20	15	23
13–14	6	Main assembly	23	23	29	29
14–16	2	Align	29	29	31	31
15–16	4	Spray and bake	19	27	23	31
16–18	1	Case up	31	31	32	32
17–20	8	Obtain packing	12	28	20	36
18–19	1	Fit knobs and dial	32	32	33	33
19–20	3	Calibrate and test	33	33	36	36
20–21	2	Pack	36	36	38	38

E = Earliest time L = Latest time

Figure 16.9 List of activities for the special waveform generator

* D. G. Malcolm, J. H. Roseboom, C. E. Clark, and W. Fazar, 'Application of a technique for R & D Program Evaluation', *Operations Research*, September–October 1959.

path analysis (C.P.A.). Again, the method will not be described here in detail, but there is available a wide range of literature on the subject, and some notable works are listed in the bibliography.

The purpose of C.P.A. is to plan the work through the medium of a *network* which shows the correct sequence of the jobs in pictorial form. The strength of the network is its ability to depict and indicate those jobs which must be performed concurrently by the use of parallel as well as series construction of the network.

An important part of C.P.A. is the listing of the jobs with their *durations* (Figure 16.9).* The durations are obtained through the application of work measurement. Often the technique is *estimating*, especially where synthetic-data values are not available. Alternatively, in some cases a second-level system of predetermined motion–times may be used to obtain the durations of the activities. A typical but simple network, based on the data given previously, is shown in Figure 16.10.

A variation on the basic C.P.A. theme is the provision of *three* weighted estimates for the durations of each activity, used in the technique known as P.E.R.T. (Program Evaluation and Review Technique). This calls for a pessimistic (P), an optimistic (O), and a most probable completion time for each activity, the time actually incorporated in the schedule being an amalgam of these times computed from the formula:

$$\text{estimated probable activity time} = \frac{P + O + 4E}{6}$$

The formula is based on the assumption that probable activity durations are distributed as *beta distributions* (see Chapter 14). These usually have a right-hand skew, and in the limit they tend to the shape of the normal distribution. The formula given is the standard one for the *mean* of a beta distribution.

16.6 APPLICATION TO REMUNERATION

The purpose of the work unit is to provide a facility for expressing the time for all jobs at a common level of performance. This ensures that equal effort is recognized as such, an essential feature of any soundly based scheme of remuneration.

A person's wage may be associated in some way with his effort or output, or it may, as in the case of day-work, be entirely unrelated to the amount of work carried out. The former means of remuneration includes the systems known as payment-by-results, and between this group of schemes and the fixed wage of day-work exist the hybrid schemes generally known as *measured day-work*. While it is true that the wage paid under measured day-work is not affected by the level of output achieved, the output is still monitored daily or weekly using the performance indices outlined earlier.

* Taken from: Dennis A. Whitmore, *Work Study and Related Management Services* (London: William Heinemann Ltd, 1968, 1970).

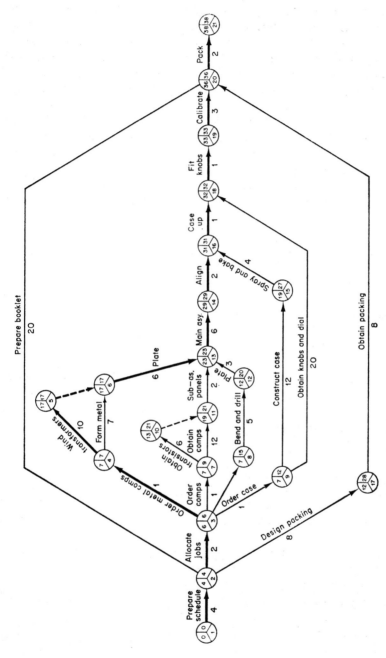

Figure 16.10 The arrow diagram for the special waveform generator

The first geared 'premium bonus' incentive schemes appeared in the late nineteenth century, and many of the systems in use currently are little different from those early schemes. In those days work measurement was not sufficiently consistent or reliable to form a sound basis for methods of payment-by-results which would pay workers fairly for their efforts. Consequently protection against 'rate-busters' was afforded by the placing of ceilings on earnings, and the development of non-linear schemes of the type introduced by Rowan. Straight proportional schemes are much more readily accepted by people, who can understand the calculations of bonuses.

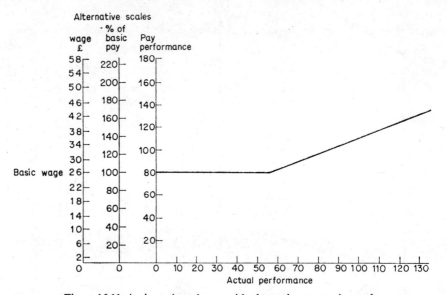

Figure 16.11 An incentive scheme, with alternative conversion scales

Figure 16.11 shows, in graphical form, a typical linear incentive-payment plan. The dependent variable (on the vertical y axis) denotes the reward for the given effort, which may be in terms of *pay performance*, or *percentage of the basic pay*, or even in actual money. The horizontal (or x) axis* carries a measure of the output attained by the worker. Expressing this in terms of actual volumes is too restrictive; it is better to use a scale common to any and all work. The performance indices have the advantage of being free from the encumbrances of specific units, and as such are the obvious choice for this purpose.

In operation, the first part of the application is the monitoring of performance by means of one of the indices described in Section 16.1. This implies that the jobs involved have been adequately covered by work measurement. Clearly it is imperative that the work measurement is reliable because the performance, and hence, in the case of incentive schemes, the worker's pay depends on its accuracy.

* *See* Chapter 6 for a description of graphical methods.

In this example the standard time for the job had been established as 4·0 s.m. In the particular week considered (38 hr) the operative had produced 510 components, equivalent to 2,040 s.m. or 34 standard hours. The resulting operator-performance was 90. The operator-performance is the most appropriate index to use because payment must be related to his effort during the time he is able to work.

The second part of the application is relating the performance to the wage to be paid. This process is shown diagrammatically in Figure 16.12. Other schemes could be devised which paid different wages for a given output merely by altering the slope (or gearing) of the line, or the position of the line, or both. The company must devise the most acceptable and appropriate relationship between performance and payment by careful selection of the parameters.

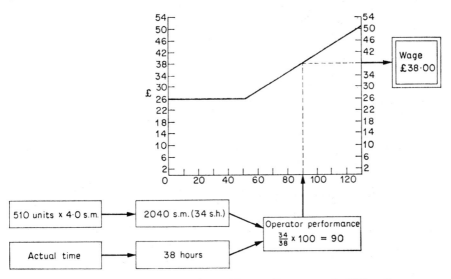

Figure 16.12 Calculating the weekly wage for an output of 510 units

Even when workers are paid on a fixed wage on a measured day-work scheme, monitoring of performance must also be based on sound work measurement. This is particularly important when the scheme is a form of graded-level day-work such as the Philips Premium Payment Plan. Payment does not vary with output, but is made at certain predetermined levels of performance. The levels usually number about six to eight, separated by five or sometimes ten performances (*see* Figure 16.13). Under these systems the worker's money is fixed, but fixed at the level determined by his average performance over an agreed period of time. Consequently any error in the standard time set by work measurement reflects in the level of wage paid, hence the importance of reliability.

A particular form of graded-level day-work was devised, for example, for use in a typing pool. At the time it was considered to be the best compromise: the reluctant management being under pressure from the staff union to introduce some form of output-related pay scheme.

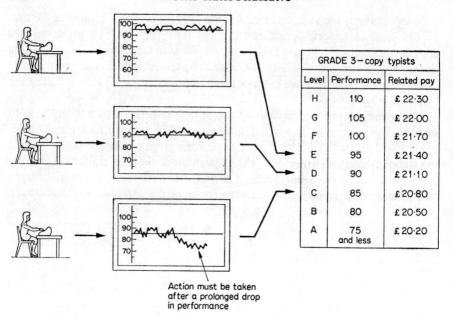

Figure 16.13 Pictorial representation of the premium payment plan (daily rates of pay)

In the usual way the operators' performances are calculated and checked to ensure that they achieve their respective payment levels. For example, suppose three typists are paid respectively on levels C, D, and E, on the scheme shown in Figure 16.13. They are required to achieve weekly performances of 85, 90, and 95 respectively in order to justify their wages, although they will receive the stated wages regardless of their actual performances. In cases of default (as in the instance of typist 3) other forms of disciplinary action than pecuniary penalty must be taken.

17 New Developments

17.1 IN WHICH DIRECTION?

In Chapter 2 the history of work measurement over the past 200 years was traced from its early primitive beginnings through to the relatively advanced methods of the present day. The chapter described how work measurement had turned a full circle, moving from elementary overall studies, through various stages of refinement which reached a peak in the 1940s, returning in the late 1960s to the more readily applied overall studies made possible by the introduction of mathematical techniques into the discipline.

In the field of P.M.T.S. the search for methods which were less time-consuming to apply led to the development of higher-level systems exemplified by such techniques as MTM-2 and MTM-3 which are described in Chapter 8. The drive was stimulated by the reluctance of managements to accept systems whose methods were laborious though admittedly precise, but whose precision was not justified in many applications. The advent of the table-top computer and time-sharing terminal made detailed P.M.T.S. once again a feasible proposition for economic work measurement.

Although it is not possible to predict the future developments in work measurement, certain trends may be identified from present-day research.

The unprecedented increase in the number of indirect jobs over the last decade was discussed in Chapter 15. The trend already apparent in the United States may be identified in the United Kingdom, with the indirect work-force increasing more rapidly than the total employees. Gradual mechanization and automation of processes and decline in manufacturing is changing the shape of labour distribution. The changing emphasis from purely manual to white-collar work demands the development of new techniques of measurement.

With so many years of work-measurement experience at shop-floor level behind them it was inevitable that practitioners should adapt the well-tried techniques to this new situation. Thus all existing methods of measurement, in spite of some novel titles, still rely on the three basic groups of techniques, namely timing, estimating, and P.M.T.S. In this age of such rapidly advancing technology the approach does seem to be somewhat old-fashioned and not a little ponderous. However, it would appear that these methods will persist into the future as far as one can foresee.

Practitioners are ill-prepared for the changing role of the worker, who is becoming less of an automaton and a more a thinking employee. To date, despite the laudable efforts of researchers to produce a system capable of measuring work of high mental content, there has been no major breakthrough and it is difficult to see how creative work will ever be measured other than retrospectively and even then measurement will probably be confined to binary-decision level.

In the near future managements will be prompted to reconsider the use of work measurement, and to ask three questions:
1. Is the right way to control people through the use of time standards?
2. If so, are there better ways of measuring their jobs?
3. If not, how *should* control be effected?

Undoubtedly the motivationalists would reply by proposing a system based on 'motivation from within', stimulated by such principles as job enrichment and increased responsibility. In doing so they would be working on the assumption that the use of work measurement is for purposes of incentive. Standards in terms of actual time will still be necessary for planning, manning, costing, and other management information.

Future systems may well integrate the two concepts of inner motivation and objective work-standards, thus allowing people to execute the work in their own way within the required framework and in consistency with the overall objective. This in turn provides them with the stimulus to eliminate ineffective work while still providing data for management purposes.

In Chapter 2 it was shown how thinking in work measurement has turned full circle with the emergence of mathematical techniques coupled with the availability of computer time. Future trends may be in four directions of:
1. Overall timing methods based on multiple-regression analysis and similar techniques (e.g. linear programming);
2. Physiological measurement of work and effort;
3. Computerized work measurement and use of comprehensive data-banks of universal standard data;
4. Analyses of jobs through the application of computerized predetermined motion–time systems.

A decline in the popularity of individual financial incentive schemes coupled with a growth in the practice of employee-motivation which includes job enrichment, flexible working hours and the like, could have the effect of switching interest from individual-performance to department or even corporate-performance measurements. In such circumstances practitioners will need the more gross over-all methods of measurement rather than the type provided by micro-

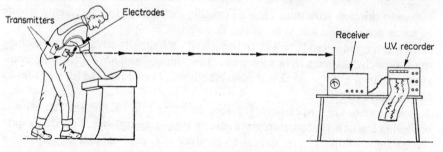

Figure 17.1 Use of telemetry in monitoring heart-rate

P.M.T.S. and time study. The mathematical techniques of work measurement could fulfil this need.

Much research has been carried out already on measuring work in terms of *effort* as opposed to time. Some workers are at present investigating the potentials of biotelemetry systems for monitoring such biological parameters as heart rate (E.C.G.), pulse, muscle activity (E.M.G.), respiration, skin resistance, temperature, and others. Electrodes attached to the body detect and transmit impulses via micro-transmitters to central recording apparatus, which monitors the activities (Figure 17.1). In most cases polarization and depolarization provide the electrical impulses from the muscles, which are picked up by the electrodes, but sometimes photocells attached to the ear-lobes detect colour differences and blood is pumped through them. In other instances thermal-sensitive devices measure temperature changes. Relationships between these parameters and the amount of work accomplished are established in order to interpret the output traces in terms of work content. Such systems are useful in measuring the work content of purely manual jobs, but timing is still necessary for machine- and process-controlled work.

Attempts have also been made to relate mental activity to measurable physical properties, including the application of the electro-encephalograph to detect 'brain-waves'. Most of the successful research has been carried out in the laboratory under carefully controlled conditions. Unfortunately many of the methods are not suitable for use under operational conditions because they employ specially contrived conditions. These include 'secondary tasks' which are given to the subject in order to load his mental capacity to its maximum. Clearly it is difficult to ask someone to carry out a secondary task while he is performing his normal work.

A certain amount of mental work can be measured by such techniques as the Wofac Mento-Factor System, and more crudely by over-all timing, but this latter method cannot take into account the mental processes involved nor 'rate' the effort put into the mental activity by the subject. Thus, with the ever-increasing importance of clerical and supervisory work, it is becoming more necessary to be able to assess the extent of mental work by more objective means.

17.2 COMPUTERIZED WORK MEASUREMENT

With the ascendance of the microcomputer at prices which everyone can afford it was only natural that software would be made available and hardware designed specifically to accommodate the techniques of work measurement.

The arrival of the compact, battery operated, microcomputer has introduced a new dimension to the capture of work study data. Typical models from the late 1970s were Datamyte*, Husky†, and Epson HX20‡. These are 'off the shelf'

*R. M. Barnes, *Motion and Time Study*, 7th Ed. (New York: Wiley, 1980), Chpt. 25.

†G. Cook, 'Manservant – Study Work Using Electronic Notebook', *Management Services*, March 1983, pp. 28-30.

‡A. R. Watson, 'Using the Epson HX20 for Time Study', *Work Study*, April 1984, pp. 43-50.

portables equipped with suitable software written for data collection, electronic time study (E.T.S.), activity sampling, and other applications. With these standard models naturally the keyboards are of the QWERTY type and not specifically designed for these purposes. Thus some coding or even full typing of ratings is necessary. The 1980s saw the specially designed E.T.S. boards which are much preferred and simpler to use, for example Tectime from Tectime Data System Ltd. of Newcastle-under-Lyme, and Teltime by Tleford Management Services Ltd. of Wolverhampton.

A tailor-made model will have a large internal memory for storing data, an internal clock which keeps very accurate time, and a special keyboard with keys engraved with the ratings so that the observer does not have to remember that '1' means 65 rating, '2' means 70 rating, and so on. A well designed keyboard will demand as few key depressions as possible. Rather than typing three keys for 'one hundred and five' rating, for example, one key should carry this figure. This is possible, as ratings are made in steps of fives between about 65 and 130, and this requires 14 keys. Keys can be dual or triple purpose, with shift-keys, which reduces the size of the keypad.

Although several types of E.T.S. study boards are available there are two basic forms, illustrated in Figure 17.2. One design (type 'A') is small enough to hold in one hand, like a calculator while the other hand is free to operate the keys. The alternative (type 'B') is one which emulates the traditional study board and is cradled on one of the arms. This has a writing surface which can be used for making notes about elements and recording other data, but not necessary, of course, for writing down the ratings (which are entered from the keyboard) nor for writing observed times (calculated automatically).

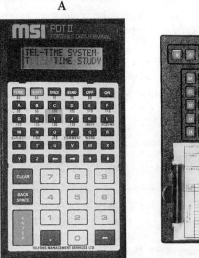
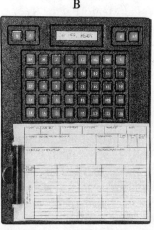

Figure 17.2

A. The TMS Portable Data Terminal (by kind permission of Telford Management Services)
B. The electronic study board (by kind permission of Tectime Data Systems Ltd)

17.21 PROCEDURE

The procedure for use using the computerized board follows that of manual time study with minor variations. The analysis of the job into elements with breakpoints is done in exactly the same way. Abridged descriptions of elements and breakpoints may be noted on a sheet to remind the observer where elements start and finish. The main difference here is that full descriptions are typed in and recorded on disk on the microcomputer rather than on observation or top sheets. With some systems this may be done on the electronic study board, but others may need data to be entered at the V.D.U. terminal. Elements are individually coded to minimize the amount of information typed in during the study.

While the basic procedure for manual and computerized time study are very similar the differences which are apparent are in the recording of the data. Manual time study requires the observer to write down the element coding, rating, and observed time. When using the computerized board the observed time is not entered as this is calculated by the electronic study board or the microcomputer.

Using a typical system the observer watches the operation of the element and rates it. If all has gone well, he presses the appropriate rating button and, at the instant of the breakpoint, another button which causes the computer to record the finishing time, and to reset its clock recording to zero to start timing the next element. In some designs the operating of the rating button performs both functions, so, of course, it must be pressed at the breakpoint. Thus all writing is eliminated; the observer merely presses the element code number button, rates the job, and presses the appropriate rating button at the breakpoint.

There needs to be provision for foreign elements, ineffective time, and any other times which are not rated for some reason, and which may be 'lost'. Facilities must be available for marking any element times which will not be included in the analysis.

Data are stored electronically in the board during the study. This may be done in either of two ways. One of these is to use the board's internal memory. This requires the memory to be large enough. Subsequently the data are downloaded to the microcomputer. Alternatively data may be dumped to micro-cassette for input later to the microcomputer.

At the end of the study the observer loads the data from the board (from its memory, or cassette) into the microcomputer, already programmed to receive the data, and runs the E.T.S. program to obtain a complete analysis.

17.22 SOFTWARE FOR ANALYSES

Printouts are only as comprehensive as the software which produces them. The choice of system will depend a lot on the sort of printout required. Print-outs vary from narrow strips which contain the bare elements of information necessary (Figure 17.3) with element descriptions summarized in up to three words, to comprehensive study sheets emulating the manual equivalents.

The advantage of the strip as used with the Epson HS20 system, is that it provides an on-the-spot readout, something which could please the trade union representatives. The main disadvantage is in the lack of information contained.

```
PACKING BA257                          PACKING
OPERATOR BL                            OPERATOR BL
OBSERVER DAW                           OBSERVER DAW
03/02/1987                             03/02/1987
TIME START = 13:35:55                  TIME START = 13:35:55
ELEMENT DESCRIPTIONS                   ELEMENT DESCRIPTIONS
1 PICK UP COMPONENT                    1 PICK UP COMPONENT
2 INSPECT                              2 INSPECT
3 WRAP                                 3 WRAP
4 PLACE IN BOX                         4 PLACE IN BOX
5 PA BOX & OBT.EMPTY                   5 PA BOX & OBT.EMPTY
CY    EL    RATE   OBS                 EL.  FREQ.   RA%   STD
NO    NO           TIME                1    1       10    .085
 1     1    100    .08                 2    1       11    .143
 1     2    100    .13                 3    1       12    .159
 1     3     95    .15                 4    1       12    .121
 1     4    100    .11                 5    1/5     12    .041
 2     1    100    .07                 TOTAL STANDARD TIME=
 2     2     85    .15                 .549
 2     3     90    .15                 CONTINGENCY % = 2
 2                                     INCLUDING CONTINGENCY
                                       TOTAL STANDARD TIME =
       1    100    .08                 .56
 4     2    100    .13
 4     3    105    .14
 4     4    105    .1
 4     5     85    .22
TIME OFF = 1.1
TIME FINISH = 13:38:39


PACKING BA257
OPERATOR BL
OBSERVER DAW
03/02/1987
TIME START = 13:35:55
ELEMENT DESCRIPTIONS
1 PICK UP COMPONENT
2 INSPECT
3 WRAP
4 PLACE IN BOX
5 PA BOX & OBT.EMPTY
EL   NO.OF   MEAN    %
NO   OBS     TIME   ERROR
 1    4      .077   5.2
 2    4      .129    .78
 3    4      .142   2.82
 4    4      .108    .93
 5    1      .187   0
TIME OFF = 1.1
TIME FINISH = 13:38:39
```

Figure 17.3 Three examples of 'strip' print-outs from the Epson HS20

Quality of print is as good as the printer being used, and most systems rely on dot matrix for draft or letter quality.

Typical print-outs for time study include study/element description listings, study record (equivalent to the observation sheet), study extension, summary sheet, and graphs of mean basic times (Figures 17.4 and 17.5).

OBSERVATION STUDY SHEET

OPERATION: INPUT DETAILS OF E120 TO COMPUTER   OBSERVER: E.J.SIMMONDS

OPERATOR: MS K MULLENS   STUDY NUMBER: 16820

DATE: 3/3/1986   START TIME: 11:41.36
                                                     FINISH TIME: 11:47.54
                                                     ELAPSED TIME: 00:06.18

ELMT	CODE	DESCRIPTION	START	FINISH	OBS.TIME	RATING	BASIC TIME
01	PU132	PICK UP INVOICE	11:41.36	11:41.49	0.13	85	0.1105
02	KEY38	KEY IN DATA FROM INV.	11:41.49	11:41.80	0.31	80	0.2480
03	PA096	PLACE ASIDE INV.TO FIL	11:41.80	11:41.90	0.10	90	0.0900
01	PU132	PICK UP INVOICE	11:41.90	11:42.04	0.14	80	0.1120
02	KEY38	KEY IN DATA FROM INV.	11:42.04	11:42.39	0.35	85	0.2975
03	PA096	PLACE ASIDE INV.TO FIL	11:42.39	11:42.50	0.11	85	0.0935
01	PU132	PICK UP INVOICE	11:42.50	11:43.63	0.13	85	0.1105
02	KEY38	KEY IN DATA FROM INV.	11:43.63	11:43.03	0.40	80	0.3200
03	PA096	PLACE ASIDE INV.TO FIL	11:43.03	11:43.12	0.09	95	0.0855
01	PU132	PICK UP INVOICE	11:43.12	11:43.26	0.14	80	0.1120
02	KEY38	KEY IN DATA FROM INV.	11:43.26	11:43.51	0.25	80	0.2000
03	PA096	PLACE ASIDE INV.TO FIL	11:43.51	11:43.62	0.11	90	0.0990
01	PU132	PICK UP INVOICE	11:43.62	11:43.77	0.15	75	0.1125
02	KEY38	KEY IN DATA FROM INV.	11:43.77	11:44.09	0.32	80	0.2560
03	PA096	PLACE ASIDE INV.TO FIL	11:44.09	11:44.19	0.10	90	0.0900
01	PU132	PICK UP INVOICE	11:44.19	11:44.32	0.13	85	0.1105
02	KEY38	KEY IN DATA FROM INV.	11:44.32	11:44.62	0.30	85	0.2550
03	PA096	PLACE ASIDE INV.TO FIL	11:44.62	11:44.74	0.12	80	0.0960
04	GET22	GET NEW SET OF INVOICE	11:44.74	11:45.25	0.51	90	0.4590
01	PU132	PICK UP INVOICE	11:45.25	11:45.37	0.12	85	0.1020
02	KEY38	KEY IN DATA FROM INV.	11:45.37	11:45.64	0.27	80	0.2160
03	PA096	PLACE ASIDE INV.TO FIL	11:45.64	11:45.75	0.11	85	0.0935
		COM:TALK TO SUPERVSR	11:45.75	11:46.98	1.23		
01	PU132	PICK UP INVOICE	11:46.98	11:47.13	0.15	75	0.1125
02	KEY38	KEY IN DATA FROM INV.	11:47.13	11:47.40	0.27	75	0.2025
03	PA096	PLACE ASIDE INV.TO FIL	11:47.40	11:47.54	0.14	70	0.0980

Figure 17.4   A typical Observation Study Sheet print-out from a microcomputer

SUMMARY SHEET

OPERATION: INPUT DETAILS OF E120 TO COMPUTER   OBSERVER: E.J.SIMMONDS
OPERATOR: MS K MULLENS   STUDY NUMBER: 16820
DATE: 3/3/1987   START TIME: 11:41.36
                                                FINISH TIME: 11:47.54

                                                ELAPSED TIME: 00:06.18
                                                FOREIGN ELMTS: 00:01.23
                                                EFFECTIVE TIME: 00:04.95

                                                OBSERVED TIME: 00:04.95
                                                DIFFERENCE: 00:00.00 = 0%

ELMT	CODE	DESCRIPTION	OBS.	BASIC.T.	FREQ.	RA%.	STD.TIME
01	PU132	PICK UP INVOICE	8	0.1103	1/1	11.00	0.1224
02	KEY38	KEY IN DATA FROM INV.	8	0.2494	1/1	12.00	0.2793
03	PA096	PLACE ASIDE INV. TO FILE	8	0.0932	1/1	11.00	0.1035
04	GET22	GET NEW SET OF INVOICES	1	0.4590	1/6	13.50	0.0868

                                                TOTAL SM   = 0.5920
                                                CONTINGENCIES = 0.0000
                                                STANDARD TIME = 0.5920

Figure 17.5 A typical Summary Sheet print-out from a microcomputer

17.23 PROS AND CONS

The main advantages of using computerized time study are as follows:

1. Data collection is far simpler because the amount of writing is reduced:
   (*a*) on certain models the ratings are entered by a single button depression,
   (*b*) no times are read off a watch, nor written down.
2. The simplified recording allows observers to concentrate more on observing.
3. Flyback error is eliminated.
4. Analysis time is drastically reduced – claims of savings of up to 85 per cent of analysis time are made by systems consultants.
5. The consequent savings could affect management services staffing levels.
6. The integrity of the study is preserved although elements can be edited.
7. Presentation of results is improved.
8. Many versions have a small alpha-numeric display to show what has been input.
9. The saving and organizing of synthetic data is simplified.

Saberi[*] has shown that E.T.S. is well received by trade union members. In his sample survey of members where E.T.S. had been installed he found that 13 per cent were sceptical about E.T.S., but 78 per cent were favourable with nine per cent enthusiastic. Many liked the idea that data could not be tampered with by observers, and the absence of clock rating. He found several cases where stop-watches had

[*] M. Saberi, *Electronic Work Quantification, a Review of Techniques*, M.B.A. dissertation, Middlesex Business School, 1986.

been banned by unions, but E.T.S. was allowed.

He also found that in the majority of cases the pay-back period was between six to 18 months. Cost savings ranged between a few thousand pounds to over £25 000 per annum, corresponding to between 10 to 30 per cent of the total annual running costs of work study departments.

17.3 OTHER USES

Several packages are made available in E.T.S. kits, and may include:

1. Activity sampling for data collection.
2. Rated activity sampling, as an alternative to time study.
3. Synthetic databanks for storing standard times.

Activity sampling data may be collected on the study board for multi-person study. Systems can accommodate around 100 operatives and a similar number of activities.

The number of observations can be several thousand. These facilities are more than adequate for one observer.

Both random and systematic times can be used, the time intervals being generated by the study board itself thus saving a considerable amount of preparation time. Most of the designed boards give an audible alarm when the observation is due. The software usually provides for ratings to be included for rated activity sampling (see Chapter 10). As an extension of time study, rated activity sampling provides standard times, computed by the terminal. The print-outs can be as in Figure 10.5 of Chapter 10.

One of the most important outcomes of E.T.S. is the compiling of a data bank of standard times. The computer's capabilities for organizing data, rapid calculation, fast retrieval, and storing of data in a very compact way, are ideal for this purpose. Computer packages are available for the building up of data banks from time study (or P.M.T.S.) data, and can be integrated with E.T.S. packages.

A work measurement data base for synthetic times, such as Tectime's FAST program will provide certain essential facilities, described below. The software permits the creation of a data bank of elemental standard times. Each user may need these times in a specifically defined format which will assist in the synthesis of job times. The software will allow the user to define the format to suit his needs. Searching and retrieval may be done by categories of work such as electrical, maintenance, clerical, and others. From the data bank elements may be extracted using code numbers and used to build up data blocks with their standard times. In turn, these may be synthesised into complete job times. Job descriptions will be printed out automatically from the data bank. Retrieval of elements from disks is very quick allowing job times to be synthesized in a few minutes.

Where elements are updated and amended it is possible to do global changes to all the jobs or data blocks which incorporate those elements. This represents a huge saving in clerical time as updating can be done in seconds rather than hours.

17.31 USE OF E.T.S.

The survey carried out by Saberi* in 1986 showed that almost all of his sample of 87 users of the two systems investigated was in favour of E.T.S.; 60 per cent claimed to be 'better off using the system', with 38 per cent who 'could not do without it', leaving just two per cent of dissatisfied customers. This is rather impressive. The results, of course, do not necessarily hold for all systems.

It is estimated that by the beinning of 1987 there were several thousand users of E.T.S. The prices of E.T.S. systems may curtail the number of potential users, but considering the relatively short pay-back period (see Section 17.23) the investment could be cost-effective.

17.4 COMPUTERIZED P.M.T.S.

For many years attempts have been made to utilize the facilities afforded by the computer in the application of work measurement. Initially these attempts were confined to merely speeding up the recording of data and calculating results.

Toward the end of 1964 the work of researchers, including Douglas Towne at the University of California, culminated in the production of a Fortran computer program which generated simple analyses from appropriately input data. Since that time research has been directed at developing a system in which the computer makes a significant contribution towards producing a complete analysis of complex situations.

Dr Douglas Towne has defined the five steps in making an analysis as:

1. Observing the task or synthesizing it mentally;
2. Determining the motions each hand should make;
3. Measuring the physical variables;
4. Determining the exact motion-classifications;
5. Looking up the times for the motions.

Towne's approach, in conjunction with the Wofac Company, was to develop a system called Wocom which required a minimum amount of input data of such simplicity that a work-measurement practitioner who was relatively inexperienced in the application of predetermined motion–time systems could perform the elementary analyses. Such input data would include the X, Y, and Z co-ordinates of the starting position and finishing position of the hands, and the co-ordinates of the workplace and material locations. Shapes and sizes of materials handled and assembled would be coded and input. The computer would 'think out' the complete and best analyses for itself, and include due allowance for simultaneous motions. The capabilities of the system are:†

1. The deductive logic of Wocom can synthesize Reach, Grasp, Move, Assemble, Release, Pre-position, and Body Motions, for any operation. Appropriate

---

*M. Saberi, *Electronic Work Quantification, A Review of Techniques*, M. B. A. dissertation, Middlesex Business School, 1986.

† Based on a paper presented by Douglas Towne to the Work-Factor Associates Conference.

logic could not be developed to deduce the type of manipulative motions, such as certain very unusual elements in Use or Index. With exception of Machine and Process Functions, however, Wocom can evaluate times for all manual and mental operations.
2. The deductive logic can produce very efficient use of hands. It cannot be developed to completely deduce the sequence of operations, except that with a small effort by the engineer operating the computer, Wocom can balance assembly lines more efficiently than the unaided engineer.
3. The program can easily calculate distances between hands and objects, and can predict the effects that each motion will have on them.
4. The program can easily look up the time to perform any defined motion.
5. The program can construct a short English phrase to describe each motion or sequence of motions (Work Segments) appearing in the output.

Simplification of the system required there to be certain data-banks stored in the program. These are four in number. One file stores the characteristics of objects and operations in specific terms of dimensions, shapes, and manipulative motions required for handling them. A second bank stores details of the workplace, while a third holds data referring to the tasks and operations. Five actions are specified, corresponding to certain MTM and Work-Factor motions: Assemble, Move, Get, Release, and Aside. Certain modifiers are included such as To, From, Array, and Nest. The last data-bank carries formulae which may be required in the analyses.

Refinements to the system include the ability to 'nest' a number of small tasks to obtain large Work Segments. It is also possible to test the effects of changing the workplace arrangement simply by changing the co-ordinates of the items to be moved, or by re-specifying the particular tools to be used. This facilitates the assessment of cost savings. Other programs are available which balance the tasks in a product-oriented line.

Data may be in the format of MTM, Work-Factor, or any other system of predetermined motion–times. All data, including the time data of the Wocom system being used, are stored in the central files referred to above, access to which may be obtained through the medium of the computer link. Thus a subscriber to the system may contact the comprehensive data-banks merely by dialling a number on his time-sharing telephone link, and at very little cost in computer time. Analyses may be made in a fraction of the time normally taken by manual conventional methods, and without the skill needed in making the manual analysis. Today's computer terminals are portable, and this makes the system very flexible, as the equipment may be operated in any department of any company from the nearest Post Office telephone.

There is still much to do in developing such systems for use in measuring indirect work, although analysis of highly repetitive work is a reality, and it appears that Wocom's Multiple-Regression capability will add to the feasibility of measuring certain non-repetitive operations.

Bibliography

GENERAL AND PRODUCTIVITY (Chapters 1, 2, and 3)

TEXTBOOKS

A. Abruzzi, *Work Measurement* (New York: Columbia University Press, 1952)

A. Abruzzi, *Work, Workers, and Work Measurement* (New York: Columbia University Press, 1956)

R. M. Barnes, *Motion and Time Study* (New York: Wiley, 1968)

C. Babbage, *On the Economy of Machinery and Manufacturers* (London: Charles Knight, 1835)

R. M. Currie, *The Measurement of Work* (London: Management Publications, 1965)

British Standards Institution, *Glossary of Terms Used in Work Study* (London: B.S.I., 1969)

W. Gomberg, *A Trade Union View of Time Study* (New York: Prentice Hall, 1955)

R. F. Hoxie, *Scientific Management and Labour* (New York: Appleton, 1935)

International Labour Office, *Introduction to Work Study* (Geneva: I.L.O., 1958)

D. W. Karger, F. H. Bayhe, *Engineered Work Measurement* (New York: Industrial Press, 1966)

J. M. Lowry, H. B. Maynard, G. J. Stegemerten, *Time and Motion Study and Formulas for Wage Incentives* (New York: McGraw-Hill 1940)

H. B. Maynard (ed), *Industrial Engineering Handbook* (New York: McGraw-Hill, 1971)

R. L. Morrow, *Motion Economy and Work Measurement* (New York: Ronald Press, 1957)

R. L. Morrow, *Time Study and Motion Economy* (New York: Ronald Press, 1946)

M. E. Mundel, *Motion and Time Study* (Englewood Cliffs: Prentice Hall, 1960)

B. W. Niebel, *Motion and Time Study* (Homewood, Ill.; Irwin, 1958)

R. Presgrave, *Dynamics of Time Study* (New York: McGraw-Hill, 1945)

W. H. Schutt, *Time Study Engineering* (New York: McGraw-Hill, 1943)

F. W. Shumard, *A Primer in Time Study* (New York: McGraw-Hill, 1940)

F. W. Taylor, *The Principles of Scientific Management* (New York: Harper, 1929)

D. A. Whitmore, *Work Study and Related Management Services* (London: Heinemann, 1970)

ARTICLES

Rt. Hon. Lord Beeching, 'People, Performance, and Profitability', *Work Study and Management Services*, January 1973
G. R. Burn, 'The Place of Work Measurement', *W.S. & M.S.*, May 1969
J. E. Chapman, 'Frederick Winslow Taylor, Father of Scientific Management', *Atlanta Economic Review*, July 1968
Contributed, 'Measuring Productivity in a Service Industry', *Work Study*, October 1969
E. Cule-Davies, 'Work Measurement Research', *W.S. & M.S.*, March 1970
B. T. Faherty, C. B. Hall, D. A. Sparks, 'The Quality of Work Measurement', *W.S. & M.S.*, June 1970
M. Fein, 'Work Measurement Today', *W.S.*, February 1973
J. R. de Jong, 'Productivity, and Productivity Services', *W.S.* October 1970
H. E. Kearsey, 'The Bedaux Work Unit Method', *W.S. & M.S.*, September 1970
B. J. Longhurst, R. G. P. Kerridge, 'The Application of Work Measurement in Industry', *W.S. & M.S.*, March 1972
W. Walton, 'A Short Review of the Theory and Practice of Work Measurement', *W.S.*, November 1968
W. A. Woeber, 'A Survey of Work Measurement Techniques', *W.S.* June, 1972

HUMAN RELATIONS (Chapter 4)

TEXTBOOKS

C. Argyris, *Organization and Innovation* (Homewood, Ill.: Irwin, 1965)
C. Argyris, *Integrating the Individual and the Organization* (New York: Wiley, 1964)
C. Argyris, *Personality and Organization* (New York: Harper & Row, 1957)
W. Bennis, *Changing Organizations* (New York: McGraw-Hill, 1966)
W. Bennis, K. Benne, R. Chin, *The Planning of Change* (New York: Holt, Rinehart & Winston, 1961)
J. A. C. Brown, *The Social Psychology of Industry* (London: Penguin Books, 1954)
R. N. Ford, *Motivation Through the Work Itself* (New York: American Management Association, 1969)
S. W. Gellerman, *Management by Motivation* (New York: A.M.A., 1969)
S. W. Gellerman, *Motivation and Productivity* (New York: A.M.A., 1963)
F. Herzberg, *Work and the Nature of Man* (Cleveland: World Publishing Co., 1966)
F. Herzberg, B. Mausner, B. Snyderman, *The Motivation to Work* (New York: Wiley, 1959)
G. W. Howells, *Human Aspects of Management* (London: Heinemann, 1969)
P. Lawrence, A. Turner, *Industry, Jobs, and the Worker* (Boston: Harvard Business School, 1964)

R. Likert, *The Human Organization* (New York: McGraw-Hill, 1967)
R. Likert, *New Patterns of Management* (New York: McGraw-Hill, 1961)
A. H. Maslow, *Motivation and Personality* (New York: Harper & Row, 1954)
D. C. McClelland, *The Achieving Society* (Princeton: Van Nostrand Co., 1961)
D. McGregor, *The Human Side of Enterprise* (New York: McGraw-Hill, 1960)
D. McGregor, *The Professional Manager* (New York: McGraw-Hill, 1967)
M. Scott Myers, *Every Employee a Manager* (New York: McGraw-Hill, 1970)
F. I. Roethlesberger, W. I. Dickson, *Management and the Worker* (Cambridge (Mass.): Harvard University Press, 1939)
A. Zaleznic, C. Christensen, F. I. Roethlesberger, *The Motivation, Productivity, and Satisfaction of Workers* (Boston: Harvard Business School, 1958)

ARTICLES

P. J. Butcher, 'The Human Side of Change', *W.S. & M.S.*, May 1970
D. Charles-Edwards, 'Work and Motivation', *W.S. & M.S.*, May 1972)
H. F. Goodwin, 'Improvement must be Managed', *American Institute of Industrial Engineering Proc. 19th Annual Institute Conference*, 1968
F. Herzberg, 'One More Time—How Do You Motivate Employees?' *Harvard Business Review*, January/February 1968
E. V. Krick, 'The Engineer and Resistance to Change', *Journal of Industrial Engineering*, January/February 1959
J. Leask, 'The Worker's Attitude to Scientific Work Measurement', *Time and Motion Study*, February 1953
J. P. Lowry, 'Industrial Relations in the U.S.A.', *W.S. & M.S.*, March 1970
A. H. Maslow, 'A Theory of Human Motivation', *Psych. Review*, Vol. 50, 1943
I. S. McDavid, 'People, Participation, and Motivation', *W.S. & M.S.*, September 1970
F. Oldfield, 'People, Participation, and Profitability', *W.S. & M.S.*, April 1970
A. Wilkinson, 'Motivation: Some Western European Experiments', *W.S. & M.S.*, May 1971

BASIC CONCEPTS (Chapter 5)

TEXTBOOKS

See under 'General and Productivity' section, p. 417.

ARTICLES

C. J. Anson, 'Accuracy of Time Study Rating', *Engineering*, March 1954
K. Bibby, A. L. Minter, 'Rating—an Objective Approach', *W.S. & M.S.*, June 1967
L. Brouha, 'Measuring Human Effort', *Mechanical Engineering* (U.S.A.), June 1958
R. F. Bruckart, 'Achieving Accuracy in Work Measurement', *T. & M.S.*, March 1954

E. S. Buffa, 'Pacing Effects in Production Lines', *J.I.E.*, November/December 1961

R. G. Carson, 'Consistency in Speed Rating', *J.I.E.*, January 1954

D. J. Desmond, 'The Statistical Approach to Time Study', *Proc. R. Stat. Soc.*, 1950

D. J. Desmond, 'The 1950 National Survey of Work Measurement', *Work Study and Industrial Engineering*, Vol. 6, 1962

N. A. Dudley, 'The Effect of Pacing on Operator Performance', *Int. J. of Production Research*, Vol. 1, No. 2, 1962

M. Fein, 'A Rational Basis for Normal in Work Measurement', *W.S.*, September 1967

H. Gershoni, 'Variations in Work Measurement Techniques', *W.S. & M.S.*, July 1968

M. E. Mundel, 'An Analysis of Time Study Rating Systems and Suggestions for a Simplified Systematic System', *J.I.E.*, Vol. 4, No. 4, February 1944

K. F. H. Murrell, 'Laboratory Studies of Repetitive Work', *Int. J. of Production Research*, Vol. 2, No. 4, 1963

K. F. H. Murrell, 'On the Validity of Work Measurement Techniques', *W.S.*, April 1968

J. W. Pike, 'Speed and Effort Rating', *T. & M.S.*, October 1964

R. J. Sury, 'The Simulation of a Paced Single Stage Work Task', *Int. J. of Production Research*, Vol. 4, No. 2, 1965

R. J. Sury, 'Comparative Study of Performance Rating Systems', *Int. J. of Production Research*, March, 1962

C. Thomas, 'Special Report on a Test of a Group of Time Study Men', *Advert. Management*, August 1953

L. F. Thomas, 'Work Measurement and the Psychology of Judgement', *W.S.*, May 1969

E. S. Valfer, 'Evaluation of a Multiple-Image Time Study Rating Film', *Proc. I.E. Inst. (Univ. of California)*, February 1952

MATHEMATICS OF MEASUREMENT (Chapter 6)

TEXTBOOKS

A. H. Bowker, G. J. Lieberman, *Engineering Statistics* (Englewood Cliffs: Prentice Hall, 1959)

A. B. Clark, R. L. Disney, *Probability and Random Processes for Engineers and Scientists* (New York: Wiley, 1970)

R. G. Coyle, *Mathematics for Business Decisions* (London: Nelson, 1971)

R. Dorfman, P. A. Samuelson, R. M. Solow, *Linear Programming and Economic Analysis* (New York: McGraw-Hill, 1953)

W. Feller, *Introduction to Probability Theory and its Applications* (New York: Wiley, 1968)

S. I. Gass, *Linear Programming* (New York: McGraw-Hill, 1969)

W. W. Garvin, *Introduction to Linear Programming* (New York: McGraw-Hill, 1960)

J. H. Heward, P. M. Steele, *Business Control through Multiple Regression Analysis* (London: Gower Press, 1972)

A. G. Holzmann, R. Glaser, H. Schaefer, *Matrices and Mathematical Programming* (Chicago: Encyclopedia Britannica, 1963)

P. Meyer, *Introduction to Probability and Statistical Applications* (Reading, Mass.: Addison-Wesley, 1965)

K. Meisels, *Primer of Linear Programming* (New York: New York University Press, 1962)

M. J. Moroney, *Facts from Figures* (London: Penguin Books, 1964)

S. B. Richmond, *Operational Research for Management Decisions* (New York: Ronald Press, 1963)

M. Simonnard, *Linear Programming* (Englewood Cliffs: Prentice Hall, 1966)

J. Tennant-Smith, *Mathematics for the Manager* (London: Nelson, 1971)

ARTICLES

D. Anderson, Linear Programming Time Estimating Functions', *J.I.E.*, March/April 1965

D. Anderson, 'A Linear Programming Criteria for Establishing Time Estimating Functions', Pennsylvania State University, 1962

R. Campbell, 'An Investigation of the Application of Linear Programming to Work Standards', University of Utah, 1961

D. Crocker, 'Intercorrelation and the Utility of Multiple Regression', *J.I.E.*, Vol. XVIII, January 1967

D. Crocker, 'Letter to the Editor', *W.S.*, March 1967

J. Chisman, 'Using Linear Programming to Determine Time Standards', *J.I.E.*, April 1966

R. S. Klein, H. J. Tait, 'Faster, Better Tooling Estimates' (M.R.A.), *J.I.E.*, December 1971, and *W.S.*, April 1972

G. W. Ladd, 'Regression Method—Visual Inspection Times', *J.I.E.*, September/October 1960

R. Mellor, 'Multiple Regression', *W.S. & M.S.*, July 1970

L. H. C. Tippett, 'Statistical Methods in Textile Research', *J. of Textile Institute Transactions*, Vol. 26, February 1935

A. Moscowitz, J. Cranney, 'Multiple Linear Regression Analysis in Industrial Engineering', *W.S.*, August 1966

M. Salem, 'Multiple Linear Regression Analysis for Work Measurement of Indirect Labor', *J.I.E.*, May 1967

H. W. Secor, 'Regression Analysis—Standards Efficiency', *J.I.E.*, January 1966

D. R. Thelwel, 'Linear Programming and Multiple Regression Analysis in Estimating Manpower Needs', *J.I.E.*, March 1967

ANALYSIS STAGE, INCLUDING ACTIVITY SAMPLING
(Chapter 7)

TEXTBOOKS

R. M. Barnes, *Work Sampling* (New York: Wiley, 1957)

R. E. Heiland, W. J. Richardson, *Work Sampling* (New York: McGraw-Hill, 1957)

See also: books under 'General and Productivity' section, p. 417.

ARTICLES

N. F. Allard, 'Work Sampling—Valuable Maintenance Aid', *Plant Engineer*, September 1968

H. P. Anderson, 'Statistiche Technieken en hun Toepassingen', Rotterdam: University Press, 1971

Anon., 'Developments in Activity Sampling', *Brit. Steel Corpn*, 1969

R. Barnes, R. Andrews, 'Performance Sampling', Univ. of California, 1955

M. Clay, 'Complex Activity Sampling', *W.S.*, March 1969

R. W. Conway, 'Some Statistical Considerations in Work Sampling', *J.I.E.*, March/April 1957

D. S. Correll, R. M. Barnes, 'Industrial Applications of the Ratio-Delay Method', *Advanced Management*, September 1950

H. O. Davidson, 'Activity Sampling and Analysis—Present State of the Theory and Practice', *A.S.M.E. Paper*, 1953

R. E. Ducharne, 'More Samples=Less Accuracy', *W.S.*, August/September 1972

N. J. Fuhro, 'How to Use Sampling to Control Man and Machine Assignments', *W.S.*, November 1972

H. Gershoni, 'Selecting a Proper Work Study Technique', *W.S.*, June 1967

G. Gustat, 'Application of Work Sampling Analysis', Proc. 10th Time Study and Methods Conference of SAM-ASME, 1955

J. T. Hay, 'What is Activity Sampling?', *W.S.*, April 1966

D. Mahaffey, 'Work Sampling—To Set Standards on Indirect Labour', *Factory Management and Maintenance*, November 1954, Vol. 112

T. R. Mistry, 'Activity Sampling Application to Crane Utilization', *W.S. & M.S.*, March 1970

G. E. McAllister, 'Random Ratio-Delay', *J.I.E.*, August 1953

T. D. Milton, 'Activity Sampling Exercise on Parking Meters and Illegally Parked Cars', *W.S.*, February 1970

A. D. Moskowitz, F. A. Burhachi, 'Work Sampling—Sample Size and Confidence Limits', *W.S. & M.S.*, January 1966

Student's Forum, 'Activity Sampling', *W.S.*, August 1966

J. B. Tabernacle, F. Wharton, 'Compilation and Application of Standard Data, Pt. 3', *W.S.*, November 1970

L. H. C. Tippett, 'Statistical Methods in Textile Research', *Shirley Institute Memoirs*, Vol. 13, November 1934

L. H. C. Tippett, 'The Ratio-Delay Technique', *T. & M.S.*, May 1953

PREDETERMINED MOTION-TIME SYSTEMS (Chapter 8)

TEXTBOOKS

W. Antis, J. M. Honeycutt Jr., G. N. Koch, *'The Basic Motions of M.T.M.'* (Pittsburgh: The Maynard Foundation, 1968)

G. E. Bailey, R. Presgrave, *Basic Motion Time-Study* (New York: McGraw-Hill, 1958)

R. M. Currie, *Simplified P.M.T.S.* (London: Business Pubs., 1964)

D. W. Karger, F. H. Bayha, *Engineered Work Measurement* (New York: Industrial Press, 1966)

H. B. Maynard, J. G. Stegemerten, J. L. Schwab, *Methods–Time Measurement* (New York: McGraw-Hill, 1962)

J. H. Quick, J. H. Dundan, J. A. Malcolm Jr, *Work-Factor Time Standards* (New York: McGraw-Hill, 1962)

A. B. Segur, *Motion Time Analysis Instruction Manual* (Oak Park, Ill.: A. B. Segur, 1946)

ARTICLES

P. M. Burman and others, 'M.T.M. and the B.S.I. Rating Scale', *W.S. & M.S.*, February 1969

H. O. Davidson, 'On Balance—The Validity of Predetermined Element Time Systems', *J.I.E.*, Vol. XIII, No. 3, May/June 1962

F. Evans, 'Algorithms for the Processing of MTM-2 Tape Data Studies', *MTM Association*, 1968

F. Evans, 'MTM-2 Motion Sequences in Mechanical Maintenance Work', *MTM Association*, 1968

F. Evans, 'MTM-2 Based Maintenance Work Measurement', *MTM Association*, 1969

F. Evans, 'Tape Data Analysis', *MTM Association*, 1970

F. Evans, 'PMTS in Perspective', *W.S. & M.S.*, October 1972

C. W. Frederick, 'On Obtaining Consistency in Application of Predetermined Time Systems', *J.I.E.*, January/February 1960

J. R. Gadd, 'MTM-2 Maintenance Data', *W.S. & M.S.*, May 1963

F. B. and L. M. Gilbreth, 'Classifying the Elements of Work', *Management and Administration*, Vol. B, August 1924

A. W. Hughes, 'General Purpose Data with Slotting Techniques', *W.S.*, July 1972

K. Knott, 'Time Standards for Meat Boning Operations', *W.S. & M.S.*, June 1972

K. Knott, 'Body Motions with Weight—A Technical Comment', *W.S.*, October 1972

E. V. Krick, 'Hyperenthusiastic and Hypercritical Writing of Predetermined Motion Times', *J.I.E.*, May/June 1958

MTM Association, 'Application of MTM-2', *W.S. & M.S.*, June 1970

R. Nanda, 'The Additivity of Elemental Times', *J.I.E.*, May 1968

G. D. Rendel, 'MTM-2 in Maintenance Engineering', *J. M.T.M.A.*, July 1968

H. Schmidtke, F. Stier, 'An Experimental Evaluation of the Validity of Predetermined Element Time Systems', *J.I.E.*, Vol. XII, No. 3, May/June 1961

P. M. Steele, 'Halving Maintenance Costs', *MTM Association*, 1969

P. M. Steele, 'MTM Data System for Office Work', *W.S. & M.S.*, July 1971

J. B. Taggart, 'Comments on "An Experimental Evaluation of Predetermined Element Time Systems"', *J.I.E.* Nov./Dec. 1961

TIME STUDY and RATED-ACTIVITY SAMPLING
(Chapters 9 and 10)

TEXTBOOKS

Anon., *Introduction to Work Study* (Geneva: International Labour Office, 1969)

P. Carroll, *Time Study Fundamentals for Foremen* (New York: McGraw-Hill, 1951)

R. M. Currie, *The Measurement of Work* (London: Management Publications, 1967)

N. A. Dudley, *Work Measurement—Some Research Studies* (London: Macmillan, 1963)

W. C. Glassey, *The Theory and Practice of Time Study* (London: Business Books, 1966)

(*See also under* General and Productivity)

ARTICLES

T. Cass, 'The Accuracy of Time Studies', *W.S. & M.S.*, February 1967

J. Connolley, 'Rated Systematic Activity Sampling', *W.S. & M.S.*, November 1967

R. Elspie, O. Metcalfe, 'A Practical Technique for Sequential Work Sampling', *T. & M.S.*, January 1962

A. Flowerdew, P. Malin, 'Systematic Activity Sampling', *Work Study and Management*, December 1963

F. B. and L. M. Gilbreth, 'Stop-Watch Time Study—An Indictment or a Defense', *Taylor Society Bulletin*, June 1921

C. Graham, 'Rated Activity Sampling', *W.S. & M.*, January 1966

F. Hillebrandt, 'Sequential Work Sampling Tests', *W.S. & M.S.*, August 1967

W. Hines and J. Moder, 'Recent Advances in Systematic Activity Sampling', *J.I.E.*, September 1965

N. G. Jones, P. M. Ghare, 'Confidence Intervals for Systematic Activity Sampling', *J.I.E.*, May/June 1964

D. V. Kanon, 'Some Criticisms of Rating Practice', *J.I.E.*, June 1970

I. Landis Haines, 'Work Sampling by Fixed Interval Studies', *J.I.E.*, July/August 1963

R. Mattson, 'G.T.T., Work Measurement through Fixed Interval Observations', *Svenska Arbersgivar Foreningen*, 1968

A. Minter, 'Activity Sampling', *W.S. & M.S.*, June 1968
R. J. Sury, 'A Survey of Time Study Rating Research', *J.I. Prod. E.*, Vol. 41, No. 1
C. Travis, 'Rated Work Sampling', *J.I.E.*, September 1970

SYNTHESIS (Chapter 11)

TEXTBOOKS

R. D. Douglass, D. P. Adams, *Elements of Nomography* (New York: McGraw-Hill, 1947)
L. I. Epstein, *Nomography* (New York: 1958, Wiley)
J. Fasal, *Nomography* (New York: F. Ungar 1968)
E. Otto, *Nomography* (New York: Pergamon Press, 1964)

ARTICLES

A. Abruzzi, 'Developing Standard Data for Predictive Purposes' *Proc. I.E. Inst.* Univ. of California, February 1952
Anon., 'The British Work Measurement Data Foundation', *W.S. & M.S.*, May 1970
P. Carroll, Jr, 'Time Standards from Standard Data', *J. Soc. for Advancement of Management*, January 1939
W. C. Cooling, 'How Good is the Standard Data', *W.S.*, March 1967
R. E. Duval, 'Accurate Time Standards in Less Time', *Management Services*, July/August 1967
D. W. Field, A. Wells, 'The LAMSAC Data Bank', *W.S. & M.S.*, August 1972
C. Holt, 'A Guide to Machine Shop Setting Time Allowances', *W.S.*, April 1969
R. J. Jones, 'Standard Data for Power Station Maintenance', *W.S. & M.S.*, October 1972
E. J. Owen, 'Synthetic Data for a Capstan Lathe', *W.S.*, September 1965
E. J. Owen, 'Synthetic Data for a light Drilling Machine', *W.S.*, February and April 1966
E. J. Owen, 'Synthetic Data for Planers, Shapers, and Slotters', *W.S.*, December 1966
J. B. Tabernacle, T. Wharton, 'The Compilation and Application of Standard Data', *W.S.*, September 1970

ALLOWANCES (Chapter 12)

TEXTBOOKS

E. Bowman, R. Fetter, *Analyses of Industrial Operations* (Homewood, Ill.: R. D. Irwin, 1959)
L. Brouha, *Physiology in Industry* (New York: Pergamon Press, 1960)
E. Grandjean, *Fitting the Task to the Man* (London: Taylor & Francis, 1969)
K. F. H. Murrell, *Ergonomics (Seminar Papers)* (London: B.P.C., 1961)
K. F. H. Murrell, *Ergonomics: Man in his Working Environment* (London: Chapman & Hall, 1965)

T. F. O'Connor, *Production and Probability* (Manchester: Mechanical World Monographs No. 65, 1952)

ARTICLES

H. Ashcroft, 'The Productivity of Several Machines under the Care of One Operator', *J. Royal Stat. Soc.*, Vol. XII, No. 1, 1950

H. S. Belding, J. F. Hatch, 'Index for Evaluating Heat Stress in Terms of Resulting Physiological Strain', *J. American Soc. of Heating and Air Conditioning Engineers*, August 1955

N. Bhatia, K. F. H. Murrell, 'An Industrial Experiment in Organized Rest Pauses', *Human Factors*, No. 11, 1969

J. Brown, 'Measurement of Fatigue', *W.S. & M.S.*, December 1967

A.R.T., 'R.A.—What is Achieved?', *W.S. & M.S.*, June 1963

D. B. Chaffin, 'Physical Fatigue Percentage: What it is—How it is Predicted', *MTM Journal*, July/August 1969

E. A. Cyrol, 'How to Determine Standard Allowances for Unavoidable Delays', *Mill & Factory*, October 1945

E. A. Cyrol, 'The Matter of Fatigue Allowances', *Mill & Factory*, July 1948

H. Freeman, W. Wright, W. Duvall, 'Machine Interference', *Mechanical Engineering*, August 1932

K. F. H. Murrell, 'The Relationship Between Work and Rest Pauses', *W.S. & M.S.*, June 1967

T. F. O'Connor, 'Compensation for Interference', *T. & M.S.*, December 1952

T. F. O'Connor, 'Interference—Does it Matter?', *T. & M.S.*, October 1952

D. Jones, 'Mathematical and Experimental Calculation of Machine Interference Time', *The Research Engineer*, January 1949

D. Jones, 'A Simple Way to Figure Machine Downtime', *Factory Management and Maintenance*, October 1946

C. E. Lewis, R. F. Scherberger, F. A. Miller, 'A Study of Heat Stress in Extremely Hot Environments', *Brit. Journal of Industrial Medicine*, Vol. 17, 1960

R. F. Lomicka, J. M. Allderige, 'Mathematical Man-machine Analyses', *J.I.E.*, May/June 1957

B. Moores, 'Relaxation Allowance', *W.S. & M.S.*, March 1968

E. A. Müller, 'The Physiological Basis of Rest Pauses in Heavy Work', *Quarterly J. of Exp. Physiology*, Vol. 38, 1953

J. Tennant Smith, 'Machine Interference and Related Problems', *W.S.*, June to October 1965

PRODUCTION STUDIES (Chapter 13)

Student's Forum, 'Production Study', *W.S.*, Sept. 1967

ESTIMATING (Chapter 14)

ARTICLES

F. C. Allen, 'Comparative Estimation', *W.S. & M.S.*, October 1966

R. Hough, 'Comparative Estimating—the Points System', *W.S. & M.S.*, June 1969

P. W. Kirby, 'Category Estimating', *W.S. & M.S.*, April 1970

R. E. Gates, 'Comparative Estimating for Indirect Work Evaluation', *W.S. & M.S.*, February 1972

MEASUREMENT AND CONTROL OF INDIRECT WORK (Chapter 15)

TEXTBOOKS

J. E. Bayhylle, *Productivity Improvements in the Office* (Bristol: Engineering Employers Federation, 1968)

S. A. Birn, R. M. Crossan, R. W. Eastwood, *Measurement and Control of Office Costs* (New York: McGraw-Hill, 1966)

J. Constable, D. Smith, *Group Assessment Programmes* (London: Business Publications, 1966)

L. C. Harmer, *Clerical Work Measurement* (H.M.S.O., C.A.S. Occasional Papers No. 9, 1968)

D. A. Whitmore, *Measurement and Control of Indirect Work* (London: Heinemann, 1970)

ARTICLES

M. E. Addison, 'The Control of Indirect Work', *Management in Action*, April 1971

Anon., 'U.M.S. in Practice', *Maintenance Engineering*, April 1966

Anon., 'B.W.D. Goes Metric', *W.S.M.S.*, May 1970

H. P. Bakkenes, 'Standards for Office Work', *MTM Jnl*, Vol. 8/1

E. H. Benge, 'Measuring White-Collar Output', *System*, June 1930

S. R. Catt, 'Benefits of C.W.M.', *O & M Bulletin*, May 1969

D. Connor, 'Self Generated Clerical Performance Reports' *J.I.E.*, July 1970

P. Carpenter, 'Productivity in the Office', *P.O.T. Jnl*, Summer 1971

T. Clarke, 'Universal Maintenance Standards', *W.S.M.*, September 1964

J. Dening, 'Work Measurement Foils Parkinson's Law', *Business*, March 1966

K. Digney, 'Work Measurement for Smaller Plants Too', *F.M.M.*, January 1955, Vol. 113

H. P. Dolan, 'Establishing Work Measurement Standards for Non-productive Labor', *The Controller*, June 1952

J. Day, 'Standard Times for Gas-powered Fork Trucks', *Modern Materials Handling*, November 1954

F. Evans, 'MTM-2 Motion Sequences in Mechanical Maintenance Work', *MTM Association*, 1969
F. Evans, 'MTM-2 Based Maintenance Work Measurement', *MTM Association* 1971
M. Fein, 'Short Interval Scheduling', *J.I.E.*, February 1972
J. Gadd, 'MTM-2 Maintenance Data', *W.S.M.S.*, May 1969
D. L. Gerber, 'Short Interval Scheduling', *Systems and Procedures Journal*, November/December 1966
A. Gilchrist, 'Introducing Management Controls into a Maintenance Department', *Chemistry and Industry*, 1968
J. S. Grant, 'Variable Factor Programming', *W.S.M.S.*, June 1968
W. S. Halson, 'Measurement and Control of Clerical Work', *Office Management*, Summer 1965
W. S. Halson, 'Improving Productivity in Offices', *W.S.M.S.*, February 1968
C. Herriott, 'Clerical Work Measurement', *O & M Bulletin*, August 1966
A. D. R. Johnson, 'V.F.P. in Indirect Work', *W.S.M.*, March 1965
R. G. Lea, 'A Fresh Approach to Clerical and Indirect Work Measurement', *W.S.*, April 1972
A. J. Levy, 'Work Measurement in the Office', *Systems and Procedures*, November 1957
H. Nance, 'Five Techniques for Measuring Clerical Work', *The Office*, May 1967
L. M. Norton, 'Control and Indirect Labour', *W.S.*, February 1967
B. Pearce, 'Group Capacity Assessment', *W.S.*, May 1970
K. Rackham, 'Group Capacity Assessment', *The Director*, April 1966
J. Rousseau, 'Basic Work Data', *W.S.M.S.*, March 1967
D. Shrubsall, 'Office Modapts', *W.S.M.S.*, August 1969
D. Shrubsall, 'Introducing Office Modapts', *W.S.M.S.*, February 1969
H. Smalley, 'Predetermined Standards for Typing', *University of Connecticut*, June 1954
J. Spring, 'Master Clerical Data', *The Office*, May 1966
P. Steele, 'MTM Data System for Office Work', *W.S.M.S.*, July 1971
P. Steele, 'Halving Maintenance Costs', *MTM Association*, 1969
A. Thorncroft, 'A Manual Approach to Office Work', *The Financial Times*, 7 August 1968
G. Turner, 'The Control of Indirect Payroll Costs', *W.S.*, October 1967
J. Wilkinson, 'How to Manage Maintenance', *Harvard Business Review*, March/April 1968
I. Williams, 'How Good are your Cleaners?' *Industrial Society*, January 1971
S. Young, 'Measurement of Indirect Work', *J.I.E.*, August 1971

THE USES OF WORK MEASUREMENT (Chapter 16)

TEXTBOOKS

M. Avery, *Time Study, Incentives and Budgetary Control* (London: Business Publications, 1964)

A. Battersby, *Network Analysis for Planning* (London: Macmillan, 1964)
J. Batty, *Standard Costing* (London: Macdonald & Evans, 1966)
P. Carroll, *Time Study for Cost Control* (New York: McGraw-Hill, 1954)
P. Carroll, *Better Wage Incentives* (New York: McGraw-Hill, 1957)
W. D. Falcon, *Value Analysis/Value Engineering* (New York: American Management Association, 1964)
N. C. Hunt, *Methods of Wage Payment, in British Industry* (London: Pitman, 1951)
G. Kourim, *Wertanalyse* (Munich: R. Oldenbourg Verlag, 1968)
K. Lockyer, *Critical Path Analysis* (London: Pitman, 1964)
J. Louden, J. Keith, J. Wayne Deegan, *Wage Incentives* (New York: Wiley, 1959)
D. G. Malcolm, A. J. Rowe, *Management Control Systems* (New York: Wiley, 1960)
R. Merriott, *Incentive Payment Schemes* (London: Staples Press, 1957)
L. D. Miles, *Techniques of Value Analysis and Engineering* (New York: McGraw-Hill, 1961)
P. Sinclair, *Budgeting* (New York: Ronald Press, 1934)
Anon., *Sixteen Case Studies in Value Analysis* (London: *British Productivity Council*, 1966)

ARTICLES

O. Altum, 'A Group Incentive Plan for the Office', *N.A.C.A.*, March 1951
K. Boaz, 'Incentives in the Warehouse', *Factory Management and Maintenance*, September 1953
R. Boyde, 'Incentive Pay for Typists', *Management Review*, March 1955
P. Brewster, 'Optimum Economic Manning', *W.S. & M.S.*, September 1968
H. Gershoni, 'Motivation and micro-method when Learning Manual Tasks', *W.S. & M.S.*, September 1971
L. S. Hill, 'Some Cost Accounting Problems in PERT Cost', *J.I.E.*, February 1966
L. S. Hill, 'Some Pitfalls in the Design and Use of PERT Networking', *J. Academy of Management*, June 1965
E. Horning, 'Wage Incentive Practices', *National Industrial Conf. Board Study in Personnel Policy*, No. 68, 1945
J. Keller, 'Incentive Pay Increased Office Production', *The Office*, March 1954
P. Matthews, C. Brechin, 'Incentives Cut Painting Costs', *Factory Management and Maintenance*, June 1941
R. P. Nadkarni, 'Can You Measure Productivity of Engineers?', *W.S. & M.S.*, November 1971
B. Payne, 'The Administration of an Incentive Plan, *Michigan Business Review*, March 1954
R. Putterbaugh, 'Incentives Applied to Order Filling', *Mill and Factory*, June 1948
D. Saxton, 'Inspectors on Piece Rates Up Quality, Output, and Earnings', *Factory Management and Maintenance*, February 1940

J. H. Van Santen, 'Method and Time Study for Mental Work', *W.S.*, September 1970

G. Webdill, M. Fletcher, C. Michan, 'Progress Forecasting based on MTM', *W.S. & M.S.*, February 1969

FUTURE DEVELOPMENTS (Chapter 17)

TEXTBOOKS

R. Mattson, *Arbetsmätning och ADB* (Stockholm: SAF Tech. Dept., 1969)

R. M. Barnes, *Motion and Time Study*, 7th Ed. (New York: Wiley, 1980)

ARTICLES

W. G. Allen, 'Decision Systems and Work Study', *W.S. & M.S.*, November 1970

Anon., 'Computerized Time Study Gaining Momentum', *Steel*, October 1963

Anon., 'IBM 1130 Work Measurement Aids—Application Description', *IBM Tech. Pub.*, H 20-0249.0—1966

E. Boepple, L. Kelly, 'How to Measure Thinking', *J.I.E.*, July 1971

Contr., 'Measuring the Human's Mental Processes', *W.S. & M.S.*, May 1968

Contr., 'Mental Process Measurement Increases Productivity', *Metalworking Production*, January 1968

Contr., 'Computerized Work Measurement', *W.S.*, February 1972

G. Cook, 'Manservant - Study Work Using Electronic Notebook', *Management Services*, March 1983, pp. 28-30

J. Heward, 'Computer System for Standard Data', *W.S. & M.S.*, September 1970

J. Heward, 'Work Study and Computers', *W.S. & M.S.*, January 1971

M. F. Mobach, 'AUTORATE—Computer Aid to Industrial Engineering', *Proc. 16th Conf. of Ind. Eng. Inst.*, 1963

R. E. Raven, 'The Frontiers of Work Study—2', *W.S.*, October 1972

W. Rohmert, W. Laurig, 'Mental Load of Radar Controllers', *W.S. & M.S.*, July 1972

R. W. Tomlinson, 'Ergonomic Methods in Work Study', *W.S. & M.S.*, November 1970

A. R. Watson, 'Using the Epson HX20 for Time Study', *Work Study*, April 1984, pp. 43-50

Index

Abnormal posture, 290
Accuracy, in activity sampling, 122
  in MTM-2, 181
  in P.M.T.S., 156
  in rated activity sampling, 251
  in time study, 242
Activities Diary Sheet, 370
Activity sampling, 78, 118
  definition, 118
  error (accuracy), 122
  number of observations, 121
  procedure, 119
Actual performance, 27
Advantages of P.M.T.S., 153
Algorithms, 172, 183, 185, 188, 387
Allowances, 42, 242, 287
  contingency, 308
  bonus increment, 43, 44
  interference, 300
  policy, 43, 44
  process, 294
  unoccupied time, 295
Analytical estimating, 325
  definition, 326
Application ratios, all systems, 46
  for P.M.T.S., 154
Apply pressure, in MTM, 171
  in MTM-2, 186
  in MTM-3, 195
Ashcroft's numbers, 306
Assignable causes, 135
Attention time, 195

Babbage, Charles, 14
Barnes, Ralph M., 21, 147, 149, 246, 290
Basic time, 42, 116
Batch control ticket, 360, 361
Bedaux, Charles E., 18, 49
Bedaux scale for rating, 49
Behaviour of practitioner, 37
Bench-mark jobs, 335, 342
Bend and arise, in MTM, 174
  in MTM-2, 188
  in MTM-3, 196
Beta distribution, 206

Binomial distribution, 77, 121
Blanket allowance, 294
Blue-collar workers, 353, 354
Body motions, in MTM, 174
Break-points, 208
British Standards Institution
  rating scale, 50

Categories, determination of, 337
Category estimating, 346
Category mid-points, 339, 342
Check times, definition, 235
Clerical Work Improvement Programme, 23, 383
Clerical work measurement, 205, 379
Comparative estimating, 335
Compilation of basic times, 116
Compute, element in Work-Factor, 204
Computers in work measurement, 414
Concentration, 291
Confidence, level of, 125
Consecutive element timing, 212
Consistency in rating, 52, 59
Consultation, 36, 248
Constant element, 106
Contingency allowance, 308
Continuous timing, 212
Control charts, 121, 136, 137, 138
Control documents, 374
  loops, 357
Correlation, 84
Costing, 390, 393, 400
Cost variance, 32, 391
Coulomb, M., 14, 15, 16
Counting instructions, 362, 373
Crank, element in MTM-2, 186
Critical path analysis, 405, 406
Cumulative timing (*see* continuous timing)

Data blocks, 386, 388
Data collection, 118, 370
Decision models (*see* algorithms)
Definitions
  activity sampling, 118

437

INDEX

Definitions—*Contd*
  analytical estimating, 326
  attention time, 295
  check time, 235
  comparative estimating, 335
  contingency allowance, 308
  counting instructions, 262
  element, 106
  estimating, 325
  governing element, 106, 295
  inside work, 295
  machine running time, 295
  outside work, 295
  predetermined motion-time system, 143
  production study, 310
  rated activity sampling, 246
  rating, 47
  relaxation allowance, 287
  selected basic time, 116
  standard data, 268
  standard performance, 50
  standard rating, 42
  synthesis, 268
  synthetic data, 268
  work cycle, 105, 295
  work measurement, 6, 40
Department performance, 27, 389
Design, 392
Desmond, D. J., 60
Diary sheet, 371, 372
Differential timing, 212
Disengage, element in MTM, 170
Dispersion, 113
Distribution of MTM motions, 179
Documentation, for category estimating, 348
  for indirect work, 370
  for synthesis, 270
  for time study, 224
  for varifactor synthesis, 367
Dry run period, 366
Dudley, N. A., 53, 60

Earned time (in minutes, or hours), 360
Effectiveness ratio, 385
Efficiency, 362
Elapsed time, 236, 240
Electrocardiograph (E.C.G.), 413
Electroencephalograph (E.E.G.), 413
Electronic time study, 211
Element breakdown, 105, 225, 228, 280
  definition, 105
  types of, 106
  for P.M.T.S., 158
Elements, 41, 208
Error assessment, in activity sampling, 122
  in MTM, 181
  in mean basic time, 242
  on number of cycles, 116
  in P.M.T.S., 156
  in rated activity samplng, 251
  in timing, 240
Estimating, definition, 325
Extension, 41, 214, 240
  graphical methods, 215
  rated activity sampling, 254
Eye movements and actions,
  in MTM, 173
  in MTM-2, 187
  in Work-Factor, 201

Factor time, 379
Fatigue allowance, 288
Fatigue, meaning of, 288
Feedback of data, 357
Film analyses, 149
Financial incentives (*see* incentives)
Flat rating, 59
Flow process chart, 369
Flowerdew and Malin, 246, 252
Flyback timing, 212, 213
Foot motions, in MTM-2, 187
Foreign element, 106
Formulae, use in synthetics, 284
Frequency of occurrence, 108, 249
Fry, T. C., 308

Galton, Sir Francis, 86
Gantt chart, 18, 361, 403
Geared schemes, 408, 409
Get, element in MTM-2, 182
Gilbreth, F. B., 21, 144
Governing element, 106, 295
Gradient error in rating, 58
Grandjean, E., 288
Graphical methods
  for number of observations, 111
  for synthetic data, 277
  for extension, 215
Grasp, element in P.M.T.S., 160
Group Capacity Assessment, 379

Handle, element in MTM-3, 194
Hawthorne Plant, 13
Heart-rate, 413
Heat and humidity, 291

INDEX

Herzberg, F., 34
Hyperbola, 215, 216

Incentives (financial), 408
Inconsistent rating, 59
Index of determination, 88, 91, 92, 283
Inside work, 295
Inspection, measurement of, 202
Intercorrelation, 93
Interference, 300

Job enrichment, 360, 412

Kymograph, 21

Levelling, 53
Level of confidence, 125
Line of best fit, 81
Linear programming, 94
Load factor, 298
Log-log rating chart, 218, 219
Log-normal distribution, 337, 349
Long-cycle work measurement, 261
Loose rating, 59
Lowry, Maynard, and Stegemerten, levelling system, 19, 53, 150

Machine element, 106
Machine interference, 300
Machine loading, 31
Machine running time, definition, 295
Macro-P.M.T.S., 143, 205
Maintenance work, 205
Management ratios, 5
Manpower productivity measure, 7
Man/year equivalent, 7
Manning tables, 378
Manual element, 106
Maslow, A., 34
Maynard, H. B., 147, 161
Mayo, G. Elton, 13
Measured daywork, 406, 409
Mental work, measurement of, 146, 198
Mento-Factor System, 146, 198
Methods of sampling, 127
Methods–Time Measurement, 22, 161
  analysis, 107, 108, 274, 275, 394, 395, 397
MTM Consortium Data, 386
MTM-2, 176
MTM-2 Maintenance Data, 386
MTM-3, 192
Micro-P.M.T.S., 41, 143

Minor job ticket, 388
Modal extension, 214, 215
Monotony, 291
Morrow, R. L., 19, 56
Motion combinations, 188
Motion Time Analysis (MTA), 145
Motivation, 36, 358, 412
Move, element in MTM, 168
Mulligan, P. B., 148, 383
Mulligan standards, 145, 148
Multi-image films, 55
Multiple regression analysis, 86, 416
Mundel's rating method, 19, 51, 54
Murrell, K. F. H., 289, 290

Nadler's rating method, 50, 56
National Coal Board, 246, 252
Needs and drives, 34
Networks, 406
Noise, effects of, 291
Nomograms, for activity sampling, 126
  for number of observations (timing), 110
  for synthesis, 285
Non-linear relationships and graphs, 281
Normal distribution, 70, 341
Number of observations, timing techniques, 109
  activity sampling, 121
  rated activity sampling, 248, 253

Objective rating, 53
Objectives of work measurement, 40
Observation record sheet, 234
  for rated activity sampling, 257
Occasional element, 106
Operations list, 371
Operator performance, 27, 389
Outside work, 295
Overall performance, 27, 389
Owen, R., 12

Parabolic curve, 123, 282
Performance, indices, 29
  reports, 374
  ratios, 27, 362
Perronet, J., 14
Personal allowance, 288
P.E.R.T., 406
Photography in P.M.T.S., 149
Physical effort, 290
Planning, 402
Position, element in MTM, 168

Posture, 290
Presentation of data, 276
Probability, use of, 112, 135
Process allowances, 294
Production study, definition, 310
   procedure, 311
Productivity, 1
Put, element in MTM-2, 184

Quick, J. H., 20, 22, 145, 149

Random attention, 304
Random numbers, 140
   samples, 127
   times, 120, 131, 248, 262
   time tables, 131
Rated activity sampling, 246
   definition, 246
Rating, 47, 236
   consistency of, 52
   errors, 57
   levels, 51
   maintenance of, 65
   equivalent of MTM, 150
   research, 52, 53
   scale (B.S.I.), 50
   scales, 49
   equivalent in Work-Factor, 150
Ratio delay, 118
Ratio of staff to analysts, 380
Reach, element in MTM, 163
Reading time (MTM), 199
React time (Work-Factor), 201
Realization, 362
Reciprate chart, 57, 62, 217
Reciprate chart for rating, 57, 62
   construction of, 57, 219
Reference period summary, 376, 377
Regrasp, element in MTM-2, 186
Regression analysis, 82
Relaxation allowance, 239, 287
   sheet for, 241
   tables of, 292
Release, element in MTM, 167
Remuneration, 32, 406
Repetitive element, 106
Reports, 374
Research, 52, 53
Resistance to change, 34

Sammie, 415
Sample means, 72
Sampling, 127
Sampling methods, 127, 250

Schedule of standard times, 375
Scheduling, 402
Scientific management, 13, 16
Scott–Mulligan Manual, 383
Security, 35
Segur, A. B., 22, 144, 145
Selection of a work measurement technique, 102
Selective element timing, 212
Self-recording, 359
Short interval scheduling, 360
Signout procedure, 382
Simplex, 94
Simultaneous motions, in MTM, 175
Slotting techniques, 335
Snap reading method, 118
Speeds of application of techniques, 46, 154
Split-action stop-watches, 211
Standard costing, 32, 389, 390, 391
Standard deviation, 71
   error, 72, 113, 121, 253, 316
   error chart, 138
   error of difference, 317
Standard performance, 27
   rating, definition, 42
   time, 26
Status, 35
Steep rating, 59
Step, element in MTM-2, 187
   element in MTM-3, 195
Stop-watches, types, 209, 210, 211
Straight line, 79, 277, 284
Stratified sampling, 127
Supervisor, his role in work measurement, 358, 360
Synthesis, definition, 268
Synthetic data, 268
   rating system, 56
Systematic sampling, 133, 250, 254

Tabulation of data, 276
Tape Data Analysis, 23, 148, 189, 386, 387
Target times, 52, 383
Task Programming, 382
Task Ticket, 382
Taylor, F. W., 16, 269
Team-work, 297
Therbligs, 144
Tight rating, 59
Time bands, 335
Time intervals for systematic sampling, 133, 248, 251

Time-measurement unit (T.M.U.), 150, 163
Timing error, 240
  procedure, 209
Tippett, L. H. C., 19
T.M.U. (*see* Time-measurement unit)
Total Productivity Index (T.P.I.), 2
Towne, D., 415
Transport, element in MTM-3, 194
  element in Therbligs, 144
Turn, element in MTM, 170

Unit of count, 362, 375
Unit of measurement (*see* Unit of production)
Unit of production, 108, 262, 361
Unit of work, 26, 29
Unoccupied time allowance, 295
Utilization ratio, 362

Validity of P.M.T.S., 155
Value analysis/engineering, 392
Variable element, 106

Variables in P.M.T.S., 160
Variance, in costing, 391
  statistical (*see* Standard deviation)
Varifactor synthesis, 353
VeFAC Programming, 381
Vibration, 291
Vigilance, 291
Visual fatigue, 291
Volume activity sheet, 371

Westinghouse leveling system, 53, 150
White-collar workers, prevalance of, 353, 354, 412
Wocom, computer system, 146, 151, 415
Work allocation, 360
Work cycle, definition, 105, 295
Work-Factor, 145
  Mento-Factor System, 143, 200
Work measurement, definition, 6, 40
Work units, 26, 29
Wright, W., 308
Writing times (in MTM), 199